Springer Series in
Nuclear
and **Particle Physics**

Springer Series in **Nuclear** and **Particle Physics**

Editors: Mary K. Gaillard · J. Maxwell Irvine · Erich Lohrmann · Vera Lüth
Achim Richter

Hasse, R. W., Myers W. D.
Geometrical Relationships of Macroscopic Nuclear Physics

Belyaev, V. B.
Lectures on the Theory of Few-Body Systems

Heyde, K. L. G.
The Nuclear Shell Model

Gitman, D. M., Tyutin I. V.
Quantization of Fields with Constraints

Sitenko, A. G.
Scattering Theory

Fradkin, E. S., Gitman, D. M., Shvartsman, S. M.
Quantum Electrodynamics with Unstable Vacuum

Aleksei G. Sitenko

Scattering Theory

Translated by O. D. Kocherga

With 32 Figures

Springer-Verlag

Berlin Heidelberg New York
London Paris Tokyo
Hong Kong Barcelona

Professor Dr. *Aleksei G. Sitenko*

Institute for Theoretical Physics, Academy of Sciences of the Ukrainian SSR
Metrologicheskaya 14b, SU-252130 Kiev, USSR

Translator:

Dr. *Olga D. Kocherga*

Institute for Theoretical Physics, Academy of Sciences of the Ukrainian SSR
SU-252130 Kiev, USSR

Title of the original Russian edition: *Teoriya Rasseyaniya.*
© Vishcha Shkola, Kiev 1975

ISBN-13 : 978-3-642-84036-4   e-ISBN-13 : 978-3-642-84034-0
DOI : 10.1007 / 978-3-642-84034-0

Library of Congress-Cataloging-in-Publication Data. Sitenko, A. G. (Alekseï Grigor'evich) Scattering theory / Aleksei G. Sitenko ; translated by O. D. Kocherga. p. cm.–(Springer series in nuclear and particle physics) Translated from the Russian. "This book is based on the course in theoretical nuclear physics that has been given by the author for some years at the T. G. Shevchenko Kiev State University. This version is supplemented and revised to include new results obtained after 1971 when the first and second editions were published"–CIP pref. 1. Scattering (Physics) I. Title. II. Series. QC794.6.S3S57   1990   539.7'58–dc20 90-9592

© Springer-Verlag Berlin Heidelberg 1991
Softcover reprint of the hardcover 1st edition 1991

2157/3140-543210 – Printed on acid-free paper

To my dear daughter
Alla

Preface

This book is based on the course in theoretical nuclear physics that has been given by the author for some years at the T. G. Shevchenko Kiev State University. This version is supplemented and revised to include new results obtained after 1971 and 1975 when the first and second editions were published.

This text is intended as an introduction to the nonrelativistic theory of potential scattering. The analysis is based on the scattering matrix concept where the relationship between the scattering matrix and observable physical quantities is considered. The stationary formulation of the scattering problem is presented; particle wave functions in the external field are obtained. A formulation of the optical theorem is given as well as a discussion on time inversion and the reciprocity theorem. Analytic properties of the scattering matrix, dispersion relations, and complex moments are analyzed. The dispersion relations for an arbitrary direction scattering amplitude are proven, and analytic properties of the amplitude in the plane of the complex cosine of the scattering angle are studied in detail. This is followed by a derivation of the dispersion relations for the scattering amplitude with respect to the momentum transfer. A special chapter is devoted to the double dispersion relations. The Mandelstam representation is applied to the analysis of the analytic properties of the scattering amplitude with respect to energy and momentum transfer. The so-called inverse scattering problem is also discussed. Separable representation of the scattering amplitude is introduced including the formulation of the Hilbert-Schmidt method. Three-particle bound states and scattering are investigated. The final chapter deals with scattering of spin possessing particles and related polarization phenomena.

Problems illustrating and supplementing the material are given at the end of each chapter. A list of general reading references is given including texts on quantum mechanics and scattering theory, as well as original papers in which the reader can find a more detailed account of some topics.

The book is suggested as a manual for students and postgraduates of physical and engineering faculties of universities and colleges, as well as instructors and research workers in theoretical nuclear physics and elementary particles.

The author is grateful to Y. A. Berezhnoi, I. S. Dotsenko, P. V. Skorobogatov, V. K. Tartakovsky, V. F. Kharchenko, and A. D. Foursat for valuable discussions and assistance.

Kiev, June 1990   *A. G. Sitenko*

Contents

1. Quantum Mechanical Description and Representations

1.1 Quantum Mechanical Description of Physical Systems

In quantum mechanics any physical quantity (dynamical variable) is associated with an operator that is, in turn, associated with a linear equation, solvable only for definite eigenvalues of the operator. The relevant solutions of the linear equation are referred to as eigenfunctions. Quantum mechanics usually deals with Hermitian operators, i.e., linear self-conjugate operators with real eigenvalues. The eigenvalues of the Hermitian operator determine probable values of the physical quantity and are described by quantum numbers. The relevant eigenfunctions describe probable states of the physical system, and usually have certain conditions imposed upon them (finiteness, unambiguity, continuity). The Hermitian operator eigenfunctions also satisfy the orthonormalization and completeness conditions.

In the general case, a physical system is described by several dynamical variables. The state of the system is associated with the wave function (state vector) $\psi_\alpha(x)$, where α is the index of the state, i.e., the set of eigenvalues of the physical quantities which determine the state or relevant quantum numbers; x is the representation index, i.e., the set of variables on which the wave function depends. The square of the wave function modulus directly determines the probability that the system is at the point x for the given state α.

The values of the wave function $\psi_\alpha(x)$ corresponding to different x may be regarded as coordinates of a complex vector $|\alpha\rangle$ (infinite-dimensional in the general case) which is referred to as the state vector of the system. A linear combination of two vectors forms a new vector. The scalar product of any two vectors $|\alpha\rangle$ and $|\alpha'\rangle$ can be defined by means of the relation

$$\langle\alpha|\alpha'\rangle \equiv \int dx\, \psi_\alpha^*(x)\, \psi_{\alpha'}(x) \quad . \tag{1.1}$$

The vector norm $\|\alpha\|$ is defined by

$$\|\alpha\| = \sqrt{\langle\alpha|\alpha\rangle} \quad . \tag{1.2}$$

The set of all state vectors of the system forms a linear vector space — the Hilbert space $\mathcal{H}$ that possesses a very important property — there exists in it a denumerable orthonormalized basis, i.e., a set of orthonormalized vectors in terms of which any vector can be expanded.

Let us consider an operator Q. Suppose its eigenfunctions $\psi_q(x)$ (q is the eigenvalue of the operator Q) form a complete orthonormalized set, i.e.,

$$\int dx\, \psi_{q'}^*(x)\, \psi_q(x) = \delta_{qq'} \quad . \tag{1.3}$$

The wave function $\psi_\alpha(x)$ can be expanded in terms of $\psi_q(x)$. We have

$$\psi_\alpha(x) = \sum_q c_\alpha^q\, \psi_q(x) \quad , \tag{1.4}$$

or

$$|\alpha\rangle = \sum_q c_\alpha^q\, |q\rangle \quad .$$

The squared modulus of the expansion coefficient c_α^q describes the probability that the quantity Q is equal to q in the state α. To define the mean value of the quantity Q in this state, we multiply each eigenvalue q by the relevant probability $|c_\alpha^q|^2$ and take the sum of all the products obtained, so that

$$\overline{Q} = \sum_q q\, |c_\alpha^q|^2 \quad . \tag{1.5}$$

The set of expansion coefficients c_α^q may be regarded as coordinates of the vector $|\alpha\rangle$ in the coordinate system determined by vectors $|q\rangle$, i.e., as the wave function of the state α in the q-representation. This becomes evident in terms of Dirac notation, in which

$$\psi_\alpha(x) \equiv \langle x|\alpha\rangle \quad , \qquad \psi_q(x) \equiv \langle x|q\rangle \quad , \quad \text{and} \quad c_\alpha^q \equiv \langle q|\alpha\rangle \quad .$$

Then relation (1.4), which describes the transition from the q- to the x-representation, may be written as

$$\langle x|\alpha\rangle = \sum_q \langle x|q\rangle \langle q|\alpha\rangle \quad , \tag{1.6}$$

and the mean value (1.5) then takes the form

$$\overline{Q} = \langle \alpha|Q|\alpha\rangle \quad . \tag{1.7}$$

As follows from (1.6), the eigenfunction of the operator Q in the x-representation, $\langle x|q\rangle \equiv \psi_q(x)$, is associated with the transformation from the q- to the x-representation. The notation $\langle x|q\rangle$ for the transformation function emphasizes the symmetry of the representation index x and the state index q. It is not difficult to verify that the inverse transformation function $\langle q|x\rangle$ is equal to the complex conjugate of the direct transformation function $\langle x|q\rangle^*$. Indeed, let us multiply (1.6) by $\langle x|q \rangle^*$ and integrate over x. Then, because the functions (1.2) are orthonormalized, we have

$$\langle q|\alpha \rangle = \int dx \, \langle x|q \rangle^* \, \langle x|\alpha \rangle \quad ,$$

from which we find, according to the definition of the transformation function, that

$$\langle q|x \rangle = \langle x|q \rangle^* \quad . \tag{1.8}$$

The transformation law for matrices follows immediately from (1.6), i.e.,

$$\langle x'|Q|x \rangle = \sum_{q,q'} \langle x'|q' \rangle \, \langle q'|Q|q \rangle \, \langle q|x \rangle \quad , \tag{1.9}$$

and the transformation is provided by the same transformation functions.

An example of the transformation function is the momentum operator eigenfunction in the coordinate representation

$$\psi_p(\boldsymbol{r}) \equiv \langle \boldsymbol{r}|\boldsymbol{p} \rangle = \exp\left(\frac{i}{\hbar}\,\boldsymbol{p}\boldsymbol{r}\right) \quad . \tag{1.10}$$

This function represents the transition from the momentum to the coordinate representation. Another example is the eigenfunction of the angular momentum operator in the momentum representation

$$\psi_{lm}(\boldsymbol{n}) \equiv \langle \boldsymbol{n}|lm \rangle = Y_{lm}(\vartheta, \varphi) \quad , \tag{1.11}$$

where $\boldsymbol{n}$ is the unit vector along the momentum direction $\boldsymbol{p}$. Function (1.11) provides the transformation from the representation given by moment values to the one given by particle motion directions. The transition from one representation to another (i.e., from one set of dynamical variables to some other set) is referred to as canonical transformation. Canonical transformations are realized by unitary operators, and the physical properties of the system are invariant with respect to these transformations.

Along with the unitary canonical transformations associated with the transitions between different sets of dynamical variables, quantum mechanics also treats unitary transformations describing the time evolution of the states. In contrast to the canonical transformations, unitary operators are time-independent, so that the change of the state with time is now described by the unitary operator action on the wave function.[1] The various approaches describing the time evolution of physical systems are called representations. The latter representations

---

[1] We remind the reader of the definition of a unitary operator. A unitary operator in the Hilbert space $\mathcal{H}$ is a linear operator U that maps the whole space onto itself and conserves the norm for all vector states $|\psi \rangle$, so that $D(U) = \mathcal{R}(U) = \mathcal{H}$ and $\| U\psi \| = \| \psi \|$. [$D(U)$ is the domain of definition, $\mathcal{R}(U)$ is the mapping domain of the operator.] The vector norm conservation condition requires that the unitary operator U must satisfy the relation $U^+U = 1$, and since $\mathcal{R}(U) = \mathcal{H}$, we have $UU^+ = 1$. The inverse to the unitary operator is $U^{-1} = U^+$. The two conditions $U^+U = 1$ and $UU^+ = 1$ completely determine the unitary operator.

dealing with the time-dependence of the wave functions must not be confused with the representations of physical quantities and states, which are determined by the choice of independent variables.

1.2 Schrödinger Representation

In the Schrödinger representation, the dynamical variables are associated with time-independent operators, the time evolution of the system being governed by the wave function variations. The time evolution of the wave function is governed by the Schrödinger equation,

$$i\hbar \frac{\partial \psi(t)}{\partial t} = H \psi(t) \quad , \tag{1.12}$$

where H is the Hamiltonian of the system. The wave function time dependence may be described by the time shift operator $R(t,0)$,

$$\psi(t) = R(t,0) \psi(0) \quad , \tag{1.13}$$

where $\psi(0)$ is the initial wave function at $t = 0$. The Schrödinger equation (1.12) yields the operator equation for $R(t,0)$,

$$i\hbar \frac{\partial R(t,0)}{\partial t} = H R(t,0) \quad , \tag{1.14}$$

with the initial condition $R(0,0) = 1$. The formal solution of (1.14) may be written as

$$R(t,0) = \exp\left(-\frac{i}{\hbar} H t\right) \quad . \tag{1.15}$$

Since H is a Hermitian operator, the time shift operator (1.15) is unitary, i.e.,

$$R^+(t,0) = R^{-1}(t,0) \quad . \tag{1.16}$$

The mean values of physical quantities depend on time. Let us define the operator $\dot{Q}$, associated with the time derivative of the physical quantity Q, to be the quantity whose mean value is equal to the time derivative of the mean value $\overline{Q}$, i.e.,

$$\overline{\dot{Q}} = \dot{\overline{Q}} \quad . \tag{1.17}$$

Making use of (1.13, 16), we find in the general case that

$$\dot{Q} \equiv \frac{dQ}{dt} = \frac{\partial Q}{\partial t} - \frac{i}{\hbar} [Q, H] \quad , \tag{1.18}$$

where $[a, b] \equiv ab - ba$. If Q does not depend explicitly on time, we have

$$\dot{Q} = -\frac{i}{\hbar}[Q, H] \quad . \tag{1.19}$$

According to (1.19), the mean value of the physical quantity is time-independent only if the relevant operator commutes with the Hamiltonian of the system.

1.3 Heisenberg Representation

Let us consider an alternative description of the physical system. Suppose the wave functions are not changed with time, while the operators associated with the dynamical variables are time-dependent. If the wave functions of this Heisenberg representation are defined in a way to ensure their coincidence with the Schrödinger wave functions at a certain time, say, $t = 0$, then

$$\psi_H = \psi_S(0) \quad . \tag{1.20}$$

According to (1.13), the transition from the Schrödinger to the Heisenberg representation is performed by means of the unitary transformation

$$\psi_H = R^{-1}(t, 0)\,\psi_S(t) \quad , \tag{1.21}$$

and the relation between the Heisenberg and the Schrödinger operators may be written as

$$Q_H(t) = R^{-1}(t, 0)\,Q_S\,R(t, 0) \quad . \tag{1.22}$$

The time differentiation of (1.22) yields, within the context of (1.14), the Heisenberg equation of motion, which governs the operator time evolution in the Heisenberg representation:

$$i\hbar\,\frac{d}{dt}\,Q_H(t) = [Q_H(t),\,H] \quad . \tag{1.23}$$

If the operator commutes with H, it is time-independent in the Heisenberg representation, too. In particular, the Hamiltonian H remains unchanged under the transition from the Schrödinger to the Heisenberg representation and vice versa. It should be noted that though (1.23) formally reproduces (1.19), these two equations have quite different meanings. Relation (1.19) defines the derivative operator $\dot{Q}$, while (1.23) describes the time evolution of the Heisenberg operator Q_H.

1.4 Interaction Representation

Let us consider a system that consists of several interacting parts. The Hamiltonian of such a system may be written as a sum

$$H = H_0 + V \quad , \tag{1.24}$$

where H_0 is the part of the Hamiltonian that is independent of the interaction between the subsystems and V represents the interaction. It is convenient to describe the time evolution of states in such systems in the interaction representation.

Let us introduce a new wave function $\psi_I(t)$ defined by

$$\psi_I(t) = R_0^{-1}(t,0)\, \psi_S(t) \quad , \tag{1.25}$$

where the shift operator $R_0(t,0)$ is a solution of

$$i\hbar \frac{\partial R_0(t,0)}{\partial t} = H_0\, R_0(t,0) \tag{1.26}$$

with the initial condition $R_0(0,0) = 1$. The formal solution of (1.26) is

$$R_0(t,0) = \exp\left(-\frac{i}{\hbar}\, H_0 t\right) \quad . \tag{1.27}$$

Differentiating (1.25) with respect to time and making use of the Schrödinger equation (1.12), one finds the equation of motion in the interaction representation to be

$$i\hbar \frac{\partial \psi_I(t)}{\partial t} = V_I(t)\, \psi_I(t) \quad , \tag{1.28}$$

where $V_I(t)$ is the interaction operator in the interaction representation defined by

$$V_I(t) = R_0^{-1}(t,0)\, V_S\, R_0(t,0) = \exp\left(\frac{i}{\hbar}\, H_0 t\right) V_S\, \exp\left(-\frac{i}{\hbar}\, H_0 t\right) \quad . \tag{1.29}$$

Any operator time-dependence in the interaction representation is described by an analogous relation:

$$Q_I(t) = \exp\left(\frac{i}{\hbar}\, H_0 t\right) Q_S \exp\left(-\frac{i}{\hbar}\, H_0 t\right) \quad . \tag{1.30}$$

Differentiating the latter yields the operator time evolution in the interaction representation given by

$$i\hbar \frac{d}{dt}\, Q_I(t) = [Q_I(t),\, H_0] \quad . \tag{1.31}$$

The interaction representation is an intermediate one, lying between the Heisenberg and Schrödinger representations. The time-dependence of operators is similar to that of the Heisenberg representation for a system with no interactions, and the time evolution of the wave function is completely determined by the interaction. In what follows, we shall omit the subscript I when writing any quantities in the interaction representation.

1.5 Time-Dependent Green Functions

The time dependence of the wave function in the Schrödinger representation may be described, instead of by the time shift operator $R(t,0)$, in terms of the retarded and advanced Green functions $G^{(+)}(t)$ and $G^{(-)}(t)$. The latter possess the advantageous property that their spectral representations may be analytically continued to the complex plane.

The retarded and advanced Green functions $G^{(+)}(t)$ and $G^{(-)}(t)$ are determined by the equation

$$\left(i\hbar \frac{\partial}{\partial t} - H\right) G^{(\pm)}(t) = \delta(t) \quad , \tag{1.32}$$

where $\delta(t)$ is the usual delta function; the initial conditions are

$$G^{(+)}(t) = 0 \quad \text{for} \quad t < 0 \quad ,$$
$$G^{(-)}(t) = 0 \quad \text{for} \quad t > 0 \quad . \tag{1.33}$$

Formal solutions of (1.32) with the initial conditions (1.33) may be represented by the operators

$$G^{(+)}(t) = \begin{cases} 0 & t < 0 \\ -\frac{i}{\hbar} \exp\left(-\frac{i}{\hbar} Ht\right) & t > 0 \end{cases} \quad ; \tag{1.34}$$

$$G^{(-)}(t) = \begin{cases} \frac{i}{\hbar} \exp\left(-\frac{i}{\hbar} Ht\right) & t < 0 \\ 0 & t > 0 \end{cases} \quad . \tag{1.35}$$

Inasmuch as neither (1.32) nor (1.33) contain operators which do not commute with H, both functions $G^{(+)}(t)$ and $G^{(-)}(t)$ commute with the Hamiltonian. This also follows from the explicit expressions (1.34, 35). Since H is Hermitian, the functions $G^{(+)}(t)$ and $G^{(-)}(t)$ satisfy the relation

$$G^{(-)}(t) = \left[G^{(+)}(t)\right]^{+} \quad . \tag{1.36}$$

The function $G^{(+)}(t)$ is unitary for $t > 0$, and $G^{(-)}(t)$ is unitary for $t < 0$.

It is not difficult to verify that if the wave function $\psi(t)$ satisfies the Schrödinger equation (1.12), then the operator $G^{(+)}$ $\left(G^{(-)}\right)$ enables one to express the wave function $\psi(t')$ for any later (earlier) instant t' in terms of its value for t:

$$\psi(t') = i\hbar G^{(+)}(t' - t) \psi(t) \quad \text{for} \quad t' > t \quad , \tag{1.37}$$

and

$$\psi(t') = -i\,\hbar\,G^{(-)}(t' - t)\,\psi(t) \quad \text{for} \quad t' < t \quad . \tag{1.38}$$

Therefore, the operator $G^{(+)}$ describes the future evolution of the wave function of the system associated with the Hamiltonian H, and $G^{(-)}$ describes the evolution in the past. (The Green functions $G^{(+)}(t)$ and $G^{(-)}(t)$ are also called propagators.)

If the Hamiltonian is of the form (1.24), then one may introduce, in the same manner as $G^{(+)}(t)$ and $G^{(-)}(t)$, the retarded and advanced Green functions $G_0^{(+)}(t)$ and $G_0^{(-)}(t)$ for the nonperturbed Hamiltonian H_0. These functions may be written in a form analogous to (1.34, 35), so that

$$G_0^{(+)}(t) = \begin{cases} 0 & t < 0 \\ -\frac{i}{\hbar}\,\exp\left(-\frac{i}{\hbar}\,H_0 t\right) & t > 0 \end{cases} \quad ; \tag{1.39}$$

$$G_0^{(-)}(t) = \begin{cases} \frac{i}{\hbar}\,\exp\left(-\frac{i}{\hbar}\,H_0 t\right) & t < 0 \\ 0 & t > 0 \end{cases} \quad . \tag{1.40}$$

Making use of (1.32, 33) for the functions $G^{(+)}(t)$ and $G^{(-)}(t)$, and the analogous equation and conditions for $G_0^{(+)}(t)$ and $G_0^{(-)}(t)$, one easily obtains

$$G^{(\pm)}(t - t') = G_0^{(\pm)}(t - t') + \int_{-\infty}^{\infty} dt''\, G_0^{(\pm)}(t - t'')\, V G^{(\pm)}(t'' - t') \; . \tag{1.41}$$

This may be regarded as the integral equation that governs the Green functions $G^{(+)}(t)$ or $G^{(-)}(t)$ for given nonperturbed Green functions $G_0^{(+)}(t)$ or $G_0^{(-)}(t)$. Insofar as the Green functions $G^{(\pm)}(t)$ and $G_0^{(\pm)}(t)$ satisfy the condition (1.33), integration in (1.41) actually extends over a finite range. Therefore, (1.41) is a Volterra integral equation and hence has a unique solution.

Problems

1.1   Find the eigenfunctions of some operator $\hat{q}$ in its proper representation.

Suppose q_0 is an eigenvalue of the operator $\hat{q}$. The eigenfunctions of $\hat{q}$ in any representation are then governed by the equation

$$\hat{q}\,\psi_{q_0} = q_0\,\psi_{q_0} \quad . \tag{1.42}$$

The action of the operator $\hat{q}$ in its proper representation reduces to multiplication by the quantity q, which is the argument of the wave function ψ_{q_0}. So q may be regarded as the independent variable of the representation under consideration.

Let us rewrite (1.42) as

$$(q - q_0)\,\psi_{q_0}(q) = 0 \quad . \tag{1.43}$$

As follows from (1.43), $\psi_{q_0}(q) = 0$ for all $q \neq q_0$. The nonzero value of the function $\psi_{q_0}(q)$ associated with $q = q_0$ may be normalized in an arbitrary manner, e.g., taken to be equal to one. Introducing the Kronecker symbol, one may write the eigenfunctions of the operator $\hat{q}$ in its proper representation as

$$\psi_{q_0}(q) = \delta_{qq_0} \quad . \tag{1.44}$$

It is evident that functions (1.44) form an orthonormalized set:

$$\sum_q \psi_{q_0}^*(q)\, \psi_{q_0'}(q) = \delta_{q_0 q_0'} \quad . \tag{1.45}$$

If the quantity q takes a continuum of values, then the Kronecker symbol is to be replaced by the delta function

$$\delta_{qq_0} \rightarrow \delta(q - q_0) \quad .$$

1.2   Derive expressions for the position and momentum operators, as well as for the corresponding eigenfunctions normalized to delta functions, in the coordinate and momentum representations.

In the coordinate representation,

$$\hat{r} = r \quad , \qquad \hat{p} = -i\hbar \frac{\partial}{\partial r} \quad ; \tag{1.46}$$

$$\psi_{r_0}(r) = \delta(r - r_0) \quad , \qquad \psi_{p_0}(r) = \frac{1}{(2\pi\hbar)^{3/2}} \exp\left(\frac{i}{\hbar}\, p_0 r\right) \quad . \tag{1.47}$$

In the momentum representation,

$$\hat{r} = i\hbar \frac{\partial}{\partial p} \quad , \qquad \hat{p} = p \quad ; \tag{1.48}$$

$$\psi_{r_0}(p) = \frac{1}{(2\pi\hbar)^{3/2}} \exp\left(-\frac{i}{\hbar}\, p\, r_0\right) \quad ,$$
$$\psi_{p_0}(p) = \delta(p - p_0) \quad . \tag{1.49}$$

The operators $\hat{r}$ and $\hat{p}$ are described by mirror expressions (with the exception of the signs) in the canonically conjugated representations.

1.3   Show that the angular momentum eigenfunctions can be described by the same expression in the coordinate and momentum representations.

Let us write equations for the moment square and projection eigenfunctions in any arbitrary representation:

$$\hat{M}^2 \psi_{lm} = \hbar^2 (l+1) \psi_{lm} \quad ,$$
$$\hat{M}_z \psi_{lm} = \hbar m \psi_{lm} \quad , \tag{1.50}$$

where l and m are the quantum numbers of the square of the angular momentum and its projection onto a chosen axis. Obviously, the quantum numbers l and

m describe the state of the physical system and do not depend on the representation. Explicit forms of the operators $\hat{M}^2$ and $\hat{M}_z$, as well as those of the eigenfunctions, depend on the choice of the representaion.

Let us find the angular momentum operator in the momentum representation. Note that the operators $\hat{r}$ and $\hat{p}$ enter the expression for $\hat{M}$ symmetrically (with the exception of the sign), i.e.,

$$\hat{M} = \hat{r} \times \hat{p} \quad , \tag{1.51}$$

and that they are of mirror form (again with the exception of the sign) in the canonically conjugated representations (1.46, 48). Substituting (1.46, 48) into (1.51), one thus finds

$$\hat{M}_r = -i\hbar r \times \frac{\partial}{\partial r} \quad , \qquad \hat{M}_p = -i\hbar p \times \frac{\partial}{\partial p} \quad , \tag{1.52}$$

i.e., the angular momentum operator is expressed in terms of p and $\partial/\partial p$ in the momentum representation in the same manner as in terms of r and $\partial/\partial r$ in the coordinate representation. The square of the angular momentum and the projection operators may be written in the spherical coordinate system as

$$\hat{M}^2 = -\hbar^2 \left\{ \frac{1}{\sin \vartheta} \frac{\partial}{\partial \vartheta} \left(\sin \vartheta \frac{\partial}{\partial \vartheta} \right) + \frac{1}{\sin^2 \vartheta} \frac{\partial^2}{\partial \varphi^2} \right\} \quad , \tag{1.53}$$

$$\hat{M}_z = -i\hbar \frac{\partial}{\partial \varphi} \quad , \tag{1.54}$$

where the angles ϑ and φ describe the direction of the vector r in the coordinate and vector p in the momentum representation, respectively.

Since the equations for the angular momentum eigenfunctions are of the same form in the coordinate and momentum representations, their solutions, associated with the same l and m, have to coincide:

$$\psi_{lm}(\boldsymbol{n}_r) = Y_{lm}(\boldsymbol{n}_r) \quad , \qquad \boldsymbol{n}_r = \frac{\boldsymbol{r}}{r} \quad ; \tag{1.55}$$

$$\psi_{lm}(\boldsymbol{n}_p) = Y_{lm}(\boldsymbol{n}_p) \quad , \qquad \boldsymbol{n}_p = \frac{\boldsymbol{p}}{p} \quad . \tag{1.56}$$

Here $Y_{lm}(\boldsymbol{n}) \equiv Y_{lm}(\vartheta, \varphi)$ is the spherical function. Thus, the angular dependence of the angular momentum eigenfunctions is similar in the momentum and coordinate representations.

2. The Scattering Matrix and Transition Probability

2.1 The Scattering Matrix

One of the crucial problems in quantum theory is the study of particle interactions. In the following the particle is regarded as a physical system localized in space by the action of intrinsic forces. A two-particle collision is accompanied either by elastic scattering or by inelastic processes (inelastic scattering and various reactions). The scattering is referred to as elastic if the collision has no effect on the intrinsic states of the particles involved, whereas inelastic scattering modifies the intrinsic state of one or both particles; reactions lead to particle redistribution, i.e., the escaping particles are not the same as the initial ones. Here we restrict ourselves to reactions producing only two particles.

The collision process may be devided into three stages. In the first stage, the particles are widely spaced and their interaction may be ignored. The second stage is reached when the particles are sufficiently close to each other for their interaction to be appreciable and the onset of some process to occur. The third stage deals with the resultant products at distances sufficient for the interaction to be again regarded as negligible. A suitable way to describe scattering and reactions is to introduce the scattering operator S, whose matrix elements form the scattering matrix. (The scattering operator was introduced by *Heisenberg* [2.1].) The scattering matrix describes the relation between the initial and final states of the system, corresponding to stages one and three just described.

Let $\psi(-\infty)$ be the initial wave function associated with the relative motion and intrinsic states of particles at $t = -\infty$, and $\psi(\infty)$ be the final wave function describing the particle system after the collision, at $t = \infty$. The scattering matrix may be defined by means of the relation

$$\psi(\infty) = S\,\psi(-\infty) \quad . \tag{2.1}$$

Probability conservation immediately yields the unitarity condition for the scattering matrix

$$S^+S = 1 \quad , \tag{2.2}$$

i.e., $S^+ = S^{-1}$. Obviously, the final wave function of the system can coincide with the initial wave function only if the state of the system is not changed. Therefore, all probable transitions in the system, i.e., all possible scattering and reaction processes, are associated with the difference between the final wave function and the initial one:

$$\psi'(\infty) \equiv \psi(\infty) - \psi(-\infty) = \mathfrak{T}\,\psi(-\infty) \quad , \tag{2.3}$$

where $\mathfrak{T}$ is the transition operator,

$$\mathfrak{T} = S - 1 \quad . \tag{2.4}$$

In the general case of an arbitrary reaction, the total Hamiltonian H may be written as a sum in either of two ways:

$$H = H_0 + V = H_0' + V' \quad , \tag{2.5}$$

where the operators H_0 and H_0' describe the relative kinetic energy and intrinsic motion of the colliding and escaping particles, respectively, and V and V' are interaction potentials for colliding and excaping particles. For scattering (both elastic and inelastic), $H_0 = H_0'$ and $V = V'$.

Now we introduce the eigenfunctions of the operators $\dot{H}_0$ and H_0', defined by

$$H_0\,\varphi_\alpha = E_\alpha\,\varphi_\alpha \qquad \text{and} \qquad H_0'\,\varphi_\beta = E_\beta\,\varphi_\beta \quad , \tag{2.6}$$

where E_α and E_β are the eigenvalues of H_0 and H_0'. The functions φ_α and φ_β form complete orthonormalized sets which satisfy the conditions

$$(\varphi_\alpha\,\varphi_{\alpha'}) = \delta_{\alpha\alpha'} \quad , \qquad (\varphi_\beta\,\varphi_{\beta'}) = \delta_{\beta\beta'} \quad . \tag{2.7}$$

Note that the systems of functions φ_α and φ_β are not mutually orthogonal for reactions in which the nature of the colliding particles is changed. In the case of scattering, the systems of functions φ_α and φ_β coincide. Since the interaction range is finite, the potentials V and V' may be disregarded in the initial $\psi(-\infty)$ and final $\psi(\infty)$ states, respectively.

Expanding the initial and final state wave functions $\psi(-\infty)$ and $\psi(\infty)$ in terms of eigenfunctions of the operators H_0 and H_0', respectively, results in

$$\psi(-\infty) = \sum_\alpha c_\alpha\,\varphi_\alpha \quad , \qquad \psi(\infty) = \sum_\beta c_\beta\,\varphi_\beta \quad . \tag{2.8}$$

Substituting (2.8) into (2.1) yields

$$c_\beta = \sum_\alpha S_{\beta\alpha}\,c_\alpha \quad , \tag{2.9}$$

where

$$S_{\beta\alpha} = (\varphi_\beta\,, S\,\varphi_\alpha) \quad . \tag{2.10}$$

If the system is in a definite state φ_{α_0} at $t = -\infty$, then the expansion coefficient c_α in (2.8) is equal to $\delta_{\alpha\alpha_0}$. The coefficient c_β of the final wave function expansion (2.8) describes the probability of finding the system in the state β after the collision. If $c_\alpha = \delta_{\alpha\alpha_0}$, then

$$c_\beta = S_{\beta\alpha_0} \quad .$$

Thus, the probability that a system with the initial state α will be found in the state β after the collision is equal to

$$|c_\beta|^2 = |S_{\beta\alpha}|^2 \quad . \tag{2.11}$$

As follows from the unitary condition (2.2) for the scattering matrix,

$$\sum_\beta |S_{\beta\alpha}|^2 = 1 \quad , \tag{2.12}$$

i.e., the sum of all probabilities is equal to one.

According to (2.3), the transition probability for the initial and final states α and β is determined by the transition matrix element $\mathcal{T}_{\beta\alpha}$, i.e.,

$$W_{\alpha \to \beta} = |\mathcal{T}_{\beta\alpha}|^2 \quad . \tag{2.13}$$

We emphasize the difference between the probability (2.11) of finding the system in the state β after the collision if it was initially in the state α, and the probability (2.13) of the transition from the state α to the state β. The two probabilities coincide only for mutually orthogonal states, i.e., if $(\varphi_\alpha, \varphi_\beta) = 0$.

The probable state of the system is called the reaction channel: α is the entrance channel, β is the exit channel. The channel is described by the energy of the relative particle motion and quantum numbers associated with particle intrinsic states. It is clear that the scattering matrix element $S_{\beta\alpha}$ describes the relation between various channels in the system. The entrance and exit channels coincide for elastic scattering ($\alpha = \beta$); if the exit channel differs from the entrance ($\alpha \neq \beta$), either inelastic scattering or a reaction occurs.

For a clear understanding of the time evolution of the process, we only consider scattering processes which do not change the nature of the particles involved in the collision ($H_0 = H_0'$ and $V = V'$). In this case, the scattering matrix may be immediately related to the time shift operator in the interaction representation.

2.2 Time Shift Operator
in the Interaction Representation

Let us introduce the time shift operator in the interaction representation $U(t, t_0)$. It transforms the wave function for t_0 into that for t; hence

$$\psi(t) = U(t, t_0)\, \psi(t_0) \quad . \tag{2.14}$$

According to (1.28), this operator is determined by the differential equation

$$i\hbar\, \frac{\partial U(t, t_0)}{\partial t} = V(t)\, U(t, t_0) \tag{2.15}$$

and the initial condition

$$U(t_0, t_0) = 1 \quad . \tag{2.16}$$

Time integration of (2.15) with the initial condition (2.16) reduces it to the integral equation

$$U(t, t_0) = 1 - \frac{i}{\hbar} \int_{t_0}^{t} dt'\, V(t')\, U(t', t_0) \quad . \tag{2.17}$$

By means of successive iteration, the solution of (2.17) may be written as an infinite series

$$\begin{aligned}
U(t, t_0) = 1 &+ \left(-\frac{i}{\hbar}\right) \int_{t_0}^{t} dt_1\, V(t_1) \\
&+ \left(-\frac{i}{\hbar}\right)^2 \int_{t_0}^{t} dt_1 \int_{t_0}^{t_1} dt_2\, V(t_1)\, V(t_2) + \ldots \\
&= \sum_{n=0}^{\infty} \left(-\frac{i}{\hbar}\right)^n \int_{t_0}^{t} dt_1 \int_{t_0}^{t_1} dt_2 \ldots \\
&\times \int_{t_0}^{t_{n-1}} dt_n\, V(t_1)\, V(t_2) \ldots V(t_n) \quad .
\end{aligned} \tag{2.18}$$

The latter expansion may be reduced to symmetric form by the chronologic ordering operator P, introduced by *Dyson* [2.2]:

$$P\big(V(t_1)\, V(t_2) \ldots V(t_n)\big) = V(t_i)\, V(t_j) \ldots V(t_r) \quad , \tag{2.19}$$
$$t_i > t_j > \ldots > t_r \quad .$$

P rearranges any product of time-dependent operators, so that the time arguments of the factors decrease from left to right.

After P has been applied to the integrand of (2.18), all the integration ranges may be extended to cover the range t_0 to t. However, since the integral is symmetric with respect to permutations of the integration variables, the result

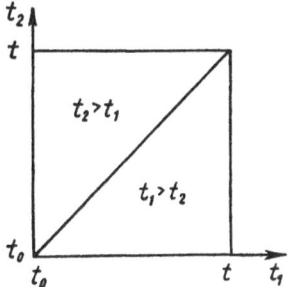

Fig. 2.1. Ranges of integration of the chronologic product of two operators

obtained is $n!$ times greater than the correct one. To show this we consider the double integral

$$I_2 = \int_{t_0}^{t} dt_1 \int_{t_0}^{t} dt_2 \, P\big(V(t_1)\, V(t_2)\big) \quad .$$

Since the chronologic product in the integrand of I_2 is

$$P\big(V(t_1)\, V(t_2)\big) = \begin{cases} V(t_1\, V(t_2) & \text{for} \quad t_1 > t_2 \\ V(t_2)\, V(t_1) & \text{for} \quad t_1 < t_2 \end{cases} \quad ,$$

the integration range in I_2 may be divided into two parts $t_1 > t_2$ and $t_1 < t_2$ (Fig. 2.1). Then

$$I_2 = \int_{t_0}^{t} dt_1 \int_{t_0}^{t_1} dt_2 \, V(t_1)\, V(t_2) + \int_{t_0}^{t} dt_2 \int_{t_0}^{t_2} dt_1 \, V(t_2)\, V(t_1) \quad .$$

Interchanging the integration variables $(t_1 \rightleftarrows t_2)$ in the second term of I_2, we find that

$$I_2 = 2! \int_{t_0}^{t} dt_1 \int_{t_0}^{t_1} dt_2 \, V(t_1)\, V(t_2) \quad .$$

The n-fold integral in (2.18) may be dealt with in a similar manner.

Thus, the time shift operator expansion may be presented in the form

$$U(t, t_0) = \sum_{n=0}^{\infty} \frac{1}{n!} \left(-\frac{i}{\hbar}\right)^{n} \int_{t_0}^{t} dt_1 \int_{t_0}^{t} dt_2 \dots \int_{t_0}^{t} dt_n$$
$$\times P\big(V(t_1)\, V(t_2) \dots V(t_n)\big) \quad . \tag{2.20}$$

The formal summation of the infinite series (2.20) yields a short expression for the time shift operator, i.e.,

$$U(t, t_0) = P \exp\left[-\frac{i}{\hbar} \int_{t_0}^{t} dt' \, V(t')\right] \quad . \tag{2.21}$$

Because the interaction operator $V(t)$ is Hermitian, the time shift operator $U(t, t_0)$ satisfies the relation

$$U^+(t, t_0) = U(t_0, t) \quad .$$

The time shift operator $U(t, t_0)$ possesses the group property

$$U(t, t') U(t', t_0) = U(t, t_0) \quad . \tag{2.22}$$

Substituting $t = t_0$ in (2.22), we obtain

$$U(t_0, t) U(t, t_0) = 1 \quad ,$$

or

$$U^+(t, t_0) = U^{-1}(t, t_0) \quad . \tag{2.23}$$

Therefore, the time shift operator $U(t, t_0)$ is unitary.

Having employed the definition (2.14) of the time shift operator in the interaction representation, the scattering matrix may be defined by means of the relation

$$S = \lim_{\substack{t_0 \to -\infty \\ t \to \infty}} U(t, t_0) \quad .$$

According to (2.21), we then have

$$S = P \exp\left[\frac{-i}{\hbar} \int_{-\infty}^{\infty} dt \, V(t)\right] \quad , \tag{2.24}$$

i.e., the scattering matrix is expressed directly in terms of the colliding particles' interaction operator in the interaction representation

$$V(t) = \exp\left(\frac{i}{\hbar} H_0 t\right) V \exp\left(-\frac{i}{\hbar} H_0 t\right) \quad . \tag{2.25}$$

The interaction $V(t)$ is usually assumed to vanish for $t \to \mp\infty$, a requirement that may be satisfied by using

$$V \to V e^{-\varepsilon t} \qquad (\varepsilon > 0) \tag{2.26}$$

in (2.25) and then approaching the limit $\varepsilon \to 0$. The substitution (2.26) implies that the interaction appears adiabatically for $t = -\infty$ and dissappears for $t = \infty$. Adiabatic appearance and disappearance of the interaction ensures unitarity of the scattering operator S.

It should be mentioned here that the S matrix may be written in the form (2.24) only in the case of scattering, i.e., when $V' = V$. Equation (2.24) does not hold for processes accompanied by particle redistribution, when the exit channel interaction V' differs from the entrance one V. The explicit S matrix form for the latter case may be obtained by simple extension. Indeed, inasmuch as the process is symmetric with respect to the entrance and exit channels, the time instant before which V is dominant and after which V' becomes important, may always be taken as zero. Therefore, one may assume the process to be governed by the entrance interaction V for the time interval $-\infty$ to 0 and by the exit interaction

V' during the time 0 to ∞. Within the context of the group property (2.22) of the time shift operator, the scattering matrix for a process accompanied by particle redistribution may be written as

$$S = P \exp\left\{ -\frac{i}{\hbar} \left[\int_0^\infty dt\, V'(t) + \int_{-\infty}^0 dt\, V(t) \right] \right\} \quad . \tag{2.27}$$

Here $V(t)$ is assumed to vanish for $t \to -\infty$, and $V'(t) \to 0$ for $t \to \infty$.

2.3 Integrals of Motion and S Matrix Diagonalization

The integrals of motion in quantum mechanics are the dynamical variables whose operators Q satisfy the condition

$$\frac{dQ}{dt} = 0 \quad . \tag{2.28}$$

In the Schrödinger representation, this reduces to

$$\frac{dQ}{dt} \equiv \frac{\partial Q}{\partial t} - \frac{i}{\hbar} [Q, H] = 0 \quad . \tag{2.29}$$

If Q does not depend on time explicitly, then (2.29) reduces to the requirement that the operator Q must commute with the Hamiltonian of the system,

$$[Q, H] = 0 \quad , \tag{2.30}$$

which, according to (2.21), means that (2.28) is satisfied in the Heisenberg representation, too. In the interaction representation, (2.30) yields $[Q, H_0] = -[Q, V]$ and, if $[Q, H_0] = 0$, then $[Q, V] = 0$. It is not difficult to verify that, by virtue of the scattering matrix definition (2.24),

$$[Q, S] = 0 \quad . \tag{2.31}$$

Operators which commute can be reduced to the diagonal form simultaneously. Therefore, taking the indices α and β to be the integrals of motion of the system q (e.g., energy, angular momentum, isotopic spin, etc.), one may transform the S matrix to the diagonal form

$$S_{\beta\alpha} = \langle \gamma' | S^q | \gamma \rangle\, \delta_{qq'} \quad , \tag{2.32}$$

where γ includes all the quantum numbers other than q. Note that some integrals of motion do not commute, therefore all the integrals of motion cannot be simultaneously represented by α.

The S matrix can be shown to be independent of the integrals of motion which are not invariant with respect to the same transformations as the Hamiltonian. For example, in the case of the field with central symmetry, the S matrix cannot depend on the projection of the total moment of the system. Indeed, S commutes with all projections of the total moment I:

$$[I, S] = 0 \quad .$$

With the projection onto the z axis being denoted by m, the commutation condition for the scattering matrix and the moment projection I_z

$$I_z S = S I_z$$

may be written in the matrix form

$$\sum_{m''} \langle m' | I_z | m'' \rangle \langle m'' \gamma' | S | \gamma m \rangle = \sum_{m''} \langle m' \gamma' | S | \gamma m'' \rangle \langle m'' | I_z | m \rangle \quad .$$

Since $\langle m'' \gamma' | S | \gamma m \rangle = \langle \gamma' | S^m | \gamma \rangle \delta_{mm'}$, we have

$$\langle m' | I_z | m \rangle \langle \gamma' | S^m | \gamma \rangle = \langle \gamma' | S^m | \gamma \rangle \langle m' | I_z | m \rangle \quad .$$

The matrix element I_z is off-diagonal in the representation under consideration; therefore, S^m does not depend on m.

The latter property of the S matrix is a direct corollary of the spatial isotropy. Indeed, $S_{\beta\alpha}$ determines the transition probability in the system and cannot depend on the choice of the coordinate system, whereas the projection of the total moment varies under coordinate system rotations. One may show in a similar manner that the scattering matrix S cannot depend on the total momentum of the system, the isotopic spin projection, etc.

2.4 Transition Probability per Unit Time

Let us find the transition probability of a system per unit time for two arbitrary states. Making use of the definition (2.4) and bearing in mind that the energy is an integral of motion, one may write the transition matrix in the form

$$\mathcal{T}_{\beta\alpha} = -2\pi i \, t_{\beta\alpha} \, \delta(E_\alpha - E_\beta) \quad , \tag{2.33}$$

where the matrix element $t_{\beta\alpha}$ corresponds to the states α and β associated with the same energy; $t_{\beta\alpha}$ is the matrix element on the energy shell. The constant factor $-2\pi i$ is included in (2.33) for the sake of convenience.

According to (2.13), the transition probability over an infinite time period, $W_{\alpha \to \beta}$, is

$$W_{\alpha\to\beta} = (2\pi)^2 \, |t_{\beta\alpha}|^2 \, \big[\delta(E_\alpha - E_\beta)\big]^2$$

$$= (2\pi)^2 \, |t_{\beta\alpha}|^2 \, \delta(E_\alpha - E_\beta)$$

$$\times \lim_{T\to\infty} \frac{1}{2\pi\hbar} \int_{-T/2}^{T/2} dt \, \exp\left[\frac{i}{\hbar}(E_\alpha - E_\beta)t\right]$$

$$= \frac{2\pi}{\hbar} \, |t_{\beta\alpha}|^2 \, \delta(E_\alpha - E_\beta) \lim_{T\to\infty} T \quad .$$

In order to find the transition probability per unit time $w_{\alpha\to\beta}$, we divide the total probability $W_{\alpha\to\beta}$ by the total transition time T:

$$w_{\alpha\to\beta} = \frac{2\pi}{\hbar} \, |t_{\beta\alpha}|^2 \, (E_\alpha - E_\beta) \quad . \tag{2.34}$$

If the final state lies in the continuum part of the spectrum, the probability that the transition will produce a state in the elementary phase volume $\Delta\Gamma_\beta$ is

$$\Delta w_{\alpha\to\beta} = \frac{2\pi}{\hbar} \, |t_{\beta\alpha}|^2 \, \delta(E_\alpha - E_\beta)\Delta\Gamma_\beta \quad . \tag{2.35}$$

After introducing the number of final states with respect to the unit energy interval,

$$\Delta\varrho_\beta = \frac{\Delta\Gamma_\beta}{\Delta E_\beta} \quad , \tag{2.36}$$

and carrying out energy integration in (2.35) with regard to the delta function, one finds the transition probability per unit time to be

$$\Delta w_{\alpha\to\beta} \equiv \int dE_\beta \, (\Delta w_{\alpha\to\beta})/\Delta E_\beta$$

$$= \frac{2\pi}{\hbar} \, |t_{\beta\alpha}|^2 \, \Delta\varrho_\beta \quad , \quad E_\beta = E_\alpha \quad . \tag{2.37}$$

The cross section $\Delta\sigma_{\alpha\to\beta}$ of the process is defined as the ratio of the transition probability per unit time $\Delta w_{\alpha\to\beta}$ to the incident flux density j_α, i.e.,

$$\Delta\sigma_{\alpha\to\beta} = \frac{\Delta w_{\alpha\to\beta}}{j_\alpha} \quad . \tag{2.38}$$

If we consider the particle scattering to be an example of a transition in a continuum, then both initial and final states α and β are associated with definite values of the relative particle momentum $p = \hbar k$ and $p' = \hbar k'$ (the rest of quantum numbers will be denoted by γ and γ'). Let the relevant relative motion functions be the plane waves

$$\varphi_k = e^{ikr} \quad \text{and} \quad \varphi_{k'} = e^{ik'r} \tag{2.39}$$

which are normalized to unit particle density. Then the incident flux density is $j_\alpha = \hbar k/\mu$ and the total energy of the system is $E_\alpha = (\hbar^2 k^2/2\mu) + E_\gamma$, where

μ is the reduced mass of the colliding particles and E_γ is the intrinsic energy of the particle. The number of final states per unit volume (the final state density) is $\Delta \Gamma_\beta = \Delta k'/(2\pi)^3$. Since $E_\beta = (\hbar^2 k'^2/2\mu') + E_{\gamma'}$, where μ' is the reduced mass and $E_{\gamma'}$ is the intrinsic energy of escaping particles, the final state density per unit energy interval is

$$\Delta \varrho_\beta = \frac{\mu' k'}{(2\pi)^3 \hbar^2} \Delta o \quad ,$$

where Δo is the elementary solid angle. After applying (2.37, 38), one obtains the differential scattering cross section $d\sigma_{\gamma \to \gamma'}$ for an elementary solid angle do,

$$d\sigma_{\gamma \to \gamma'} = \frac{\mu \mu'}{4\pi^2 \hbar^4} \frac{k'}{k} \left| \langle k' \gamma' | t | \gamma k \rangle \right|^2 do \quad , \tag{2.40}$$

where $\langle k' | t | k \rangle$ is the transition operator matrix element over the states (2.39); the quantities k and k' are related by the energy conservation law

$$\frac{\hbar^2 k^2}{2\mu} + E_\gamma = \frac{\hbar^2 k'^2}{2\mu'} + E_{\gamma'} \quad . \tag{2.41}$$

It is sometimes convenient to employ an alternative normalization of the states (2.39), namely, to normalize the wave functions to the delta function of energy. For the wave functions φ_{En} and $\varphi_{E'n'}$ (where E and E' are particle kinetic energies, n and n' are unit vectors along the directions of k and k'), the normalization condition may be written as

$$\int d\mathbf{r} \, \varphi_{E'n'}^* \varphi_{En} = \delta(E - E') \, \delta(n - n') \quad . \tag{2.42}$$

The functions φ_{En} differ from φ_k only by a normalization factor,

$$\varphi_{En} = C \varphi_k \quad . \tag{2.43}$$

Since

$$\int d\mathbf{r} \, \varphi_{k'}^* \varphi_k = (2\pi)^3 \, \delta(k - k')$$

and

$$\delta(k - k') = \frac{1}{k^2} \frac{dE}{dk} \delta(E - E') \, \delta(n - n') \quad ,$$

we find

$$C^2 = \frac{\mu k}{8\pi^3 \hbar^2} \quad . \tag{2.44}$$

Introducing the matrix element of the transition operator t over the states φ_{En} and $\varphi_{E'n'}$, $\langle n'E' | t | En \rangle$, we rewrite the scattering cross section (2.40) as

$$do_{\gamma \to \gamma'} = \frac{(2\pi)^4}{k^2} \langle n'\gamma'|t|\gamma n\rangle^2 \, do \quad . \tag{2.45}$$

The indices E and E' may be omitted in the matrix element since the energies E and E' satisfy (2.41).

2.5 Integral Equation for the t Matrix

In order to derive an integral equation for the transition operator t, we employ the scattering matrix expansion in a power series of the interaction. According to (2.18, 23), the scattering matrix may be written as a series

$$S = \sum_{n=0}^{\infty} S^{(n)} \quad ,$$

$$S^{(n)} = \left(-\frac{i}{\hbar}\right)^n \int_{-\infty}^{\infty} dt_1 \int_{-\infty}^{t_1} dt_2 \cdots \tag{2.46}$$

$$\times \int_{-\infty}^{t_{n-1}} dt_n \, V(t_1) \, V(t_2) \cdots V(t_n) \quad ,$$

with the interaction $V(t)$ being given by (2.25, 26). Expansion (2.46) may be rewritten in the matrix form,

$$S_{\beta\alpha} = \sum_{n=0}^{\infty} S_{\beta\alpha}^{(n)} \quad , \tag{2.47}$$

and we assume the states α and β to be the eigenstates of the nonperturbed Hamiltonian H_0.

In the zeroth approximation, we have

$$S_{\beta\alpha}^{(0)} = \delta_{\alpha\beta} \quad . \tag{2.48}$$

The first expansion term is equal to

$$S_{\beta\alpha}^{(1)} = -\frac{i}{\hbar} \int_{-\infty}^{\infty} dt \, \exp\left[\frac{i}{\hbar}(E_\beta - E_\alpha)t - \varepsilon|t|\right] \langle \beta|V|\alpha\rangle$$

$$= -2\pi i \, \delta(E_\alpha - E_\beta) \langle \beta|V|\alpha\rangle \quad . \tag{2.49}$$

Since there are no singularities in (2.49), the parameter ε may immediately tend to zero. Formula (2.49) describes the first-order matrix element of the traditional perturbation theory.

Having introduced the intermediate states γ, we present the second expansion term as

$$S_{\beta\alpha}^{(2)} = \left(-\frac{i}{\hbar}\right)^2 \sum_{\gamma} \int_{-\infty}^{\infty} dt_1 \, \exp\left[\frac{i}{\hbar}(E_\beta - E_\alpha)t_1\right]$$

$$\times \int_{-\infty}^{t_1} dt_2 \, \exp\left[\frac{i}{\hbar}(E_\gamma - E_\alpha)t_2 - \varepsilon|t_2|\right] \langle \beta|V|\gamma\rangle \langle \gamma|V|\alpha\rangle \quad . \tag{2.50}$$

Since

$$\int_{-\infty}^{t_1} dt_2 \, \exp\left[\frac{i}{\hbar}(E_\gamma - E_\alpha)t_2 - \varepsilon|t_2|\right] = \frac{\hbar}{i} \frac{\exp\left[(i/\hbar)(E_\gamma - E_\alpha)t_1\right]}{E_\gamma - E_\alpha - i\varepsilon} \qquad (2.51)$$

[to prove (2.51) for $t_1 > 0$, the identity

$$(x - i\varepsilon)^{-1} = (x + i\varepsilon)^{-1} + 2\pi i \, \delta(x)$$

is required], we have

$$S_{\beta\alpha}^{(2)} = -2\pi i \, \delta(E_\alpha - E_\beta) \sum_\gamma \frac{1}{E_\alpha - E_\gamma + i\varepsilon} \langle\beta|V|\gamma\rangle \langle\gamma|V|\alpha\rangle$$

$$= -2\pi i \, \delta(E_\alpha - E_\beta) \left\langle \beta \left| V \frac{1}{E_\alpha - H_0 + i\varepsilon} V \right| \alpha \right\rangle \quad . \qquad (2.52)$$

The procedure for the calculation of the matrix elements of other expansion terms in (2.46) is analogous, and the resultant scattering matrix may be easily shown to be

$$S_{\beta\alpha} = \delta_{\beta\alpha} - 2\pi i \, (E_\alpha - E_\beta) \langle\beta|t|\alpha\rangle \quad , \qquad (2.53)$$

where the operator t of the transition on the energy shell is given by the infinite series

$$t = V + V \frac{1}{E - H_0 + i\varepsilon} V + V \frac{1}{E - H_0 + i\varepsilon} V \frac{1}{E - H_0 + i\varepsilon} V + \dots \; . (2.54)$$

This infinite series may be easily proved to be equivalent to the integral equation

$$t = V + V \frac{1}{E - H_0 + i\varepsilon} t \quad , \qquad \varepsilon \to 0 \quad . \qquad (2.55)$$

We have already shown that the transition operator matrix elements completely determine the transition probabilities in the system. Therefore, the scattering problem may be formally reduced to solving the integral equation for the transition operator t.

2.6 Transformation of the Scattering Matrix. Cross Sections

Let us consider in detail relations between the scattering matrix and quantities which can be observed experimentally, such as cross sections and angular distributions. We have already shown that the squared modulus of the scattering matrix element $S_{\beta\alpha}$ is related to the transition probability between the initial state α and the final one β. Taking α and β as the angles associated with particle motion directions, one finds that the squared modulus of $S_{\beta\alpha}$ characterizes the probability of observing particles moving along a given direction.

However, it is often easier to find the scattering matrix in a representation other than the one directly associated with the quantities measured. Then the conservation laws lead to some restrictions on the form of the scattering matrix. If the sets α and β contain quantum numbers that correspond to the integrals of motion, then the scattering matrix is diagonal with respect to these. Therefore, if we are interested in the nature of the angular distribution determined by the conservation laws, the scattering matrix must be transformed from the representation given by the quantum numbers of the integrals of motion to the one associated with the angles.

The transformation of the S matrix given by some set of quantum numbers into the S matrix given by another set is determined by the general formula (1.9). In particular, the transition from the angular momentum representation to the one given by the particle escape angles is described by the expression

$$\langle n'|S|n\rangle = \sum_{lml'm'} \langle n'|l'm'\rangle \, \langle l'm'|S|lm\rangle \, \langle lm|n\rangle \quad , \tag{2.56}$$

where the transformation functions are

$$\langle n'|l'm'\rangle = Y_{l'm'}(n') \quad ,$$

$$\langle lm|n\rangle = Y_{lm}^*(n) \quad .$$

Since the S matrix is diagonal,

$$\langle l'm'|S|lm\rangle = S^l \, \delta_{ll'} \, \delta_{mm'} \quad ,$$

and since in the case of a centrally symmetric field the elements S^l do not depend on the quantum number m of the projection of the angular momentum, one may employ the addition formula for spherical functions,

$$\sum_{m=-l}^{l} Y_{lm}^*(n) \, Y_{lm}(n') = \frac{2l+1}{4\pi} \, P_l(\cos\vartheta) \tag{2.57}$$

(ϑ is the angle between the vectors n and n'). As a result, we obtain

$$\langle n'|S|n\rangle = \frac{1}{4\pi} \sum_{l} (2l+1) \, S^l \, P_l(\cos\vartheta) \quad , \tag{2.58}$$

and, if the transition is inelastic, we have

$$\langle n'\beta|S|\alpha n\rangle = \frac{1}{4\pi} \sum_{l} (2l+1) \, S^l \, P_l(\cos\vartheta) \quad , \tag{2.59}$$

where α and β are all other quantum numbers describing the initial and final states.

The relation between the transition matrix t and the scattering matrix S on the energy shell is

$$S = 1 - 2\pi i t \quad . \tag{2.60}$$

Thus, within the context of (2.45), we find that

$$d\sigma_{\alpha\to\beta} = \frac{1}{4k^2} \left| \sum_l (2l+1) \left(\delta_{\alpha\beta} - S^l_{\beta\alpha} \right) P_l(\cos\vartheta) \right|^2 do \quad . \tag{2.61}$$

Formula (2.61) describes the angular distribution of the reaction products for an arbitrary transition from the state α into the one β.

Taking $\beta = \alpha$, we obtain from (2.61) the differential cross section of the elastic scattering,

$$d\sigma_e = \frac{1}{4k^2} \left| \sum_l (2l+1)(1 - S^l_{\alpha\alpha}) P_l(\cos\vartheta) \right|^2 do \quad . \tag{2.62}$$

To find the elastic scattering integral cross section, we integrate (2.62) over the angles, then

$$\sigma_e = \frac{\pi}{k^2} \sum_{l=0}^{\infty} (2l+1) \left| 1 - S^l_{\alpha\alpha} \right|^2 \quad . \tag{2.63}$$

When deriving (2.63), we employed the orthonormalization condition for the Legendre polynomials

$$\int do\, P_l(\cos\vartheta) P_{l'}(\cos\vartheta) = \frac{4\pi}{2l+1} \delta_{ll'} \quad . \tag{2.64}$$

Integrating (2.61) over the angles and carrying out summation over all the final states β other than the initial one ($\beta \neq \alpha$), we obtain the total reaction cross section

$$\sigma_r = \frac{\pi}{k^2} \sum_{l=0}^{\infty} (2l+1) \sum_{\beta\neq\alpha} \left| S^l_{\alpha\alpha} \right|^2 \quad . \tag{2.65}$$

The unitarity condition for the scattering matrix, (2.12), enables one to express the reaction cross section σ_r in terms of the scattering matrix diagonal element $S^l_{\alpha\alpha}$:

$$\sigma_r = \frac{\pi}{k^2} \sum_{l=0}^{\infty} (2l+1) \left(1 - \left| S^l_{\alpha\alpha} \right|^2 \right) \quad . \tag{2.66}$$

Thus, knowledge of the scattering matrix diagonal element $S^l_{\alpha\alpha}$ is sufficient for obtaining both elastic scattering and reaction (all inelastic processes) cross sections σ_e and σ_r.

According to (2.63, 66), the elastic scattering and reaction cross sections σ_e and σ_r may be written as sums of partial cross sections $\sigma_e^{(l)}$ and $\sigma_r^{(l)}$ which, respectively, describe elastic scattering and reactions in the state with definite angular momentum l, i.e.,

$$\sigma_e = \sum_l \sigma_e^{(l)} \quad , \qquad \sigma_e^{(l)} = \frac{\pi}{k^2}(2l+1)\left|1 - S_{\alpha\alpha}^l\right|^2 \quad ; \tag{2.67}$$

$$\sigma_r = \sum_l \sigma_r^{(l)} \quad , \qquad \sigma_r^{(l)} = \frac{\pi}{k^2}(2l+1)\left(1 - \left|S_{\alpha\alpha}^l\right|^2\right) \quad . \tag{2.68}$$

The reaction cross section (2.66) vanishes if the scattering matrix diagonal element modulus is equal to one, $\left|S_{\alpha\alpha}^l\right| = 1$, i.e., if

$$S_{\alpha\alpha}^l = \exp(2\mathrm{i}\delta_l) \quad , \tag{2.69}$$

where δ_l is a real function of energy, referred to as the scattering phase shift at infinity. In this case, only elastic scattering occurs. The relevant cross section is expressed directly in terms of the scattering phase shift:

$$\sigma_e = \sum_l \sigma_e^{(l)} \quad ,$$

$$\tag{2.70}$$

$$\sigma_e^{(l)} = \frac{4\pi}{k^2}(2l+1)\sin^2\delta_l \quad .$$

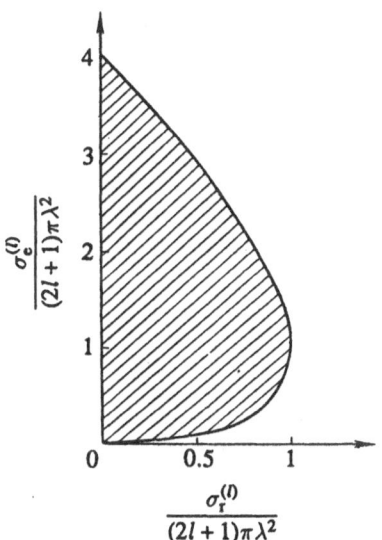

Fig. 2.2. Range of allowed partial cross sections $\sigma_e^{(l)}$ and $\sigma_r^{(l)}$

Since $\sin^2\delta_l \leq 1$, the partial cross section does not exceed the unitary limit,

$$\sigma_e^{(l)} \leq \frac{4}{k^2}(2l+1) \quad ,$$

which is attained only if the scattering phase shift is equal to an odd multiple of $\pi/2$.

With the inclusion of inelastic transitions, $\left|S_{\alpha\alpha}^l\right| < 1$ and the reaction cross section is not equal to zero. Elastic scattering always occurs, too, since then

$\left|1 - S_{\alpha\alpha}^l\right|$ is not zero and thus the elastic scattering cross section does not vanish. Figure 2.2 shows the range of allowed partial cross sections $\sigma_e^{(l)}$ and $\sigma_r^{(l)}$.

Problems

2.1   Find the relation between the differential scattering cross sections for a two-particle collision in the laboratory coordinate system and the center-of-mass system.

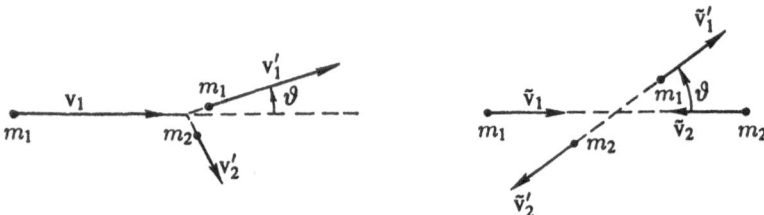

Fig. 2.3. Scattering angles in the laboratory and center-of-mass coordinate systems

Suppose a particle with mass m_1 and velocity v_1 in the laboratory coordinate system (l.s.) collides with a stationary particle of mass m_2. Let us denote the particle velocities after the collision by v_1' and v_2'. Then the scattering angle in the l.s. is the angle ϑ between the vectors v_1' and v_2' (Fig. 2.3). Since $v_2 = 0$, the velocity of the center-of-mass system (c.m.s.) with respect to the l.s. is given by

$$u = \frac{m_1}{m_1 + m_2}\, v_1 \quad . \tag{2.71}$$

The particle velocities before the collision in the c.m.s. are

$$\tilde{v}_1 = \frac{m_2}{m_1 + m_2}\, v_1 \quad , \qquad \tilde{v}_2 = \frac{m_1}{m_1 + m_2}\, v_1 \quad , \tag{2.72}$$

and the relative velocity of the colliding particles is equal to v_1:

$$\tilde{v}_1 - \tilde{v}_2 = v_1 \quad . \tag{2.73}$$

By virtue of energy conservation, the c.m.s. particle velocities maintain the same absolute values after the collision but may change their directions, i.e.,

$$|\tilde{v}_1'| = |\tilde{v}_1| \quad , \qquad |\tilde{v}_2'| = |\tilde{v}_2| \quad , \tag{2.74}$$

and, due to momentum conservation, the final particle velocities have opposite directions. We denote the angle between $\tilde{v}_1'$ and $\tilde{v}_1$ as $\tilde{\vartheta}$, which is the scattering angle in the c.m.s.

In order to find the relation between the scattering angles in the l.s. and c.m.s., we take the projections of the vector equation

$$v_1' = \tilde{v}_1' + u$$

parallel and perpendicular to v_1. The ratio of the equations obtained yields

$$\tan \vartheta = \frac{\sin \tilde{\vartheta}}{\gamma + \cos \tilde{\vartheta}} \quad , \tag{2.75}$$

where γ is the mass ratio, $\gamma = m_1/m_2$. If the particle masses are equal, then $\gamma = 1$, so that (2.75) is simplified to

$$\vartheta = \tfrac{1}{2} \tilde{\vartheta} \quad , \tag{2.76}$$

i.e., in the collision of two particles with equal masses, the scattering angle in the c.m.s. is twice the scattering angle in the l.s.

By virtue of the Lorentz invariance of the total cross section, we have

$$\sigma(\vartheta)\, do = \tilde{\sigma}(\tilde{\vartheta})\, d\tilde{o} \quad , \tag{2.77}$$

where $do = \sin \vartheta\, d\vartheta\, d\varphi$ and $d\tilde{o} = \sin \tilde{\vartheta}\, d\tilde{\vartheta}\, d\tilde{\varphi}$. Taking $\tilde{\varphi} = \varphi$, we find within the context of (2.75) that

$$\sin \vartheta\, d\vartheta = \frac{1 + \gamma \cos \tilde{\vartheta}}{\left(1 + 2\gamma \cos \tilde{\vartheta} + \gamma^2\right)^{3/2}} \sin \tilde{\vartheta}\, d\tilde{\vartheta} \tag{2.78}$$

Thus we obtain the relation between the l.s. and c.m.s. scattering cross sections $\sigma(\vartheta)$ and $\tilde{\sigma}(\tilde{\vartheta})$,

$$\sigma(\vartheta) = \frac{\left(1 + 2\gamma \cos \tilde{\vartheta} + \gamma^2\right)^{3/2}}{1 + \gamma \cos \tilde{\vartheta}}\, \tilde{\sigma}(\tilde{\vartheta}) \quad . \tag{2.79}$$

Formula (2.79) is simplified for $\gamma = 1$, i.e., if the particles involved in the collision have equal masses:

$$\sigma(\vartheta) = 4 \left(\cos \frac{\tilde{\vartheta}}{2} \right) \tilde{\sigma}(\tilde{\vartheta}) \quad . \tag{2.80}$$

2.2 Derive the Lorentz tranformation law for the matrix elements of the operator S for states with fixed energies and momenta.

Suppose the system K moves with respect to the system K' with constant velocity u. The wave function ψ is transformed into ψ' by transition from K to K':

$$\psi' = L \psi \quad , \tag{2.81}$$

where L is a unitary operator that produces the Lorentz transformation. The operators are transformed according to the law

$$S' = LSL^{-1} \quad , \tag{2.82}$$

and the Lorentz-invariance of the scattering operator implies that

$$S' = S \quad . \tag{2.83}$$

However, the choice of the initial and final states may result in changes of the matrix elements of S under the transition from K to K'. Let the initial and final states ψ_α and ψ_β be determined in K by given particle energies and momenta,

$$\alpha \equiv p_1 , E_1 ; \; p_2 , E_2 ; \; \ldots \quad ,$$

$$\tag{2.84}$$

$$\beta \equiv \tilde{p}_1 , \tilde{E}_1 ; \; \tilde{p}_2 , \tilde{E}_2 ; \; \ldots \quad .$$

The states $\psi_{\alpha'}$ and $\psi_{\beta'}$ are described in K' by the relevant quantities with primes:

$$\alpha \equiv p'_1 , E'_1 ; \; p'_2 , E'_2 ; \; \ldots \quad ,$$

$$\tag{2.85}$$

$$\beta \equiv \tilde{p}'_1 , \tilde{E}'_1 ; \; \tilde{p}'_2 , \tilde{E}'_2 ; \; \ldots \quad .$$

Particle momenta and energies in K' are related to those in K by the Lorentz transformations. Taking the direction of the vector u for the z axis, one may write the latter transformations as

$$p'_x = p_x \quad , \qquad p'_y = p_y \quad ,$$

$$\tag{2.86}$$

$$p'_z = \frac{p_z - uE/c^2}{\left(1 - \beta^2\right)^{1/2}} \quad , \qquad E' = \frac{E - up_z}{\left(1 - \beta^2\right)^{1/2}} \quad ,$$

where $\beta = u/c$ (c is the light velocity). Differentiating the first three equalities and bearing in mind that $dE = v \, dp$ (v is the particle velocity), we find the elementary volume in the momentum space to be

$$dp' = \frac{1 - uv/c^2}{\left(1 - \beta^2\right)^{1/2}} dp \quad . \tag{2.87}$$

Since $p = Ev/c^2$, the fourth equality of (2.86) may be rewritten as

$$E' = \frac{1 - uv/c^2}{\left(1 - \beta^2\right)^{1/2}} E \quad . \tag{2.88}$$

Dividing (2.87) by (2.88), one finds the quantity dp/E to be invariant with respect to the Lorentz transformations.

Let us consider the probability that the system undergoes a transition from some phase volume Δ into some other one $\tilde{\Delta}$, i.e.,

$$P_{\Delta\tilde{\Delta}} = \sum_{\alpha\beta} |S_{\beta\alpha}|^2 \quad . \tag{2.89}$$

Though Δ and $\tilde{\Delta}$ may be modified in different Lorentz systems, the transition probability (2.89) obviously cannot depend on the choice of the Lorentz system, and is invariant. Taking $dp/(2\pi\hbar)^3$ to be the elementary phase volume in momentum space, one may write the transition probability (2.89) as

$$P_{\Delta\tilde{\Delta}} = \int_{\Delta\tilde{\Delta}} \frac{d\mathbf{p}_1}{(2\pi\hbar)^3} \frac{d\mathbf{p}_2}{(2\pi\hbar)^3} \cdots \frac{d\tilde{\mathbf{p}}_1}{(2\pi\hbar)^3} \frac{d\tilde{\mathbf{p}}_2}{(2\pi\hbar)^3} |\langle\beta|S|\alpha\rangle|^2 \quad , \tag{2.90}$$

or

$$P_{\Delta\tilde{\Delta}} = \int_{\Delta\tilde{\Delta}} \frac{d\mathbf{p}_1}{(2\pi\hbar)^3 E_1} \frac{d\mathbf{p}_2}{(2\pi\hbar)^3 E_2} \cdots \frac{d\tilde{\mathbf{p}}_1}{(2\pi\hbar)^3 \tilde{E}_1} \frac{d\tilde{\mathbf{p}}_2}{(2\pi\hbar)^3 \tilde{E}_2} \cdots$$
$$\times \left| \left(\tilde{E}_1 \tilde{E}_2 \ldots \right)^{1/2} \langle\beta|S|\alpha\rangle \left(E_1 E_2 \ldots \right)^{1/2} \right|^2 \quad . \tag{2.91}$$

The quantities $d\mathbf{p}_1/E_1$, $d\mathbf{p}_2/E_2$, etc. are invariant, therefore, $(\tilde{E}_1\tilde{E}_2\ldots)^{1/2}$ $\langle\beta|S|\alpha\rangle(E_1 E_2\ldots)^{1/2}$ is also invariant under the Lorentz tranformation. Thus, the matrix element of the operator S in the system K' is related to the matrix element of S in the system K by

$$S'_{\beta\alpha} = \frac{(E_1 E_2 \ldots \tilde{E}_1 \tilde{E}_2 \ldots)^{1/2}}{(E'_1 E'_2 \ldots \tilde{E}'_1 \tilde{E}'_2 \ldots)^{1/2}} S_{\beta\alpha} \quad . \tag{2.92}$$

3. Stationary Scattering Theory

3.1 The Scattering Amplitude

In the previous chapter, we considered time development of the scattering process and defined the scattering matrix in terms of the unitary time shift operator in the interaction representation. The problem may also be formulated in such a way that the description is stationary rather than time-dependent, and the wave function of the system at large distances from the scatterer is assumed to be a superposition of the incident and scattered waves.

First of all, we consider elastic scattering which does not change intrinsic states of the particles involved in the collision. In the center-of-mass system, the two-particle scattering problem is treated as scattering of a particle with reduced mass μ in the field $V(r)$ of a stationary force center. The Hamiltonian of the system is

$$H = H_0 + V \quad ,$$

where $H_0 = (-\hbar^2/2\mu)\,\Delta$ is the kinetic energy of relative motion. The scattering wave function is governed by the Schrödinger equation

$$H\psi = E\psi \tag{3.1}$$

for positive energies ($E > 0$), with the boundary conditions which require that the solution at large distances from the scatterer must be a sum of the incident and scattered waves.

In order to find the solution with the required asymptotics, we rewrite the Schrödinger equation in the form

$$(E - H_0)\psi = V\psi \tag{3.2}$$

and treat the right-hand part as a given function. Then the general solution of (3.2) may be written as

$$\psi(r) = \varphi(r) + \int dr' \, G_0(E; r - r') \, V(r') \, \psi(r') \quad , \tag{3.3}$$

where φ is the general solution of (3.2) with vanishing right-hand part, i.e.,

$$(E - H_0)\varphi = 0 \quad , \tag{3.4}$$

and $G_0(E; r - r')$ is the Green function that satisfies the inhomogeneous equation with a point source

$$(E - H_0)\, G_0(E; r - r') = \delta(r - r') \quad . \tag{3.5}$$

The first term in (3.3) may be regarded as the incident wave that describes the free motion of the particle. The second term is associated with the scattered wave which can be either converging or diverging (it depends on the choice of the Green function).

If the eigenfunctions of the equation (3.4) are plane waves

$$\varphi_{k'}(r) = e^{ik'r} \quad , \qquad E' = \frac{\hbar^2 k'^2}{2\mu} \quad , \tag{3.6}$$

then the Green function is described by the integral

$$G_0(E; r - r') = \int \frac{dk'}{(2\pi)^3} \frac{\exp\left[ik(r - r')\right]}{E - E'} \quad . \tag{3.7}$$

Integrating (3.7) over the angle variables yields

$$G_0(E; r - r') = \frac{\mu}{2\pi^2 i \hbar^2} \frac{1}{|r - r'|} \int_{-\infty}^{\infty} dk'\, k' \frac{\exp(ik'|r - r'|)}{k^2 - k'^2} \quad , \tag{3.8}$$

where $E = \hbar^2 k^2/2\mu$. Integration over k may be carried out by means of the residue method; the boundary conditions determine the procedure of encircling the poles $k' = \pm k$ of the integrand. In order to obtain a diverging wave at large distances, we take the integration contour in the plane of the complex variable k' as shown in Fig. 3.1a. Since the integral over the circle of infinitely large radius vanishes, the integral in (3.8) is equal to $2\pi i$ times the residue at the only pole enclosed by the integration contour. Thus we obtain

$$G_0^{(+)}(E; r - r') = -\frac{\mu}{2\pi \hbar^2} \frac{\exp(ik|r - r'|)}{|r - r'|} \quad . \tag{3.9}$$

In order to calculate the Green function associated with the converging wave, we take the integration contour as given in Fig. 3.1b. Then

$$G_0^{(-)}(E; r - r') = -\frac{\mu}{2\pi \hbar^2} \frac{\exp(-ik|r - r'|)}{|r - r'|} \quad . \tag{3.10}$$

Another way to find how to encircle the poles when calculating the Green functions, is to substitute $E + i0$ and $E - i0$ for E in the denominator of (3.7) for the Green functions associated with diverging and converging waves, respectively. Then we have

$$G_0^{(\pm)}(E; r - r') = \int \frac{dk}{(2\pi)^3} \frac{\exp[ik(r - r')]}{E - E' \pm i0} \quad , \tag{3.11}$$

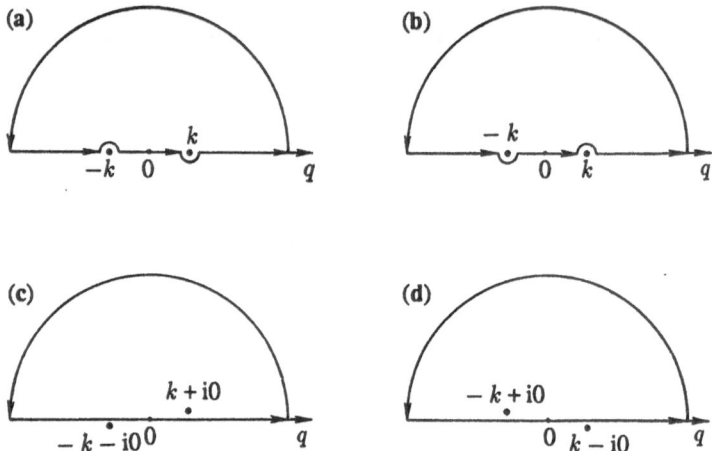

Fig. 3.1. Integration contours for the Green functions $G_0^{(+)}(E; r - r')$ and $G_0^{(-)}(E; r - r')$

and the integration contour which consists of the real axis and the semicircle of infinite radius in the upper half-plane, contains either the pole $k' = k + iO$ in the case of $G_0^{(+)}$, or the pole $k' = -k + iO$ in the case of $G_0^{(-)}$.

Taking φ in (3.3) to be the plane wave corresponding to the incident particle with momentum k and the Green function to be the converging wave (3.9), we rewrite (3.3) as

$$\psi_k^{(+)}(r) = \varphi_k(r) - \frac{\mu}{2\pi\hbar^2} \int dr' \frac{\exp(ik|r - r'|)}{|r - r'|} V(r') \psi_k^{(+)}(r') . \qquad (3.12)$$

This integral equation determines the wave function of the scattering problem. It is equivalent to the Schrödinger equation with boundary conditions.

At large distances, the Green function (3.9) may be approximated by the expression

$$G_0^{(+)}(E; r - r') \rightarrow -\frac{\mu}{2\pi\hbar^2} \frac{\exp(ikr)}{r} e^{-ik'r'} \quad (r \rightarrow \infty) , \qquad (3.13)$$

where $k' = (r/r)k$. Therefore, if the potential is nonzero within a finite region, then the wave function asymptotic is given by

$$\psi_k^{(+)}(r) \rightarrow \varphi_k(r) + f(k, k') \frac{\exp(ikr)}{r} \quad (r \rightarrow \infty) , \qquad (3.14)$$

where

$$f(k, k') = -\frac{\mu}{2\pi\hbar^2} \int dr \, e^{-ik'r} V(r) \psi_k^{(+)}(r) . \qquad (3.15)$$

The coefficient $f(k, k')$ before the diverging wave is usually referred to as the scattering amplitude. Thus, k' must be regarded as the final particle momentum.

According to (3.15), the scattering amplitude $f(k, k')$ depends on the energy of relative motion, the angle between the vectors k and k', and the scattering potential.

Once the wave function is known, it is a simple matter to calculate the differential scattering cross section. The radial density of the scattered particle flux at large distances is

$$j_r = \frac{i\hbar}{2\mu}\left(\psi\,\frac{\partial\psi^*}{\partial r} - \psi^*\,\frac{\partial\psi}{\partial r}\right) \underset{r\to\infty}{\longrightarrow} \frac{\hbar k}{\mu r^2}\left|f(k, k')\right|^2 \quad . \tag{3.16}$$

Having divided the number of particles scattered into the elementary solid angle do per unit time $dN = j_r r^2\, do$ by the incident flux density $j_0 = \hbar k/M$, we find the cross section to be

$$d\sigma_e = |f(k, k')|^2\, do \quad . \tag{3.17}$$

Thus, the differential scattering cross section per unit solid angle is directly determined by the squared modulus of the scattering amplitude.

3.2 The Lippmann–Schwinger Equation

It is convenient to rewrite the results obtained in a symbolic form that is suitable for extensions to many-particle systems. Let us consider a quantum mechanical system of several interacting particles. We assume the Hamiltonian to be a sum of two parts

$$H = H_0 + V \quad ,$$

where H_0 describes the nonperturbed motion of the system and V is the interaction which vanishes if the distance between the interacting parts of the system is sufficiently large.

In the stationary formulation, the scattering problem reduces to the Schrödinger equation

$$(E - H)\psi = 0 \tag{3.18}$$

with fixed boundary conditions, the energy E of the system being positive. At infinity, the solution ψ must be a sum of the incident wave φ, governed by the nonperturbed equation

$$(E - H_0)\varphi = 0 \quad , \tag{3.19}$$

and the diverging scattered wave.

The solution of (3.18) with the above mentioned boundary conditions formally may be written as

$$\psi^{(+)} = \varphi + G_0^{(+)}(E)\, V\, \psi^{(+)} \quad , \tag{3.20}$$

where $G_0^{(+)}(E)$ is the Green function of the nonperturbed equation (3.19),

$$G_0^{(+)}(E) = \frac{1}{E - H_0 + iO} \quad . \tag{3.21}$$

The choice of the integration contour enclosing the pole of (3.21) follows from the requirement that the asymptotics of $\psi^{(+)}$ must contain a diverging scattered wave.

The formal solution (3.20) of the Schrödinger equation (3.18) is an integral equation

$$\psi^{(+)} = \varphi + \frac{1}{E - H_0 + iO} \, V \, \psi^{(+)} \quad , \tag{3.22}$$

which is referred to as the Lippmann–Schwinger equation [3.1].

The equation

$$\psi_\varepsilon = \varphi + \frac{1}{E - H_0 + i\varepsilon} \, V \, \psi_\varepsilon \tag{3.23}$$

has a unique solution for any finite ε. Ambiguity could occur if the homogeneous equation obtained from (3.23) for $\varphi = 0$ had a solution. However, the eigenvalues corresponding to the decreasing solutions of the homogeneous equation (in the asymptotic region for which $V \to 0$) must be real. Therefore, the equation

$$\psi_\varepsilon = \frac{1}{E - H_0 + i\varepsilon} \, V \tag{3.24}$$

has no solutions for any finite ε.

The scattering wave function whose asymptotic at infinity is a sum of the incident and converging scattered waves, satisfies the equation in which the integration contour must be taken in another way, i.e.,

$$\psi^{(-)} = \varphi + \frac{1}{E - H_0 - iO} \, V \, \psi^{(-)} \quad , \tag{3.25}$$

because it is determined by the Green function

$$G_0^{(-)}(E) = \frac{1}{E - H_0 - iO} \quad . \tag{3.26}$$

Equations (3.22, 25) are written in symbolic form. To write them explicitly, one has to expand the function $V\psi$ in terms of the eigenfunctions of the nonperturbed Hamiltonian H_0. This results in

$$\psi_\alpha^{(\pm)} = \varphi_\alpha + \sum_\beta \frac{\varphi_\beta (\varphi_\beta , V \, \psi_\alpha^{(\pm)})}{E_\alpha - E_\beta \pm iO} \quad . \tag{3.27}$$

Thus, the Green function is given by

$$G_0^{(\pm)}(E) \ldots = \sum_\beta \frac{\varphi_\beta(\varphi_\beta, \ldots)}{E - E_\beta \pm iO} \quad . \tag{3.28}$$

This relation determines the expansion of the Green function $G_0^{(\pm)}$ in terms of the complete set of eigenfunctions of the operator H_0. In the coordinate representation, we have

$$\begin{aligned} G_0^{(\pm)}(E; \boldsymbol{r}, \boldsymbol{r}') &\equiv \langle \boldsymbol{r}' | G_0^{(\pm)}(E) | \boldsymbol{r} \rangle \\ &= \int \frac{d\boldsymbol{k}'}{(2\pi)^3} \frac{\varphi_{\boldsymbol{k}'}(\boldsymbol{r}) \varphi_{\boldsymbol{k}'}^*(\boldsymbol{r}')}{E - E' \pm iO} \quad , \end{aligned} \tag{3.29}$$

which is in agreement with (3.11). It is not difficult to verify that (3.27) in the coordinate representation reproduces (3.12).

If the energies of relative motion are negative, then the solutions of the Schrödinger equation (3.18) are associated with the bound states of the system. We employ the Green function (3.12) and thus reduce the Schrödinger equation for the bound states to the homogeneous integral equation

$$\psi = G_0^{(\pm)}(E) V \psi \quad . \tag{3.30}$$

In the general case, the Lippmann–Schwinger equation (3.20) has no unambiguous solutions if the system possesses bound states. Indeed, the energy levels

$$\varepsilon = P^2 / 2M + E$$

(P is the total momentum, M is the total mass, E is the energy of relative motion) are multiply degenerate because each state is associated with a certain energy distribution between the intrinsic motion and motion of the system as a whole. Since existence of a bound state infers that the homogeneous equation (3.30) has a solution for fixed ε, the solution of the nonhomogeneous equation (3.20) becomes ambiguous. For a two-particle system, the Lippmann–Schwinger equation has an unambiguous solution only in the c.m.s., for which $P = 0$ and energy degeneration does not occur ($\varepsilon = E$).

3.3 The Möller Operators Ω_+ and Ω_-

Each eigenfunction φ of the Hamiltonian H_0 associated with fixed energy $E > 0$ can be put in correspondence with the eigenfunctions $\psi^{(+)}$ and $\psi^{(-)}$ of the total Hamiltonian H associated with the same energy. This is done by the so-called Möller operators Ω_+ and Ω_-. To introduce them, we employ the definition (2.21) of the time shift operator in the interaction representation. The time shift operator is described by the integral equation

$$U(t, t_0) = 1 - \frac{i}{\hbar} \int_{t_0}^t dt' \, V(t') \, U(t', t_0) \quad . \tag{3.31}$$

Within the context of (3.31), the operator $\Omega_+ \equiv U(0, -\infty)$ tranforms the eigenstate φ of the Hamiltonian H_0 into the eigenstate $\psi^{(+)}$ of the total Hamiltonian H, so that

$$\Omega_+ \varphi = \psi^{(+)} \quad . \tag{3.32}$$

The analogous operator $\Omega_- \equiv U(0, \infty)$ transforms φ into the state $\psi^{(-)}$, i.e.,

$$\Omega_- \varphi = \psi^{(-)} \quad . \tag{3.33}$$

Having applied the operator equality (3.31) to the state φ for $t_0 = -\infty$, we obtain

$$\psi(t) = \varphi - \frac{i}{\hbar} \int_{-\infty}^{t} dt' \, \exp\left(\frac{i}{\hbar} H_0 t'\right) V$$

$$\times \exp\left[-\varepsilon |t'| - \frac{i}{\hbar} H_0 t'\right] \psi(t') \quad , \qquad \varepsilon \to 0 \quad .$$

We put $t = 0$ and apply the successive iteration method. Then

$$\psi(0) = \left(1 + \frac{1}{E - H_0 + i\varepsilon} V + \frac{1}{E - H_0 + i\varepsilon} V \frac{1}{E - H_0 + i\varepsilon} V + \dots\right) \varphi$$

$$= \varphi + \frac{1}{E - H_0 + i\varepsilon} V \psi(0) \quad , \qquad \varepsilon \to 0 \quad .$$

Comparing this equation for $\psi(0)$ to the Lippmann–Schwinger equation, we find that

$$\psi(0) = \psi^{(+)} \quad . \tag{3.34}$$

Thus, we have shown that

$$\psi^{(+)} = U(0, -\infty) \, \varphi \qquad \text{or} \qquad \psi^{(+)} = \Omega_+ \varphi \quad . \tag{3.35}$$

It is a simple matter to show in a similar manner that

$$\psi^{(-)} = U(0, \infty) \, \varphi \qquad \text{or} \qquad \psi^{(-)} = \Omega_- \varphi \quad . \tag{3.36}$$

The Möller operators Ω_+ and Ω_- are the limiting values of the operator $U(t, t_0)$ for $t_0 = \mp\infty$ and therefore, in general are isometric operators. An isometric operator, in contrast to a unitary one, does not necessarily map the whole Hilbert space onto itself. Any unitary operator is isometric, but the inverse assertion is not valid. The isometric operators satisfy the relation

$$\Omega^+ \Omega = 1 \quad . \tag{3.37}$$

However,

$$\Omega \Omega^+ \neq 1 \quad . \tag{3.38}$$

The states $\psi^{(+)}$ and $\psi^{(-)}$ usually are referred to as the scattering states, the related spaces are $\mathcal{R}_+$ and $\mathcal{R}_-$.

The eigenfunctions φ_α associated with the continuum spectrum of the operator H_0 form a complete orthonormalized set,

$$(\varphi_\alpha, \varphi_{\alpha'}) = \delta_{\alpha\alpha'} \quad . \tag{3.39}$$

The eigenfunctions $\psi_\alpha^{(+)}$ or $\psi_\alpha^{(-)}$, produced by the passage to the limit (3.35) or (3.36), also form orthonormalized sets associated with the continuum spectrum of the operator H which is assumed to coincide with that of H_0, i.e.,

$$\begin{aligned}
\left(\psi_\alpha^{(\pm)}, \psi_{\alpha'}^{(\pm)}\right) &= \left(\Omega_\pm\varphi_\alpha, \Omega_\pm\varphi_{\alpha'}\right) \\
&= \left(\varphi_\alpha, \Omega_\pm^+\Omega_\pm\varphi_{\alpha'}\right) \\
&= \left(\varphi_\alpha, \varphi_{\alpha'}\right) = \delta_{\alpha\alpha'} \quad .
\end{aligned} \tag{3.40}$$

The discrete levels lying below the boundary of the continuum spectrum of the operator H correspond to the bound states ψ_n. We denote the subspace formed by the bound states by $\mathcal{B}$. The Hilbert space for H is the direct sum of $\mathcal{B}$ and the subspace $\mathcal{R}$ of all state vectors orthogonal to the bound states ($\mathcal{B} \perp \mathcal{R}$). If the potential V entering the Hamiltonian H decreases with $r \to \infty$ more rapidly than r^{-3}, and its singularity for $r \to 0$ is weaker than r^{-2}, then asymptotic completeness occurs: the space of vectors $\psi^{(+)}$ coincides with the space of $\psi^{(-)}$ and coincides with the space of all state vectors, orthogonal to the bound states, i.e.,

$$\mathcal{R}_+ = \mathcal{R}_- = \mathcal{R} \quad . \tag{3.41}$$

The results of the scattering theory are valid provided that all these requirements are satisfied.

The scattering operator S is expressed in terms of the Möller operators by means of the relation

$$S = \Omega_-^+ \, \Omega_+ \quad . \tag{3.42}$$

Indeed,

$$S \equiv U(\infty, -\infty) = U(\infty, 0)\, U(0, -\infty) \quad ,$$

and since

$$U(\infty, 0) = U^+(0, \infty) \quad ,$$

we immediately obtain (3.42).

Equation (3.42) directly ensures unitarity of the scattering operator S; Ω_+ and Ω_- are isometric operators which map $\mathcal{H}$ onto the subspace $\mathcal{R}$ of the scattering states. This means that Ω_+ is a linear operator which maps $\mathcal{H}$ onto $\mathcal{R}$ conserving the norm, and Ω_-^+ is a linear operator which conserves the norm and maps $\mathcal{R}$ back onto $\mathcal{H}$. Therefore, the operator S is linear, conserves the norm, and maps $\mathcal{H}$ onto $\mathcal{H}$, i.e., it is unitary. This proof is essentially based on the asymptotic completeness assumption, which infers that Ω_+ and Ω_- map $\mathcal{H}$ onto the same actual range $\mathcal{R}$ ($\mathcal{R}_+ = \mathcal{R}_- = \mathcal{R}$).

We employ (3.42) to show that the scattering operator S commutes with the Hamiltonian H_0, i.e.,

$$[S, H_0] = 0 \quad . \tag{3.43}$$

To prove this, we make use of the so-called transfer relations for the Möller operators,

$$H\Omega_\pm = \Omega_\pm H_0 \quad . \tag{3.44}$$

[To verify the validity of (3.44), one has to apply (3.44) to the state φ and make use of the definitions (3.32, 33)]. Applying the transfer relations twice, it is not difficult to show that S commutes with H_0:

$$SH_0 = \Omega_-^+ \Omega_+ H_0 = \Omega_-^+ H \Omega_+ = H_0 \Omega_-^+ \Omega_+ = H_0 S \quad .$$

Since the Möller operators are isometric and $\Omega^+ \Omega = 1$, the transfer relations may be rewritten as

$$\Omega_\pm^+ H \Omega_\pm = H_0 \quad . \tag{3.45}$$

Thus Ω_+ and Ω_- may be interpreted as the operators which transform the total Hamiltonian H into the free Hamiltonian H_0. This implies that in the general case the Möller operators cannot be unitary since otherwise the operators H and H_0 would have the same spectrum by virtue of (3.45). Actually, however, the system with the Hamiltonian H possesses bound states, whereas the system with H_0 has only continuum states. The number of bound states may serve to estimate the deviation of the Möller operators from unitarity:

$$\Omega\Omega^+ = 1 - \Lambda_d \quad . \tag{3.46}$$

Here Λ_d is the projection operator to the subspace of bound states of the operator H. Sometimes, it is called the unitarity deficiency. If H has no bound states, then both Ω_+ and Ω_- are unitary operators.

The scattering operator S is definite and unitary for the complete set of vectors φ_α. The matrix elements of the scattering operator, $S_{\beta\alpha} = (\varphi_\beta, S\varphi_\alpha)$, i.e., the scattering matrix, can be immediately expressed in terms of the Heisenberg wave functions $\psi_\alpha^{(+)}$ and $\psi_\alpha^{(-)}$:

$$S_{\beta\alpha} = \left(\psi_\beta^{(-)}, \psi_\alpha^{(+)}\right) \quad . \tag{3.47}$$

To verify this relation, one has to substitute (3.42) for S and make use of (3.32, 33).

Sometimes it is convenient to introduce the scattering operator S' defined by

$$\psi^{(+)} = S' \psi^{(-)} \quad , \tag{3.48}$$

i.e.,

$$\Omega_+ = S'\Omega_- \quad . \tag{3.49}$$

The operator S' is definite and unitary only for the space formed by the continuum spectrum functions $\psi^{(\pm)}$.

We multiply (3.49) by Ω^+ and make use of (3.46) to obtain

$$\Omega_+\Omega_-^+ = S'(1 - \Lambda_d) \quad . \tag{3.50}$$

Since the matrix elements of S' contain the continuum spectrum functions, the operator S' may be interpreted as

$$S' = \Omega_+\Omega_-^+ \quad . \tag{3.51}$$

It is not difficult to show that the matrix elements of the operators S' over the states $\psi^{(\pm)}$ coincide with those of the operators S over the states φ, so that

$$\begin{aligned}
S'_{\beta\alpha} &\equiv \left(\psi_\beta^{(+)}, S'\,\psi_\alpha^{(+)}\right) = \left(\psi_\beta^{(-)}, S'\,\psi_\alpha^{(-)}\right) \\
&= \left(\psi_\beta^{(-)}, \psi_\alpha^{(+)}\right) = \left(\varphi_\beta, S\varphi_\alpha\right) = S_{\beta\alpha} \quad .
\end{aligned} \tag{3.52}$$

The unitary condition for S' may be written as

$$S'^+ S' = S' S'^+ = 1 - \Lambda_d \quad . \tag{3.53}$$

As distinct from S, the scattering operator S' commutes with the total Hamiltonian of the system,

$$[S', H] = 0 \quad . \tag{3.54}$$

This relation may be verified within the context of the definition (3.51) and the transfer relations (3.44).

3.4 The Green Functions G_0 and G

The solutions of (3.22, 25) may be expressed in terms of the asymptotic function φ by means of the Green function of the equation (3.18)

$$G^{(\pm)}(E) = \frac{1}{E - H \pm i0} \quad . \tag{3.55}$$

We define the Green functions $G_0(z)$ and $G(z)$ of the complex argument z as

$$G_0(z) = \frac{1}{z - H_0} \quad , \tag{3.56}$$

$$G(z) = \frac{1}{z - H} \quad . \tag{3.57}$$

(These functions are usually called resolvents of the operators H_0 and H). Evidently,

$$G_0^{(\pm)}(E) = G_0(E \pm i0) \quad ,$$

$$(3.58)$$

$$G^{(\pm)}(E) = G(E \pm i0) \quad .$$

The Green functions $G(z)$ and $G_0(z)$ are related by the equation

$$G(z) = G_0(z) + G_0(z) V G(z) \quad .$$

$$(3.59)$$

It may be easily verified that the solution of (3.22) may be written in the form

$$\psi^{(+)} = \lim_{\varepsilon \to 0} i\varepsilon G(E + i\varepsilon) \varphi \quad .$$

$$(3.60)$$

Indeed, multiplying (3.59) for $z = E + i\varepsilon$ by $i\varepsilon$, applying it to the function φ, and making use of the relation $\lim_{\varepsilon \to 0} i\varepsilon G_0(E + i\varepsilon) \varphi = \varphi$, one obtains (3.20). Equation (3.60) may be written in another form:

$$\psi^{(+)} = \varphi + \frac{1}{E - H + i0} V \varphi \quad .$$

$$(3.61)$$

An analogous procedure reduces the solution of (3.25) to

$$\psi^{(-)} = \varphi + \frac{1}{E - H - i0} V \varphi \quad .$$

$$(3.62)$$

The Green functions $G^{(\pm)}(E)$ may be expanded in terms of the complete set of eigenfunctions of the operator H in the same way as $G_0^{(\pm)}(E)$ were expanded in terms of eigenfunctions of the operator H_0. We denote the eigenfunctions of H associated with bound and continuum states as ψ_n and ψ_β, respectively. Suppose these functions form a complete set, so that

$$\sum_n \psi_n(\psi_n, \ldots) + \sum_\beta \psi_\beta(\psi_\beta, \ldots) = 1 \ldots \quad ,$$

$$(3.63)$$

then, having applied the operator $G^{(\pm)}$ to this equality, one obtains the expansion

$$G^{(\pm)}(E) \ldots = \sum_n \frac{\psi_n(\psi_n, \ldots)}{E - E_n} + \sum_\beta \frac{\psi_\beta(\psi_\beta, \ldots)}{E - E_\beta \pm i0} \quad ,$$

$$(3.64)$$

with the eigenvalues E_n and E_β being negative and positive, respectively.

We rewrite (3.63, 64) in the coordinate representation for the simplest case of a particle in the external field. It is convenient to take $\psi_k^{(+)}(\boldsymbol{r})$ or $\psi_k^{(-)}(\boldsymbol{r})$ to be the continuum spectrum wave functions. Then the completeness condition takes the form

$$\sum_n \psi_n(r)\,\psi_n^*(r') + \int \frac{dq}{(2\pi)^3}\, \psi_q(r)\,\psi_q^*(r') = \delta(r - r') \quad , \tag{3.65}$$

and the Green function expansion is given by

$$G^{(\pm)}(E;\, r,\, r') \equiv \langle r' \,|\, G^{(\pm)}(E) |\, r \rangle$$

$$= \sum_n \frac{\psi_n(r)\,\psi_n^*(r')}{E - E_n} + \int \frac{dq}{(2\pi)^3}\, \frac{\psi_q(r)\,\psi_q^*(r')}{E - E_q \pm iO} \quad . \tag{3.66}$$

In contrast to the free motion Green function $G_0^{(\pm)}(E;\, r, r')$, the complete Green function $G^{(\pm)}(E;\, r, r')$ describing the motion in the field $V(r)$ depends on the coordinates r and r' rather than on their difference $r - r'$.

It is not difficult to verify within the context of the explicit expression (3.66) that the Green function $G^{(\pm)}(E;\, r, r')$ indeed satisfies the Schrödinger equation (3.18) with a point source in the right-hand part, i.e.,

$$(E - H)\, G^{(\pm)}(E;\, r, r') = \delta(r - r') \quad . \tag{3.67}$$

Let us show that the Green functions $G_0^{(\pm)}(E)$ and $G^{(\pm)}(E)$ are related to the time-dependent Green functions $G_0^{(\pm)}(t)$ and $G^{(\pm)}(t)$ (Chap. 1) by a simple Fourier tranformation. We expand the time-dependent Green functions (1.34, 35) and (1.39, 40) in the Fourier integrals and introduce the Fourier components

$$G_0^{(\pm)}(E) = \int_{-\infty}^{\infty} dt\, \exp\left(\frac{i}{\hbar} Et\right) G_0^{(\pm)}(t) \quad ,$$

$$\tag{3.68}$$

$$G^{(\pm)}(E) = \int_{-\infty}^{\infty} dt\, \exp\left(\frac{i}{\hbar} Et\right) G^{(\pm)}(t) \quad .$$

In the case of retarded Green functions $G_0^{(+)}(t)$ and $G^{(+)}(t)$, integration in (3.68) actually extends from $t = 0$ to $t = \infty$. Hence, in order to ensure that the integrals converge, we introduce in the integrand the factor $\exp(-\varepsilon t)$ with $\varepsilon > 0$ and allow the parameter ε to tend to zero as soon as integration is carried out. Similarly, in the case of advanced Green functions $G_0^{(-)}(t)$ and $G^{(-)}(t)$, we introduce in the integrand the factor $\exp(\varepsilon t)$, since in this case integration extends from $t = -\infty$ to $t = 0$. As a result, the Fourier components of the time-dependent Green functions are described by the expressions (3.21), (3.26) and (3.55).

For conservative systems, the interaction potential V is time-independent. It is not difficult to show that in this case the Fourier transform of (1.41) reproduces (3.59) with $z = E \pm iO$.

The general definition (3.55) may serve to express the Green function of a system that consists of noninteracting subsystems in terms of the Green functions of individual subsystems. For example, if the system consists of two noninteracting parts, then

$$H = H_1 + H_2$$

and the Green function of the whole system is expressed in terms of individual Green functions $G_1^{(\pm)}(E)$ and $G_2^{(\pm)}(E)$ in the following way:

$$G^{(\pm)}(E) = \mp \frac{1}{2\pi i} \int_{-\infty}^{\infty} dE' \, G_1^{(\pm)}(E') \, G_2^{(\pm)}(E - E') \quad . \tag{3.69}$$

The Green functions for more general cases may be obtained in the same manner.

3.5 The Scattering Amplitude and the Transition Matrix

The Lippmann–Schwinger equation (3.22) and the integral equation (2.55) for the transition operator t enable one to derive the general relation between the elastic scattering amplitude $f(k, k')$ and the transition matrix t. Let us multiply the left- and right-hand parts of (3.22) by V from the left and apply the operator equation (2.55) to the wave function φ:

$$V \psi^{(+)} = V \varphi + V \frac{1}{E - H_0 + iO} V \psi^{(+)} \quad ,$$

$$t \varphi = V \varphi + V \frac{1}{E - H_0 + iO} t \varphi \quad .$$

A comparison of these equations shows the relation between the solution of the Lippmann–Schwinger equation and the transition operator t, i.e.,

$$V \psi^{(+)} = t \varphi \quad , \tag{3.70}$$

and then the solution of the Lippmann–Schwinger equation may be written in the form

$$\psi^{(+)} = \varphi + \frac{1}{E - H_0 + iO} t \varphi \quad . \tag{3.71}$$

In the coordinate representation, the second term in the right-hand part of (3.71) at large distances is a diverging scattered wave. The coefficient before the latter is just the scattering amplitude

$$f = -\frac{\mu}{2\pi\hbar^2} (\varphi', t \varphi) \quad , \tag{3.72}$$

or

$$f(k, k') = -\frac{\mu}{2\pi\hbar^2} \langle k' | t | k \rangle \quad . \tag{3.73}$$

This expression for the scattering amplitude also may be derived by substituting (3.70) in (3.15).

The elastic scattering amplitude expressed in terms of the matrix elements of the transition operator t over the states (2.43) is given by

$$f(k, k') = -\frac{4\pi^2}{k} \langle n'|t|n \rangle \quad . \tag{3.74}$$

We make use of the relation (2.60) between the t matrix and the S matrix on the energy shell taking into account that the scattering matrix is diagonal in the angular momentum representation. Then the elastic scattering amplitude (3.74) may be written as

$$f(k, k') = \frac{i}{2k} \sum_l (2l + 1)(1 - S_l) P_l(\cos \vartheta) \quad , \tag{3.75}$$

where ϑ is the scattering angle.

It is sometimes convenient to use the so-called partial scattering amplitudes f_l which may be defined as coefficients of the expansion

$$f(k, k') = \sum_l (2l + 1) f_l P_l(\cos \vartheta) \tag{3.76}$$

and thus are given by

$$f_l = \frac{i}{2k}(1 - S_l) \quad . \tag{3.77}$$

If only elastic scattering occurs, $|S_l| = 1$, then the partial amplitudes may be written in terms of the scattering phase shifts as

$$f_l = \frac{1}{k} \exp(i\delta_l) \sin \delta_l \quad . \tag{3.78}$$

Note that the transition matrix t may be related to the Green function G. Indeed, comparing the operator equations (2.55) and (3.59), one finds that

$$t G_0 = VG \quad . \tag{3.79}$$

Thus, if the Green function $G(E)$ of the system is known, then the scattering amplitude can be found from (3.73, 79).

3.6 Inelastic Scattering and Reactions

In the preceeding sections we considered only elastic scattering. The Lippmann–Schwinger formalism enables one to describe inelastic scattering and reactions, too.

In the Lippmann–Schwinger equation

$$\psi^{(+)} = \varphi + \frac{1}{E - H_0 + i0} V \psi^{(+)} \quad , \tag{3.80}$$

the second term in the right-hand part describes, along with elastic scattering, inelastic scattering and reactions associated with the transitions into other channels. Making use of the operator equality

$$\frac{1}{A} = \frac{1}{B} \left\{ 1 + (B - A) \frac{1}{A} \right\} \quad ,$$

where $A \equiv E - H_0 + iO$ and $B = E - H_0' + iO$, one easily obtains an equivalent form of (3.80),

$$\psi^{(+)} = \tilde{\varphi} + \frac{1}{E - H_0' + iO} V' \psi^{(+)} \quad , \tag{3.81}$$

where

$$\tilde{\varphi} \equiv \varphi - \frac{1}{E - H_0' + iO} (H_0 - H_0') \varphi = 0 \quad .$$

Actually, $\tilde{\varphi}$ must be regarded as the limiting value

$$\tilde{\varphi} = \lim_{\varepsilon \to 0} \frac{i\varepsilon}{E - H_0' + i\varepsilon} \varphi \quad .$$

We expand it in terms of the complete set of states φ_β of the Hamiltonian H_0' to obtain

$$\tilde{\varphi} = \lim_{\varepsilon \to 0} \sum_\beta \frac{i\varepsilon}{E - H_0' + i\varepsilon} \varphi_\beta(\varphi_\beta, \varphi) \quad .$$

Because the scalar product (φ_β, φ) is always finite, and since only an infinitesimal interval of values $E_\beta \simeq E$ contributes to the sum, the component $\tilde{\varphi}$ tends to zero as $\varepsilon \to 0$.

Equation (3.80) is suitable for calculating the scattered wave asymptotics in the entrance channel for $r \to \infty$, whereas (3.81) enables one to find the scattered wave asymptotics in the channel V' for $r' \to \infty$. The coefficient before the diverging wave in the channel V' is usually called the reaction amplitude. Having calculated the scattered wave asymptotics for $r' \to \infty$ from (3.81), one finds the reaction amplitude to be

$$f' = -\frac{\mu'}{2\pi\hbar^2} \left(\varphi', V' \psi^{(+)} \right) \quad , \tag{3.82}$$

where μ' is the reduced mass of the system in the exit channel. This amplitude is associated with the transition from the state φ into φ' (φ' is the eigenfunction of the Hamiltonian H_0').

The reaction amplitude (3.82) contains the solution $\psi^{(+)}$ of the Lippmann–Schwinger equation. This function describes the particles' relative motion and intrinsic states in all the channels, and only its asymptotic for $r \to \infty$ (r is the relative distance between the particles in the entrance channel) reproduces the function φ describing the free particle motion in the entrance channel. It is clear that φ' describes the intrinsic states and relative motion of noninteracting particles in the exit channel. Let k and k' be the relative momenta of incident and escaping particles, respectively, and γ and γ' be the quantum numbers associated with the entrance and exit channels. Then (3.82) may be rewritten as

$$f(k\gamma; k'\gamma') = -\frac{\mu'}{2\pi\hbar^2} \left(\varphi_{k',\gamma'} , V' \psi_{k,\gamma}^{(+)} \right) \quad , \tag{3.83}$$

$$(\hbar^2 k^2/2\mu) + E_\gamma = (\hbar^2 k'^2/2\mu') + E_{\gamma'} \quad .$$

Formula (3.83) also describes inelastic particle scattering; in this case $V' = V$ and $\mu' = \mu$.

The reaction amplitudes may be expressed in terms of the transition matrix t in the same manner as the scattering amplitude. Note that if the exit channel is different from the entrance channel then, instead of (3.70), the relation

$$V'\psi^{(+)} = t\,\varphi \tag{3.84}$$

holds, which can be easily derived from (3.81). Then the equation for t may be written in the form of (2.55) or as

$$t = V' + V' \frac{1}{E - H_0' + iO} t \tag{3.85}$$

By virtue of (3.84), the reaction amplitude (3.82) reduces to

$$f' = -\frac{\mu'}{2\pi\hbar^2} (\varphi', t\,\varphi) \tag{3.86}$$

The relation obtained formally reproduces the scattering amplitude (3.72) with μ and φ' being the reduced mass and the wave function in the exit channel.

Having introduced the initial and final state quantum numbers k, γ and k', γ', we can rewrite (3.86) as

$$f(k\gamma; k'\gamma') = -\frac{\mu'}{2\pi\hbar^2} \langle k', \gamma'|t|\gamma, k\rangle \quad , \tag{3.87}$$

or as

$$f(k\gamma; k'\gamma') = -\sqrt{\frac{\mu'}{\mu}} \frac{(2\pi)^2}{(kk')^{1/2}} \langle n', \gamma'|t|\gamma, n\rangle \tag{3.88}$$

If the perturbation theory is applicable in the range under consideration, then we can substitute the potential V or V' instead of the transition operator t in (3.86) or (3.87). Note that the matrix elements of V and V' over the states φ and φ', for which $E = E'$, are equal. This immediately follows from the relation

$$V = H - H_0 + H_0' - H_0 + V' \rightarrow V' \tag{3.89}$$

Having calculated the scattered particle flux density from (3.81), one finds the differential reaction cross section to be

$$d\sigma_{\gamma\rightarrow\gamma'} = \frac{\mu}{\mu'} \frac{k'}{k} \left| f(k\gamma; k'\gamma') \right|^2 do \quad , \tag{3.90}$$

or

$$d\sigma_{\gamma \to \gamma'} = \frac{(2\pi)^4}{k^2} \left| \langle n'\gamma'|t|\gamma n \rangle \right|^2 do \quad . \tag{3.91}$$

Thus, the cross section per unit angle is equal to the squared modulus of the reaction amplitude, multiplied by the ratio of the escape velocity of the reaction products to the relative velocity of the colliding particles.

3.7 Born Approximation and Perturbation Theory

In the general case, deriving scattering or reaction amplitudes requires knowledge of the solution of the Lippmann–Schwinger integral equation. However, if the interaction can be regarded as a weak perturbation, then the integral equation may be solved by successive approximations. Let us calculate the elastic scattering amplitude in terms of perturbation theory.

The exact wave function of the scattering problem in the coordinate representation, $\psi_k(r)$, is governed by the equation

$$\psi_k(r) = \varphi_k(r) + \int dr' \, G_0(r, r') \, V(r') \, \psi_k(r') \quad , \tag{3.92}$$

where $\varphi_k(r)$ is the incident plane wave and $G_0(r, r')$ is the retarded Green function (3.9). Regarding the potential energy V as a perturbation, we can write the solution of (3.92) as a series

$$\psi_k(r) = \psi_k^{(0)}(r) + \psi_k^{(1)}(r) + \psi_k^{(2)}(r) + \ldots \quad , \tag{3.93}$$

where each nth expansion term is proportional to the potential raised to the nth power. Successive iteration yields

$$\psi_k^{(0)}(r) = \varphi_k(r) \quad ,$$

$$\psi_k^{(1)}(r) = \int dr_1 \, G_0(r, r_1) \, V(r_1) \, \varphi_k(r_1) \quad , \tag{3.94}$$

$$\psi_k^{(2)}(r) = \int\int dr_1 \, dr_2 \, G_0(r, r_2) \, V(r_2) \, G_0(r_2, r_1) \, V(r_1) \, \varphi_k(r_1) \quad ,$$

and so on.

With the incident wave ψ_k being substituted in (3.15) for the exact wave function φ_k, we find the elastic scattering amplitude in the first, or Born, approximation to be

$$f_B(k, k') = -\frac{1}{2\pi\hbar^2} \int dr \, e^{-ik'r} \, V(r) \, \varphi_k(r) \quad . \tag{3.95}$$

With a plane wave substituted for φ_k, the scattering amplitude reduces to

$$f_B(k, k') = -\frac{\mu}{2\pi\hbar^2} \int dr\, e^{iqr} V(r) \quad , \tag{3.96}$$

where q is the particle momentum transfer,

$$q = k - k' \quad . \tag{3.97}$$

The absolute value of the vector q is

$$q = 2k \sin\frac{\vartheta}{2} \quad , \tag{3.98}$$

where ϑ is the scattering angle (the angle between the vectors k' and k). The scattering cross section in the Born approximation is given by the expression

$$d\sigma_B = \frac{\mu^2}{4\pi^2\hbar^2} \left| \int dr\, e^{iqr} V(r) \right|^2 do \quad . \tag{3.99}$$

According to (3.96), the elastic scattering amplitude in the Born approximation is determined by the Fourier component of the scattering field and depends only on the absolute value q of the momentum transfer. The Born amplitudes of the direct and inverse scattering processes, i.e., the processes with interchanged initial and final momenta, can easily be shown to satisfy the relation

$$f_B(k, k') = f_B^*(k', k) \quad . \tag{3.100}$$

If the field is centrally symmetric, then we integrate over the angles θ, φ in (3.96) $(r = r, \theta, \varphi)$ and the amplitude takes the form

$$f_B(k, k') = -\frac{2\mu}{\hbar^2} \int_0^\infty dr\, r\, \frac{\sin qr}{q} V(r) \quad . \tag{3.101}$$

Note that for $\vartheta = 0$ (i.e., $q = 0$), the integral in (3.101) diverges at infinity if the potential decreases as or slower than r^{-3}.

The angular dependence of the Born scattering amplitude is determined by particle energy. For sufficiently low particle energies, one may put $\exp(iqr) = 1$ in (3.96) and then scattering is isotropic and energy-independent. If the particle velocity is sufficiently large, then scattering is highly anisotropic and is directed mainly in the forward direction. The amplitude appreciably differs from zero only for angles $\vartheta \leq 1/kR$, where R is the range of action of the potential.

The corrections to the Born approximation (3.95) may be calculated within the context of expansions (3.93, 94). In the general case, the elastic scattering amplitude may be written as a perturbation series

$$f(k, k') = \sum_{n=1}^\infty f^{(n)}(k, k') \quad , \tag{3.102}$$

where

$$f^{(n)}(k, k') = -\frac{\mu}{2\pi\hbar^2} \int\!\!\int \dots \int d\pmb{r}_1 \, d\pmb{r}_2 \dots d\pmb{r}_n \, \varphi_{k'}^*(\pmb{r}_n) V(\pmb{r}_n)$$
$$\times \, G_0(\pmb{r}_n, \pmb{r}_{n-1}) V(\pmb{r}_{n-1}) \dots$$
$$\times \, G_0(\pmb{r}_2, \pmb{r}_1) V(\pmb{r}_1) \varphi_k(\pmb{r}_1) \; . \tag{3.103}$$

The first term in (3.102) corresponds to the Born approximation; the other terms are corrections. According to (3.103), the nth order correction to the amplitude $f^{(n)}(k, k')$ is determined by the integral whose integrand consists of the following factors (from right to left): the wave function $\varphi_k(\pmb{r}_1)$, associated with the incident particle with the momentum k; then the potential V alternating with the Green function G_0; lastly, the wave function $\varphi_{k'}^*(\pmb{r}_n)$ describing the scattered particle with the momentum k'.

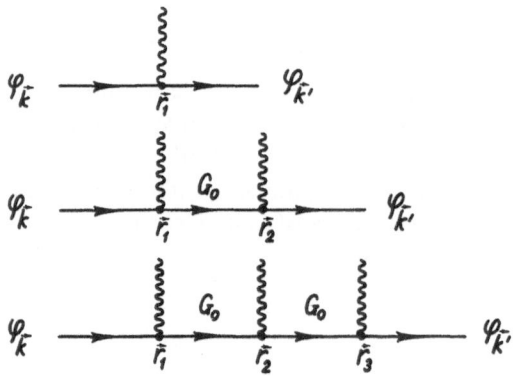

Fig. 3.2. Graphical representation of lowest-order elastic scattering amplitudes

Each amplitude $f^{(n)}(k, k')$ may be described by a simple graphical representation whose elements are associated with those of the amplitude and from which the latter can be easily reproduced. Figure 3.2 shows such graphs for the simplest elastic scattering amplitudes of the lowest orders. The initial wave function $\varphi_k(\pmb{r}_1)$ describing the incident particle with the momentum k is associated with the line of the graph that goes from infinity to the point $\pmb{r}_1$; the vertex at the point $\pmb{r}_1$ corresponds to the interaction potential $V(\pmb{r}_1)$; the line that connects the neighbor vertices $\pmb{r}_1$ and $\pmb{r}_2$ arises from the Green function $G_0(\pmb{r}_2, \pmb{r}_1)$, etc.; the line from the last vertex $\pmb{r}_n$ to infinity corresponds to the final state wave function $\varphi_{k'}^*(\pmb{r}_n)$.

The graphs give a sketch of the scattering process. The incident particle is described by the line that goes from infinity; it is scattered by the potential $V(\pmb{r}_1)$ at the point $\pmb{r}_1$, then gets to the point $\pmb{r}_2$, is scattered again, and so on. The particle motion between two subsequent scattering events is described by the Green function. For this reason, the latter is referred to as the scattering function or the propagator. The number of scattering events undergone by the particle determines the amplitude order. Having been scattered at the point $\pmb{r}_n$ for the last time, the particle goes to infinity.

In order to reproduce the amplitude from the graph, one has to write the factors associated with the graph elements and integrate over the coordinates of all the vertices. Having multiplied the result by $-\mu/2\pi\hbar^2$, one obtains the amplitude $f^{(n)}(k, k')$. It is clear that the nth order amplitude is associated with a n-vertex graph. The graph approach appreciably simplifies the calculation procedure for the high-order corrections of the perturbation theory.

The scattering picture becomes even more clear in the momentum representation of perturbation theory. The amplitude correction of the nth order with respect to the perturbation may be written as

$$f^{(n)}(k, k') = -\frac{\mu}{2\pi\hbar^2} \iint \cdots \int \frac{dk_1}{(2\pi)^3} \frac{dk_2}{(2\pi)^3} \cdots \frac{dk_{n-1}}{(2\pi)^3}$$

$$\times V_{k'k_{n-1}} \frac{1}{E - E_{k_{n-1}} + i0} V_{k_{n-1}k_{n-2}} \cdots$$

$$\times V_{k_2 k_1} \frac{1}{E - E_{k_1} + i0} V_{k_1 k} \quad . \tag{3.104}$$

The simplest amplitude graphs in the momentum representation are shown in Fig. 3.3. The propagators are described by the expressions

$$\frac{1}{E - E_{k_i} + i0} \quad ,$$

and the graph vertices are associated with the Fourier components of the potential

$$V_{k'k} = \int dr \, \exp[i(k - k')r] \, V(r) \quad . \tag{3.105}$$

The inner lines correspond to the intermediate momenta $k_1, k_2, \ldots, k_{n-1}$, over which integration must be carried out. The incoming and outgoing lines of the graph are associated with the initial and final particle momenta k and k'.

Note that formula (3.104) may also be derived from the expansion (2.54) within the context of the relation (3.73).

Now let us discuss the applicability of the successive iteration approach. According to (3.94), the first correction to the wave function may be written as

$$\psi_k^{(1)}(r) = -\frac{\mu}{2\pi\hbar^2} \int dr' \frac{\exp(ikr')}{r'} V(r - r') \varphi_k(r - r') \quad . \tag{3.106}$$

Evidently, expansion (3.93) is valid if

$$|\psi_k^{(1)}| \ll |\psi_k^{(0)}| \quad . \tag{3.107}$$

We assume the range R of the potential V to be finite, i.e., that the potential is appreciably nonzero only within the region with linear dimensions R. If the particle energy is sufficiently low for the quantity kR to be smaller than or on the order of unity, then the factor $\exp(ikr')$ in the integrand of (3.106) is insignificant and the first correction to the wave function is on the order of

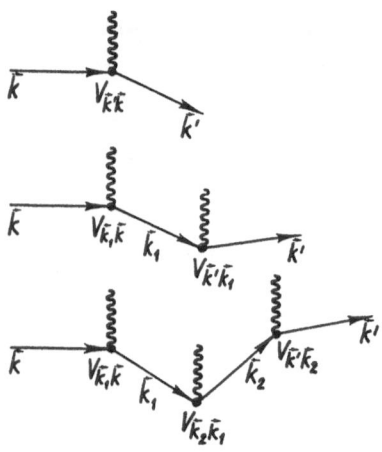

Fig. 3.3. Simplest graphical representation of elastic scattering amplitudes in the momentum representation

$$\psi_k^{(1)} \simeq \frac{\mu}{\hbar^2} R^2 \overline{V} \psi_k^{(0)} \quad , \tag{3.108}$$

where $\overline{V}$ is the effective potential. Substituting (3.108) into (3.107) yields the applicability condition for the perturbation theory:

$$\overline{V} \ll \frac{\hbar^2}{\mu R^2} \quad , \qquad kR \lesssim 1 \quad . \tag{3.109}$$

This infers that for the perturbation theory to be applicable, the effective potential must be small compared to the particle kinetic energy localized in the domain of potential action.

If particle energy is sufficiently high, then $kR \gg 1$; the integral in (3.106) is small by virtue of the oscillating factor $\exp(ikr')$ in the integrand. Let us estimate the correction (3.106) for this case. Since $\psi_k^{(0)}(r) = \exp(ikr)$, (3.106) may be rewritten as

$$\psi_k^{(1)}(r) = -\frac{\mu}{2\pi\hbar^2} \psi_k^{(0)}(r) \int dr' \frac{\exp(-ikr')}{r'} V(r - r') e^{-ikr'} \quad . \tag{3.110}$$

Since the potential is a slowly varying function at distances on the order of k^{-1}, the mean value of the potential can be put before the integral in the right-hand part of (3.110) and it is not difficult to carry out integration over angles to obtain

$$\psi_k^{(1)}(r) \simeq -\frac{\mu}{\hbar^2} \psi_k^{(0)}(r) \int_0^\infty dr' \frac{1 - \exp(2ikr')}{ik} \overline{V(r - r')} \quad . \tag{3.111}$$

We disregard the fast oscillating function $\exp(2ikr')$ which is much smaller than unity and make use of the relation

$$\int_0^\infty dr' \overline{V(r - r')} \simeq R\overline{V} \quad .$$

Then we have

$$\psi_k^{(1)} \simeq \frac{\mu}{\hbar^2} \frac{R}{k} \overline{V} \psi_k^{(0)} \quad . \tag{3.112}$$

Substituting (3.112) into (3.107), we find the applicability condition of the perturbation theory to be

$$\overline{V} \ll \frac{\hbar^2}{\mu R^2} kR \quad , \qquad kR \gg 1 \quad . \tag{3.113}$$

This condition is considerably less stringent than (3.109).

3.8 High Energy Approximation

Let us find the high energy elastic scattering amplitude for the case when the Born approximation is inapplicable. If $\overline{V}$ is not small compared to $\hbar^2/\mu R^2$, then the Born approximation is valid if the condition (3.113) is satisfied, i.e., if energy is sufficiently high, so that $kR \gg 1$. Note that the requirement that the interaction $\overline{V}$ must be much smaller than the energy $E = \hbar^2 k^2/2\mu$ is insufficient for the Born approximation to hold. It is not difficult to verify that for $kR \gg 1$, a range of rather high energies $E \gg \overline{V}$ may exist, for which the Born approximation is invalid. This occurs if the condition (3.113) is violated, i.e., if

$$\overline{V} \underset{\sim}{>} \frac{\hbar^2}{\mu R^2} kR \quad ,$$

or

$$\overline{V}^2 \underset{\sim}{>} \frac{\hbar^2}{\mu R^2} E \quad . \tag{3.114}$$

The range in question may be immediately found from (3.114); we have

$$\left(\overline{V}^2 \Big/ \frac{\hbar^2}{\mu R^2} \right) \underset{\sim}{>} E \gg \overline{V} \quad . \tag{3.115}$$

This range exists provided that

$$\overline{V} \gg \frac{\hbar^2}{\mu R^2} \quad . \tag{3.116}$$

In order to calculate the scattering amplitude in the energy range (3.115), we first derive an approximate expression for the wave function ψ_k. If the energy E is much higher than the potential energy V at any point of the space, then the coordinate dependence of the wave function must be approximately similar to that for the free motion. Hence, the wave function ψ_k can be approximately written as

$$\psi_k = \Phi \, e^{ikr} \quad , \tag{3.117}$$

where $\varPhi$ is a function of coordinates which varies much slower than the factor $\exp(ikr)$. We take the z axis along the direction of the free particle motion. Substituting (3.117) into the Schrödinger equation (3.2) and retaining in $\Delta \psi_k$ only the terms in which the factor $\exp(ikz)$ is differentiated at least once, we obtain the equation

$$2ik \frac{\partial \varPhi}{\partial z} = \frac{2\mu}{\hbar^2} V \varPhi \quad ,$$

from which we obtain

$$\psi_k(r) = \exp\left[ikz - \frac{i}{\hbar v} \int_{-\infty}^{z} dz' \, V(x, y, z')\right] \quad . \tag{3.118}$$

Making use of the general formula (3.15), we find the scattering amplitude to be

$$f(k, k') = -\frac{\mu}{2\pi\hbar^2} \int dr \, e^{iqr} \, V(r) \, \exp\left[-\frac{i}{\hbar v} \int_{-\infty}^{z} dz' \, V(x, y, z')\right] . \tag{3.119}$$

Because scattering at small angles (with small momentum transfer $q \ll k$) is considerable at high energies, the vector q in (3.119) may be assumed to be perpendicular to the initial wave vector k, i.e., to lie in the plane (x, y). We introduce the cylindrical coordinates $r = (b, z)$ and carry out integration over z in (3.119). Thus we obtain the scattering amplitude given by

$$f(k, k') = \frac{ik}{2\pi} \int db \, e^{iqb} \left\{ 1 - \exp\left[-\frac{i}{\hbar v} \int_{-\infty}^{\infty} dz \, V(b, z)\right] \right\}$$

$$= ik \int_{0}^{\infty} db \, b J_0(qb) \left\{ 1 - \exp\left[-\frac{i}{\hbar v} \int_{-\infty}^{\infty} dz \, V(b, z)\right] \right\} . \tag{3.120}$$

This formula holds under the condition

$$kR \gg 1 \quad , \tag{3.121}$$

where R is the range of the potential V.

The momentum transfer q is related to the scattering angle ϑ as usual,

$$q = 2k \sin \frac{\vartheta}{2} \quad .$$

It is clear that the scattering amplitude (3.120) is nonzero only for small angles ($\vartheta \simeq 1/kR \ll 1$). Thus, the high energy scattering acquires diffraction nature.

Formula (3.120) can be derived from the scattering amplitude expansion in terms of partial components (3.75). Indeed, if the condition (3.121) is satisfied, then the scattering occurs in the range of small angles ($\vartheta \ll 1$) and mainly large l contribute to the expansion (3.75). Then one may apply the approximate expression for the Legendre polynomials,

$$P_l(\cos \vartheta) \simeq J_0(l\vartheta) \quad .$$

Going in (3.75) from summation over l to integration over the impact parameter $b = l\lambda$, and bearing in mind that

$$S_l \rightarrow S(b) = \exp\left[-\frac{i}{\hbar v}\int_{-\infty}^{\infty} dz\, V(b, z)\right] \quad,$$

one obtains formula (3.120).

Problems

3.1 Find the Green function $G(E; r, r')$ for a system of two interacting particles under the assumption that the energy spectrum of the system consists of a single discrete level E_0 associated with the bound state $\varphi_0(r)$ and a continuum of positive energies E_k which correspond to unbound states $\varphi_k(r)$, i.e.,

$$\varphi_0(r) = \sqrt{\frac{\alpha}{2\pi}}\,\frac{e^{-\alpha r}}{r} \quad, \qquad E_0 = \frac{\hbar^2\alpha^2}{2\mu} \quad;$$

$$\text{(3.122)}$$

$$\varphi_k(r) = e^{ikr} - \frac{1}{\alpha + ik}\,\frac{e^{ikr}}{r} \quad, \qquad E_k = \frac{\hbar^2 k^2}{2\mu} \quad.$$

The functions (3.112) occur in the case of zero-range particle interaction. In the zero-range approximation, the interaction is described by the single parameter α which determines both the ground state energy and the scattering amplitude for unbound states. The limiting value $\alpha \rightarrow \infty$ corresponds to the case of non-interacting particles.

The functions $\varphi_k(r)$ and $\varphi_{k'}(r)$ associated with different wave vectors k and k' are orthogonal,

$$\int dr\, \varphi_k(r)\, \varphi_{k'}^*(r) = (2\pi)^3\, \delta(k - k') \quad. \tag{3.123}$$

The functions $\varphi_k(r)$ are also orthogonal to the wave function $\varphi_0(r)$ describing the bound states, so that

$$\int dr\, \varphi_k(r)\, \varphi_0(r) = 0 \quad. \tag{3.124}$$

The function $\varphi_0(r)$ is normalized by the condition

$$\int dr\, \varphi_0^2(r) = 1 \quad. \tag{3.125}$$

It may be easily verified that the set of functions (3.122) satisfy the completeness condition

$$\varphi_0(r)\,\varphi_0(r') + \int \frac{dq}{(2\pi)^3}\,\varphi_q(r)\,\varphi_q^*(r') = \delta(r - r') \quad. \tag{3.126}$$

According to the general definition (3.66), the total Green function $G(E; \boldsymbol{r}, \boldsymbol{r}')$ for a system with the above mentioned energy spectrum may be written as

$$G(E; \boldsymbol{r}, \boldsymbol{r}') \equiv \langle \boldsymbol{r}' | G(E) | \boldsymbol{r} \rangle$$

$$= \frac{\varphi_0(r) \, \varphi_0(r')}{E - E_0} + \int \frac{d\boldsymbol{q}}{(2\pi)^3} \frac{\varphi_q(\boldsymbol{r}) \, \varphi_q^*(\boldsymbol{r}')}{E - E_q + i0} \quad . \tag{3.127}$$

We employ the explicit expressions (3.122) and carry out integration over the angles. The result is

$$G(E; \boldsymbol{r}, \boldsymbol{r}') = -\frac{\mu}{2\pi\hbar^2} \left\{ \frac{\exp(ik|\boldsymbol{r} - \boldsymbol{r}'|)}{|\boldsymbol{r} - \boldsymbol{r}'|} - \frac{2\alpha}{\alpha^2 + k^2} \frac{\exp[-\alpha(r + r')]}{rr'} \right.$$

$$\left. - \frac{1}{\pi rr'} \int_{-\infty}^{\infty} dq \, q \, \frac{\exp[iq(r + r')]}{(q - i\alpha)(k^2 - q^2 + i0)} \right\} \quad ,$$

$$E \equiv \frac{\hbar^2 k^2}{2\mu} \quad .$$

In order to calculate the integral, we pass to the complex plane and take the integration contour as in (3.11). Thus we find the Green function to be given by

$$G(E; \boldsymbol{r}, \boldsymbol{r}') = -\frac{\mu}{2\pi\hbar^2} \left\{ \frac{\exp(ik|\boldsymbol{r} - \boldsymbol{r}'|)}{|\boldsymbol{r} - \boldsymbol{r}'|} - \frac{1}{\alpha + ik} \frac{\exp[ik(r + r')]}{rr'} \right\}$$

$$E = \frac{\hbar^2 k^2}{2\mu} \quad . \tag{3.128}$$

A comparison of this expression to (3.9) shows that the second term within the curly braces in (3.128) describes the interaction effect. For $\alpha \to \infty$, (3.128) reduces to the Green function for noninteracting particles.

The long-distance asymptotic of the Green function (3.128) is given by

$$G(E; \boldsymbol{r}, \boldsymbol{r}') \underset{r \to \infty}{\to} -\frac{\mu}{2\pi\hbar^2} \frac{\exp(ikr)}{r} \varphi_{-\boldsymbol{k}'}(\boldsymbol{r}') \quad , \qquad \boldsymbol{k}' \equiv \frac{\boldsymbol{r}}{r} k \quad . \tag{3.129}$$

[cf. (3.13)].

3.2 Calculate the Green function for a system of three noninteracting particles.

Let us denote the particle coordinates and masses by $\boldsymbol{r}_1$, $\boldsymbol{r}_2$, $\boldsymbol{r}_3$ and m_1, m_2, m_3, respectively. We transform to the c.m.s. and introduce the relative coordinates

$$\boldsymbol{x} = \boldsymbol{r}_1 - \frac{m_2 \boldsymbol{r}_2 + m_3 \boldsymbol{r}_3}{m_2 + m_3} \quad , \qquad \boldsymbol{r} = \boldsymbol{r}_2 - \boldsymbol{r}_3 \quad . \tag{3.130}$$

The Hamiltonian of the system may be written as

$$H_0 = -\frac{\hbar^2}{2m} \nabla_{\boldsymbol{x}}^2 - \frac{\hbar^2}{2\mu} \nabla_{\boldsymbol{r}} \quad , \tag{3.131}$$

where m and μ are the reduced masses,

$$m = \frac{m_1(m_2 + m_3)}{m_1 + m_2 + m_3} \quad , \qquad \mu = \frac{m_2 m_3}{m_2 + m_3} \quad . \tag{3.132}$$

According to the general definition (3.21), the Green function in the coordinate representation $G_0(E)$ is determined by

$$\langle x', r' | G_0(E) | r, x \rangle = \int \frac{dp}{(2\pi)^3} \int \frac{dq}{(2\pi)^3}$$
$$\times \frac{\exp\left[ip(x - x') + iq(r - r') \right]}{E - E_p - E_q + iO} \quad , \tag{3.133}$$
$$E_p = \frac{\hbar^2 p^2}{2m} \quad , \qquad E_q = \frac{\hbar^2 q^2}{2\mu} \quad .$$

After integration in the space of the vector q and over the angle variables of the vector p is performed, we obtain

$$\langle x', r' | G_0(E) | r, x \rangle = \frac{i}{8\pi^3} \frac{\mu}{\hbar^2} \frac{1}{|r - r'|} \frac{1}{|x - x'|} \int_{-\infty}^{\infty} dp\, p$$
$$\times \exp\left[i \sqrt{\frac{2}{\hbar^2}(E - E_p + iO)} |r - r'| \right.$$
$$\left. + ip|x - x'| \right] \quad . \tag{3.134}$$

This formula is in agreement with (3.69) where the Green functions of the variables x and r are defined as in (3.9).

We introduce the dimensionless quantities x and y, defined by

$$x \equiv \sqrt{\frac{2m}{\hbar^2} E} \; |x - x'| \quad , \qquad y \equiv \sqrt{\frac{2\mu}{\hbar^2} E} \; |r - r'| \quad ,$$

and the dimensionless integration variable ξ,

$$\xi = p \Big/ \sqrt{\frac{2m}{\hbar^2} E} \quad .$$

Then

$$\langle x', r' | G_0(E) | r, x \rangle = \frac{1}{2\pi^3} (m\mu)^{3/2} \frac{E^2}{\hbar^6} \frac{1}{xy} \tag{3.135}$$
$$\times \frac{\partial}{\partial x} \int_{-\infty}^{\infty} d\xi \, \exp\left(-y \sqrt{\xi^2 - 1 - iO} + ix\xi \right) .$$

To calculate the integral in (3.135), it is convenient to introduce complex variables ξ. Since $\xi = \pm(1 + iO)$ are the branch points, it is necessary to cut the complex ξ plane from $-\infty$ to -1 and from 1 to ∞. (The left cut lies somewhat lower than the real axis, the right one lies above it). We deform the integration contour as shown in Fig. 3.4 and take the value of the root lying on the upper edge of the cut with the opposite sign. Thus we have

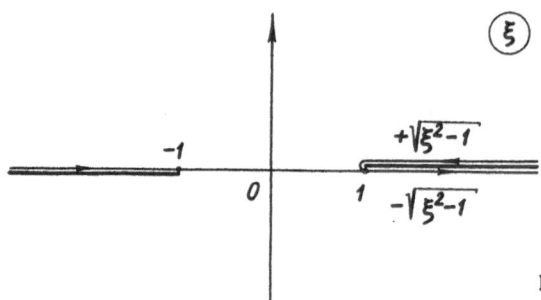

Fig. 3.4. Cuts and integration contour in the complex ξ plane for (3.135)

$$I = \int_1^\infty d\xi \, \exp\left(i x \xi\right) \left[\exp\left(-y \sqrt{\xi^2 - 1}\right) - \exp\left(y \sqrt{\xi^2 - 1}\right)\right] \quad .$$

This expression easily reduces to the integral representation of the Hankel function, so that

$$I = i\pi \, \frac{y}{\sqrt{x^2 + y^2}} \, H_1^{(1)}\left(\sqrt{x^2 + y^2}\right) \quad .$$

Because

$$\frac{\partial}{\partial x} \, \frac{y}{\sqrt{x^2 + y^2}} \, H_1^{(1)}\left(\sqrt{x^2 + y^2}\right) = -\frac{xy}{x^2 + y^2} \, H_2^{(1)}\left(\sqrt{x^2 + y^2}\right) \quad ,$$

we obtain

$$\langle \boldsymbol{x}', \boldsymbol{r}' | G_0(E) | \boldsymbol{r}, \boldsymbol{x} \rangle = -\frac{i}{2\pi^2} \, \frac{(m\mu)^{3/2}}{\hbar^6} \, \frac{E^2}{x^2 + y^2} \, H_2^{(1)}\left(\sqrt{x^2 + y^2}\right) , \quad (3.136)$$

where

$$x^2 + y^2 = \frac{2E}{\hbar^2} \left\{ m(\boldsymbol{x} - \boldsymbol{x}')^2 + \mu(\boldsymbol{r} - \boldsymbol{r}')^2 \right\} \quad .$$

Equation (3.136) determines the Green function for a system of three noninteracting particles.

3.3 Find the Green function for two noninteracting particles in the high energy approximation.

Suppose the energy of relative particle motion is $E = p^2/2m$ ($\boldsymbol{p} = \hbar \boldsymbol{k}$ is the relative momentum). The Hamiltonian $H_0 = \hat{p}^2/2m$, where $\hat{\boldsymbol{p}}$ is the momentum operator, may be written as

$$H_0 = \frac{p^2}{2m} - \frac{1}{2m} \, (\boldsymbol{p} + \hat{\boldsymbol{p}})\,(\boldsymbol{p} - \hat{\boldsymbol{p}})$$

$$\equiv \frac{p^2}{2m} - \frac{1}{m} \, \boldsymbol{p}\,(\boldsymbol{p} - \hat{\boldsymbol{p}}) + \frac{1}{2m} \, (\boldsymbol{p} - \hat{\boldsymbol{p}})^2 \quad . \quad (3.137)$$

The high energy approximation implies disregarding the last term in the right-hand part of (3.137), i.e., instead of the Hamiltonian H_0 we introduce the so-called eikonal Hamiltonian $\tilde{H}_0$:

$$\tilde{H}_0 = \frac{p^2}{2m} - \frac{1}{m} p(p - \hat{p}) \quad . \tag{3.138}$$

This substitution is justified only for diffractive scattering, with only small values of the momentum transfer being important. In the high energy approximation, instead of energy conservation one considers the conservation of the momentum projection on the direction of k, i.e.,

$$pk = \text{const} \quad . \tag{3.139}$$

Because the Hamiltonian H_0 depending on the square of the momentum operator $\hat{p}$ is replaced by the eikonal Hamiltonian that is linear with respect to $\hat{p}$, the scattering problem is simplified. The eikonal Hamiltonian $\tilde{H}_0$ depends only on the momentum component along the direction of the initial momentum k, which infers that the high energy approximation entirely disregards transverse motion.

Having substituted the eikonal Hamiltonian $\tilde{H}_0$ for the nonperturbed Hamiltonian H_0 in the expression (3.21) for the Green function, we thus define the eikonal Green function $\tilde{G}_0(E)$ as

$$\langle p' | \tilde{G}_0(E) | p \rangle = \frac{(2\pi\hbar)^3}{(1/m)\, p\, (p - \hat{p}) + iO}\, \delta(p - p') \quad . \tag{3.140}$$

In the momentum representation, the functions G_0 and $\tilde{G}_0$ are only slightly different within a small solid angle about the direction of k. If the incident particle energy is sufficiently high, this range of angles is of the greatest interest since only here is the high energy scattering amplitude considerable.

The eikonal Green function in the coordinate representation is given by [3.2]

$$\langle x' | \tilde{G}_0(E) | x \rangle = -i\, \frac{m}{\hbar^2 k}\, \exp\big[ik(z' - z)\big]\, \theta(z' - z)\, \delta(b' - b) \quad , \tag{3.141}$$

where $\theta(z)$ is the Heaviside function. The delta function $\delta(b' - b)$ (b and b' are the components of the vectors x and x' in the plane perpendicular to the vector k) appears in the eikonal Green function $\tilde{G}_0$ because the transverse particle motion is disregarded.

Substituting $\tilde{G}_0$ for G_0 reduces the integral equation (3.20) or (2.55) to an ordinary first-order differential equation; its solution is given by

$$
\langle x' | t | x \rangle = -i\, \frac{\hbar^2 k}{m}\, \exp\big[ik(z' - z)\big]
$$
$$
\times \frac{d}{dz'} \left\{ \theta(z' - z) \right.
$$
$$
\left. \times \frac{d}{dz}\, \exp\left[-i\, \frac{m}{\hbar^2 k}\right] \int_z^{z'} dz''\, V(b, z'') \right\} \delta(b' - b) \quad . \tag{3.142}
$$

For $k_z = k'_z = k$ (on the energy shell), the matrix element (3.142) takes the form

$$\langle b'|t|b \rangle \equiv \langle k'_z = k, b' \,|t|\, b, k_z = k \rangle$$
$$= -\mathrm{i}\,\frac{\hbar^2 k}{m}\,\omega(b)\,\delta(b' - b) \quad, \tag{3.143}$$

where $\omega(b)$ is the so-called profile function,

$$\omega(b) = 1 - \exp\left[-\mathrm{i}\,\frac{m}{\hbar^2 k}\,\int_{-\infty}^{\infty} dz\,V(b, z)\right] \quad. \tag{3.144}$$

Making use of (3.143), we find the high energy elastic scattering amplitude to be given by the expression

$$f(k, k') = \frac{\mathrm{i}k}{2\pi}\,\int db\,\exp[\mathrm{i}(k - k')b]\,\omega(b) \quad, \tag{3.145}$$

which reproduces (3.120).

3.4 Find the amplitude of high energy particle scattering by a system of several bound particles.

Suppose V is the energy of interaction of the incident particle with each constituent of the scatterer, and H' is the intrinsic Hamiltonian of the scatterer. Particle scattering by a many-body system is described by the transition operator T which is governed by the equation

$$T = V + V G'_0 T \quad, \tag{3.146}$$

where the Green function G'_0 describes both probe-target relative motion and the intrinsic motion within the scatterer. With g' being the Green function responsible for the intrinsic motion in the scatterer, the Green function G'_0 may be written as

$$G'_0(E) = -\frac{1}{2\pi\mathrm{i}}\,\int_{-\infty}^{\infty} dE'\,G_0(E - E')\,g'(E') \quad, \tag{3.147}$$

where G_0 is defined by (3.21). In the high energy approximation, G_0 must be replaced in (3.147) by $\tilde{G}_0$.

In the simplest case of a two-particle scatterer, the intrinsic Hamiltonian is

$$H' = -\frac{\hbar^2}{2}\,\Delta + U(r) \quad, \tag{3.148}$$

where $U(r)$ is the potential describing the interaction of the scatterer constituents, r is the relative distance between the particles, μ is the reduced mass. Suppose we know both the spectrum of eigenvalues E_f of the Hamiltonian (3.148) (in the general case it consists of discrete levels and a continuum) and the eigenfucntions $\varphi_f(r)$. Then the intrinsic Green function g' may be written in the form

$$\langle r'|g'(E)|r\rangle = \sum_f \frac{\varphi_f^*(r')\,\varphi_f(r)}{E - E_f + i0} \quad, \tag{3.149}$$

where the summation over f includes summation over the discrete levels and integration over the continuum part.

If particle motion within the scatterer is disregarded, then the spectrum of eigenvalues of (3.148) degenerates to the single value $E_f = 0$. Then, since the set of eigenfunctions $\varphi_f(r)$ is complete, we find from (3.149) that

$$\langle r'|g'(E)|r\rangle = \frac{1}{E + i0}\,\delta(r' - r) \quad. \tag{3.150}$$

The Green function entering (3.149, 150) is written in the coordinate representation. It is not difficult to pass from (3.149, 150) to any other representation.

Substituting (3.150) into (3.147) and replacing G_0 by $\tilde{G}_0$, we thus find that

$$\tilde{G}_0' = \tilde{G}_0(E) \quad. \tag{3.151}$$

This relation is also valid for the particle scattered by a target consisting of three or more particles if the intrinsic motion in the scatterer is ignored. Relation (3.151) (the freezing condition for the scatttering particles) is necessary for the diffraction approximation.

If only two-particle interactions are taken into account, the potential V may be written as a sum of two potentials associated with the interaction between the incident particle and each constituent of the scatterer, i.e.,

$$V = \sum_i V(x - x_i) \quad. \tag{3.152}$$

(The vector x describes the relative distance between the incident particle and the center of mass of the scatterer; individual particle coordinate vectors x_i are expressed in terms of intrinsic coordinates). Substituting (3.151) for G_0' in (3.146) and making use of (3.152), we find the transition operator in the diffraction approximation. In particular, on the energy shell (for $k_z = k_z' = k$) we have

$$\langle b', r'\,|T|\,r, b\rangle = -\frac{\hbar^2 k}{m}\,\omega_{(N)}(b, r)\,\delta(r' - r)\,\delta(b' - b) \quad, \tag{3.153}$$

where $\omega_{(N)}(b, r)$ is the total profile function which depends on the individual particle profile functions $\omega_i(b - b_i)$, so that

$$\omega_{(N)}(b, r) = 1 - \prod_{i=1}^N \{1 - \omega_i(b - b_i)\} \tag{3.154}$$

(r is the set of coordinates of the intrinsic degrees of freedom of the scatterer). As follows from (3.153), the amplitude of particle scattering by a system of bound particles is given by [3.3, 4]

$$\mathcal{F}_{of}(q) = \frac{ik}{2\pi} \int db\, e^{iqb} \left[\varphi_f(r),\, \omega_{(N)}(b, r)\, \varphi_0(r) \right] \quad, \tag{3.155}$$

where $\varphi_0(r)$ and $\varphi_f(r)$ are the intrinsic wave functions of the scatterer before and after the collision. The profile function $\omega_{(N)}$ depends on the quantities which determine individual particle scattering amplitudes. Therefore, relations (3.154, 155) describe general relationship between the amplitudes of particle scattering by a many-body system and by individual constituents of this system.

4. Particle Wave Functions in the External Field

4.1 Partial Wave Expansion

Let us consider in detail the problem of particle scattering in the central field, assuming that the potential V depends only on the modulus of the distance r. We have already mentioned that this problem reduces to solving the Schrödinger equation for positive energies,

$$\{H_0 + V(r) - E\} \psi_k = 0 \quad , \tag{4.1}$$

with the boundary condition at large distances that the solution must be a sum of the incident plane and diverging scattered waves, i.e.,

$$\psi_k \underset{r \to \infty}{\longrightarrow} \exp(i\boldsymbol{k}\boldsymbol{r}) + f(\boldsymbol{k}, \boldsymbol{k}') \frac{\exp(ikr)}{r} \quad , \qquad E = \frac{\hbar^2 k^2}{2\mu} \quad . \tag{4.2}$$

Since $H_0 = -(\hbar^2/2\mu)\Delta$, (4.1) may be rewritten as

$$\{\Delta - V_r(r) + k^2\} \psi_k = 0 \quad , \tag{4.3}$$

where $V_r(r) \equiv 2\mu V(r)/\hbar^2$ will be referred to as the reduced potential. The Laplace operator in the spherical coordinate system is

$$\Delta = \frac{1}{r^2} \frac{\partial}{\partial r} \left(r^2 \frac{\partial}{\partial r} \right) - \frac{L^2}{r^2} \quad , \tag{4.4}$$

where L^2 is the operator of the angular momentum squared,

$$L^2 = -\left\{ \frac{1}{\sin \vartheta} \frac{\partial}{\partial \vartheta} \left(\sin \vartheta \frac{\partial}{\partial \vartheta} \right) + \frac{1}{\sin^2 \vartheta} \frac{\partial^2}{\partial \varphi^2} \right\} \quad . \tag{4.5}$$

The radial part of the Laplace operator may be described by several expressions:

$$\frac{1}{r^2} \frac{\partial}{\partial r} \left(r^2 \frac{\partial}{\partial r} \right) \equiv \frac{\partial^2}{\partial r^2} + \frac{2}{r} \frac{\partial}{\partial r} \equiv \frac{1}{r} \frac{\partial^2}{\partial r^2} (r \ldots) \quad . \tag{4.6}$$

Because central potentials are spherically symmetric, the angular momentum is an integral of motion. The states associated with different angular momenta take part in the scattering process independently. Therefore, it is convenient to write the solution $\psi_k(r)$ of (4.3) as a superposition of partial waves associated with definite values of the angular momentum. Taking the z axis to be directed along the incident particle momentum $\boldsymbol{k}$, we can write the partial wave expansion of the wave function $\psi_{\boldsymbol{k}}(\boldsymbol{r})$ as

$$\psi_{\boldsymbol{k}}(\boldsymbol{r}) \sim \sum_l \psi_{l,k}(r)\, P_l(\cos\vartheta) \quad , \tag{4.7}$$

where $P_l(\cos\vartheta)$ is the eigenfunction of the operator of squared angular momentum $\boldsymbol{L}^2$, i.e.,

$$\boldsymbol{L}^2 P_l(\cos\vartheta) = l(l+1)\, P_l(\cos\vartheta) \quad . \tag{4.8}$$

The expansion (4.7) may be easily transformed to an arbitrary coordinate system within the context of the addition rule (2.57) for spherical functions. Substituting (4.7) into (4.3) and making use of (4.8), we obtain a differential equation for the radial functions $\psi_{l,k}(r)$,

$$\left\{ \frac{1}{r^2} \frac{d}{dr}\left(r^2 \frac{d}{dr} \right) - \frac{l(l+1)}{r^2} - V_r(r) + k^2 \right\} \psi_{l,k}(r) = 0 \quad . \tag{4.9}$$

First of all we consider equation (4.9) for $V_r(r) = 0$, i.e.,

$$\left\{ \frac{1}{r^2} \frac{d}{dr}\left(r^2 \frac{d}{dr} \right) - \frac{l(l+1)}{r^2} + k^2 \right\} \psi_{l,k}^0(r) = 0 \quad . \tag{4.10}$$

The two independent solutions of (4.10) can be taken to be the spherical Bessel function $j_l(kr)$ and the spherical Neumann function $n_l(kr)$, or the first- and second-order spherical Hankel functions $h_l^{(+)}(kr)$ and $h_l^{(-)}(kr)$. The lowest-order spherical functions are given by

$$j_l(x) = \sqrt{\frac{\pi}{2x}}\, J_{l+1/2}(x) \quad ,$$

$$n_l(x) = \sqrt{\frac{\pi}{2x}}\, N_{l+1/2}(x) \quad , \tag{4.11}$$

$$h_l^{(\pm)}(x) = j_l(x) \pm i n_l(x) \quad ,$$

where $J_{l+1/2}(x)$ and $N_{l+1/2}(x)$ are the Bessel and Neumann functions of half-integer orders. For small x, the approximate expressions

$$j_l(x) \to \frac{x^l}{(2l+1)!!} \quad , \qquad n_l(x) \to -\frac{(2l-1)!!}{x^{l+1}} \quad , \qquad x \ll 1 \tag{4.12}$$

occur. If x is very large, then the following asymptotic formulas hold:

$$j_l(x) \to \frac{\sin[x - (l\pi/2)]}{x} \quad ,$$

$$n_l(x) \to -\frac{\cos[x - (l\pi/2)]}{x} \quad , \tag{4.13}$$

$$h_l^{(\pm)}(x) \to \mp i \frac{\exp\{\pm i[x - (l\pi/2)]\}}{x} \quad , \qquad x \gg 1 \quad .$$

The general solution of (4.10) may be written as a superposition of two independent solutions j_l and n_l, or $h_l^{(+)}$ and $h_l^{(-)}$.

Now let us study equation (4.9) for $V_r(r) \neq 0$ under the assumption that $V_r(r)$ is a monotonic function of r. Here two possibilities exist. If $V_r(r) > 0$ for any r, then the potential describes repulsion; if $V_r(r) < 0$ for any r, the potential is attractive.

In the case of attractive forces, the system can possess bound states associated with a sequence of discrete energy levels. We denote the relevant energies as $E_n = -\left(\hbar^2 \kappa_n^2\right)/2\mu < 0$, where n is the level number. As follows from (4.9), the bound state wave function asymptotic for large r is

$$\psi_n(r) \to \beta_n \frac{\exp(-\kappa_n r)}{r} \quad , \qquad r \to \infty \quad , \tag{4.14}$$

where β_n is a constant.

We assume that the potential $V_r(r)$ decreases faster than r^{-1} as r grows, i.e., its asymptotic for $r \to \infty$ is

$$V_r(r) \sim \frac{1}{r^{1+\varepsilon}} \quad , \qquad \varepsilon > 0 \quad . \tag{4.15}$$

If this condition is satisfied, the asymptotic of the wave function $\psi_{l,k}$ [solution of (4.9)] can be written as a linear combination of independent solutions of (4.10). Indeed, potential variations are rather smooth at large distances from the center, hence the wave function asymptotic may be calculated in the quasiclassical approximation. For the S-case, we have

$$\psi_{0,k}(r) \sim \frac{1}{r} \sin \int_0^r dr \sqrt{k^2 - V_r(r)} \quad . \tag{4.16}$$

We take the distance r' to be sufficiently large for the condition $k^2 \gg V_r(r')$ to be satisfied. Then, for any $r \geq r'$, we have

$$\sqrt{k^2 - V_r(r)} \simeq k - \frac{V_r(r)}{2k} \quad .$$

Having divided the integration range in (4.16) into two parts, 0 to r' and r' to r, we rewrite the function (4.16) as

$$\psi_{0,k} \sim \frac{1}{r} \sin(kr + \delta_0) \quad , \tag{4.17}$$

where

$$\delta_0 = -kr' + \int_0^{r'} dr \sqrt{k^2 - V_r(r)} - \frac{1}{2k} \int_{r'}^r dr \, V_r(r) \quad . \tag{4.18}$$

If for large distances $(r \to \infty)$

$$V_r(r) = \frac{b}{r^{1+\varepsilon}} \quad (\varepsilon > 0) \quad ,$$

then

$$\lim_{r \to \infty} \int_{r'}^{r} dr \, V_r(r) = \frac{b}{\varepsilon r'^{\varepsilon}} \quad .$$

Therefore, the phase shift (4.18) tends to a finite value as $r \to \infty$, and then (4.17) is a superposition of the solutions of equation (4.10). If $\varepsilon = 0$ (the case of Coulomb field $V_r(r) = b/r$), then

$$\int_{r'}^{r} dr \, V_r(r) = b \ln \frac{r}{r'} \quad . \tag{4.19}$$

In this case, the phase shift (4.18) increases as $\ln r$ with growing r. It is not difficult to verify that this conclusion also holds for the phase shifts with arbitrary l.

The potential $V_r(r)$ is spherically symmetric and monotonic, therefore, it can have singularities only at $r = 0$. If the potential is attractive, we assume that at $r = 0$, $V_r(r) \sim r^q$, $q \geq -2$. If this condition is not satisfied, then the energy of the system has no lower limit, i.e., the ground state energy is $-\infty$. If the potential is repulsive, there is no need to impose any restrictions on the nature of the singularity at $r = 0$.

Equation (4.9) must be supplemented with the boundary condition for $\psi_{l,k}$ which follows from (4.2). In order to derive it, we expand the incident plane wave in terms of partial waves,

$$e^{ikr} = 4\pi \sum_{lm} i^l j_l(kr) \, Y_{lm}^*(n) \, Y_{lm}(n') \quad , \tag{4.20}$$

where n and n' are the unit vectors directed along k and r. The proportionality coefficient in (4.7) may be written by analogy to (4.20), to ensure that

$$\psi_k(r) = \sum_{lm} i^l \, \psi_{l,k}(r) \, Y_{lm}^*(n) \, Y_{lm}(n') \quad . \tag{4.21}$$

Making use of the Möller operator definition, we obtain a relation for the functions (4.20, 21),

$$\psi_k(r) = \Omega_+ e^{ikr} \quad .$$

According to (4.13), the asymptotic of the plane wave radial part expansion is given by

$$\psi_{l,k}^0(r) \equiv 4\pi j_l(kr) \underset{r \to \infty}{\to} 4\pi \frac{\sin[kr - (l\pi/2)]}{kr} \quad , \tag{4.22}$$

or

$$\psi_{l,k}^0(r) \underset{r \to \infty}{\to} \frac{2\pi i}{kr} \left\{ \exp\left[-i\left(kr - \frac{l\pi}{2}\right)\right] - \exp\left[i\left(kr - \frac{l\pi}{2}\right)\right] \right\} \quad . \tag{4.23}$$

The first term within the curly braces describes the spherical wave that converges to the center, the second term is the diverging spherical wave.

Substituting (4.23) into (4.2) and expanding the scattering amplitude in terms of partial waves (3.104), we find the required asymptotic of the solution of (4.9) to be

$$\psi_{l,k}(r) \underset{r\to\infty}{\longrightarrow} \frac{2\pi i}{kr} \left\{ \exp\left[-i\left(kr - \frac{l\pi}{2}\right)\right] - S_l \exp\left[i\left(kr - \frac{l\pi}{2}\right)\right] \right\} \quad . \tag{4.24}$$

We see that far from the center the radial wave function $\psi_{l,k}$, similarly to $\psi^0_{l,k}$, is a superposition of spherical waves converging to and diverging from the center. Comparing (4.24) to (4.23) shows that the particle-field interaction modifies only the diverging wave amplitude, whereas the converging wave amplitude in (4.24) reproduces the relevant amplitude in (4.23) and is governed by the incident flux density. The amplitude variation of the wave diverging from the scattering center is described by the diagonal element of the scattering matrix S_l corresponding to the angular momentum quantum number l.

In general, the matrix elements S_l are complex. If only elastic scattering occurs, then S_l is a function of the real scattering phase shift δ_l, i.e.,

$$S_l = \exp(2i\delta_l) \quad . \tag{4.25}$$

Because the exponential is periodic, (4.25) describes the scattering phase shift ambiguously, with an uncertainty of $n\pi$, where n is an arbitrary integer. Usually one describes the scattering phase shift by an expression that vanishes as $k \to \infty$.

Substituting (4.25) into (4.24) yields the radial function asymptotic in the form

$$\psi_{l,k}(r) \underset{r\to\infty}{\longrightarrow} 4\pi \exp(i\delta_l) \frac{\sin\left[kr - (l\pi/2) + \delta_l\right]}{kr} \quad . \tag{4.26}$$

Comparing (4.26) to (4.22), we find that interaction is responsible for the phase shift δ_l in the asymptotic of the radial partial wave, which does not occur in the free wave asymptotic.

The relation between the radial functions $\psi_{l,k}(r)$ and $\psi^0_{l,k}(r)$ is determined by the Möller matrix diagonal element in the momentum representation, so that

$$\psi_{l,k}(r) = \Omega^{(l)}_+ \quad , \qquad \psi^0_{l,k}(r) \equiv 4\pi \, \Omega^{(l)}_+ \, j_l(kr) \quad . \tag{4.27}$$

If the potential range is finite, $V_r(r) = 0$ for $r > R$, then the exact solution of (4.9) off the interaction region (for $r > R$) may be written as

$$\psi_{l,k}(r) = a^{(-)}_l h^{(-)}_l(kr) + a^{(+)}_l h^{(+)}_l(kr) \quad , \qquad r > R \quad . \tag{4.28}$$

Substituting the asymptotics for $h^{(-)}_l$ and $h^{(+)}_l$ and comparing (4.28) to (4.24) for large r, one finds that the coefficient before $h^{(-)}_l$ in (4.28) is determined by the incident flux density,

$$a^{(-)}_l = 2\pi \quad ,$$

and the coefficient before $h_l^{(+)}$ is a function of the scattering matrix,

$$a_l^{(+)} = S_l a_l^{(-)} \quad . \tag{4.29}$$

Thus, the solution of (4.9) off the interaction region depends on the known functions $h_l^{(-)}$ and $h_l^{(+)}$ and the scattering matrix:

$$\psi_{l,k}(r) = 2\pi \left\{ h_l^{(-)}(kr) + S_l h_l^{(+)}(kr) \right\} \quad , \qquad r > R \quad . \tag{4.30}$$

This solution may also be expressed in terms of the functions $j_l(kr)$ and $n_l(kr)$ and the scattering phase shift δ_l, so that

$$\psi_{l,k}(r) = 4\pi \, e^{i\delta_l} \cos \delta_l \left\{ j_l(kr) - \tan \delta_l \, n_l(kr) \right\} \quad , \qquad r > R \quad . \tag{4.31}$$

If one knows the solution in the inner region, then the scattering phase shifts can be found from the continuity conditions for the wave functions and their derivatives on the interaction region boundary. The solution in the inner region ($r < R$) may be obtained by numerical integration of (4.9); for some potentials, it can be found explicitly. We denote the solution of (4.9) in the inner region by $\tilde{\psi}_{l,k}(r)$. The continuity conditions for the wave function and its derivative on the interaction region boundary ($r = R$) impose analogous conditions on the logarithmic derivatives, i.e.,

$$\frac{\tilde{\psi}'_{l,k}}{\tilde{\psi}_{l,k}} = \frac{\psi'_{l,k}}{\psi_{l,k}} \quad \text{at} \quad r = R \quad . \tag{4.32}$$

According to (4.31), the right-hand part of this equality may be expressed in terms of the scattering phase shift δ_l.

Relation (4.32) may be regarded as an equation for the scattering phase shift. If the wave function in the inner region is known, then the left-hand part of (4.32) is a given function of the potential and energy and hence the relevant scattering phase shift, in principle, can be calculated from this equation.

4.2 Square Well Potential

Let us consider particle scattering by a square well potential. This problem has an exact solution. Suppose the potential is

$$V_r(r) = \begin{cases} -V_{r0} \, , & r < R \quad , \\ 0 \, , & r > R \quad , \end{cases} \tag{4.33}$$

where V_{r0} is the depth ($V_{r0} > 0$) and R is the width of the potential well.

Equation (4.9) in the inner region $r < R$ takes a form similar to (4.10),

$$\left\{ \frac{1}{r^2} \frac{d}{dr} \left(r^2 \frac{d}{dr} \right) - \frac{l(l+1)}{r^2} + k_0^2 \right\} \tilde{\psi}_{l,k}(r) = 0 \quad , \qquad r < R \quad , \tag{4.34}$$

where $k_0^2 = V_{r0} + k^2$. Therefore, the solution of (4.34) that is finite at the point $r = 0$ may be written as

$$\tilde{\psi}_{l,k}(r) = C_l\, j_l(k_0 r) \quad , \qquad r < R \quad , \tag{4.35}$$

where C_l is a constant.

The solution in the outer region $r > R$ is

$$\psi_{l,k}(r) = A_l\, j_l(kr) + B_l\, h_l^{(+)}(kr) \quad , \qquad r > R \quad , \tag{4.36}$$

where A_l is determined by the incident wave normalization, B_l may be obtained from the boundary condition at infinity. Thus we have

$$A_l = 4\pi \quad , \qquad B_l = 4\pi i \sin\delta_l \exp(i\delta_l) \quad . \tag{4.37}$$

To find C_l, we put $\tilde{\psi}_{l,k}$ to be equal to $\psi_{l,k}$ at the point $r = R$, then

$$C_l = 4\pi \exp(i\delta_l) \cos\delta_l \left[j_l(kr) - \tan\delta_l\, n_l(kR) \right] j_l^{-1}(k_0 R) \quad . \tag{4.38}$$

Equating the logarithmic derivatives of $\tilde{\psi}_{l,k}$ and $\psi_{l,k}$ at $r = R$ yields an equation for the phase shift δ_l:

$$k_0\, \frac{j_l'(k_0 R)}{j_l(k_0 R)} = k\, \frac{j_l'(kR) - \tan\delta_l\, n_l'(kR)}{j_l(kR) - \tan\delta_l\, n_l(kR)} \quad . \tag{4.39}$$

Solving this equation with respect to $\tan\delta_l$, we find that

$$\tan\delta_l = \frac{j_l'(kR) - D_l\, j_l(kR)}{n_l'(kR) - D_l\, n_l(kR)} \quad , \tag{4.40}$$

where $D_l \equiv (k_0/k)\big(j_l'(k_0 R)/j_l(k_0 R)\big)$. Formula (4.40) gives the scattering phase shift as a function of the potential well parameters (depth V_{r0} and width R) and the incident particle energy.

It is not difficult to verify that in the case of a finite-range potential, a finite number of particles take part in the scattering. Indeed, without interaction, a particle with angular momentum M would pass at distance M/p from the scatterer. For the partial wave to be scattered, this distance must be shorter than the force range R. Assuming that $M^2 = \hbar^2 l(l+1)$ and $p = \hbar k$, we obtain the scattering criterion to be

$$\sqrt{l(l+1)} \le kR \quad . \tag{4.41}$$

Therefore, the scattering phase shift δ_l is not equal to zero only if the condition (4.41) is satisfied.

If the incident particle energies are low, the wavelength being greater than the force range, then $kR < 1$ and interaction occurs only in the S-state with $l = 0$.

Substituting $j_0(x) = (\sin x)/x$ and $n_0(x) = -(\cos x)/x$, we obtain for the S-scattering

$$\tilde{\psi}_{0,k}(r) = C_0 \frac{\sin k_0 r}{k_0 r} \quad , \qquad r < R \quad ,$$

$$(4.42)$$

$$\psi_{0,k}(r) = 4\pi \, \exp(i\delta_0) \frac{\sin(kr + \delta_0)}{kr} \quad , \qquad r > R \quad ,$$

and equation (4.39) for the scattering phase shift reduces to

$$k_0 \cot k_0 R = k \cot(kR + \delta_0) \quad . \tag{4.43}$$

Let us consider the low energy limiting case, $k^2 \ll V_{r0}$ and $kR \ll \delta_0$. Then (4.43) may be rewritten as

$$k \cot \delta_0 = -\frac{1}{a_0} \quad , \tag{4.44}$$

where $-a_0^{-1} \equiv V_{r0}^{1/2} \cot(V_{r0}^{1/2} R)$. The quantity a_0 is usually referred to as the scattering length.

The total scattering cross section is expressed in terms of the scattering phase shift as given by (2.70),

$$\sigma_e = \frac{4\pi}{k^2} \sin^2 \delta_0 \quad .$$

Within the context of (4.44), we thus obtain

$$\sigma_e = \frac{4\pi a_0^2}{1 + a_0^2 k^2} \quad . \tag{4.45}$$

Thus, in the limiting case of zero energy ($k \to 0$), the scattering cross section tends to the finite value $\sigma_e = 4\pi a_0^2$.

The repulsive square well potential

$$V_r(r) = \begin{cases} V_{r0} \, , & r < R \quad , \\ 0 \, , & r > R \quad , \end{cases} \tag{4.46}$$

may be considered in the same manner. In this case, the radial function for $r < R$ is described by equation (4.34) with $k_0^2 = k^2 - V_{r0}$. If $k^2 < V_{r0}$, then $k_0^2 = -\kappa^2 < 0$ ($\kappa^2 \equiv V_{r0} - k^2 > 0$). Considering only the S-state, we can write the solution in the inner region as

$$\tilde{\psi}_{0,k}(r) = C_0 \frac{e^{\kappa r} - e^{-\kappa r}}{\kappa r} \quad , \qquad r < R \quad . \tag{4.47}$$

In the outer region ($r > R$), the solution is given by the second relation (4.42). The joint condition for the solutions yields an equation for the phase shift:

$$\kappa \coth \kappa R = k \cot(kR + \delta_0) \quad . \tag{4.48}$$

For the limiting case $V_{r0} \to \infty$ which describes scattering by an impenetrable sphere of radius R, we have

$$\delta_0 = -kR \quad . \tag{4.49}$$

For very low energies, $kR \ll 1$, the scattering cross section is

$$\sigma_e = 4\pi R^2 \quad . \tag{4.50}$$

This cross section is four times greater than the geometric cross section of the impenetrable sphere.

Problem

Show that for sufficiently deep potential wells $(V_{r0}R^2 > 1)$, the system has no bound states if the scattering length is negative.

Let us consider a zero energy particle scattered by a sufficiently deep potential well. The equation for the radial function off the interaction region may be written as

$$\frac{d^2}{dr^2} r \psi_0(r) = 0 \quad , \qquad r > R \quad ,$$

and we immediately obtain

$$r \psi_0(r) = C(r - a) \quad , \qquad r > R \quad , \tag{4.51}$$

where C and a are constants. This equation determines a straight line that crosses the r axis at the point a. Comparing (4.51) to (4.42) for $k \to 0$, we easily find that

$$a = -\lim_{k \to 0} (\tan \delta_0)/k \equiv a_0 \quad ,$$

i.e., a reproduces the scattering length (4.44). The scattering cross section is given by the expression

$$\sigma_e = 4\pi a_0^2 \quad , \qquad k \to 0 \quad .$$

Once the zero energy cross section is known, we can find the absolute value of the scattering length, but not its sign. The scattering length can be both positive and negative. The r-dependence of the wave function for zero energies is shown in Fig. 4.1. The wave function behavior within the interaction region $r < R$ is almost energy-independent at low energies, therefore, it remains such for bound states with moderate binding energies. As follows from Fig. 4.1, if the scattering length is positive, the wave function in the inner region can be joined together with the exponentially decreasing solution in the outer region that is associated with a bound state. If the scattering length is negative, the wave function in the inner region is such that it cannot be joined to the exponentially decreasing solution in the outer region, i.e., bound states do not occur in the system. Thus, if $a > 0$, then the system can possess bound states; if $a < 0$, the system has no bound states.

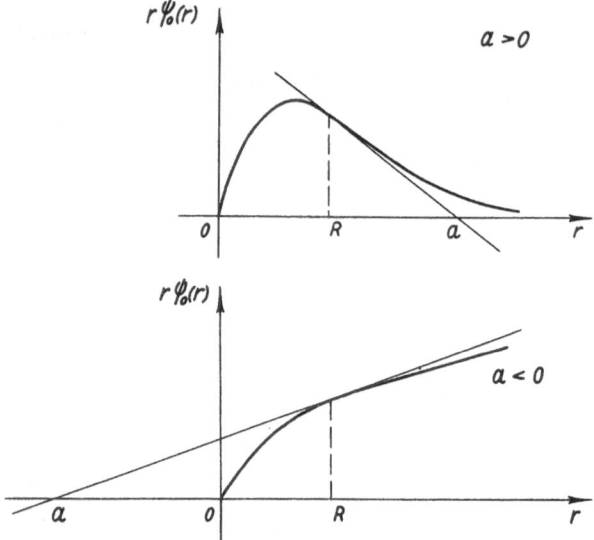

Fig. 4.1. Zero-energy radial wave function $r\,\psi_0(r)$ as a function of r for positive and negative values of the scattering length a

4.3 Coulomb Field

With further applications in view, we consider the wave functions for the Coulomb field

$$V(r) = \frac{Ze^2}{r} \quad . \tag{4.52}$$

Equation (4.9) with the Coulomb potential takes the form

$$\left\{\frac{1}{r^2}\frac{d}{dr}\left(r^2\frac{d}{dr}\right) - \frac{l(l+1)}{r^2} - \frac{2\xi k}{r} + k^2\right\}\psi_{l,k}(r) = 0 \quad , \tag{4.53}$$

where $\xi = Ze^2/\hbar v$ is the so-called Coulomb parameter ($v = \hbar k/\mu$ is the particle velocity). We write the wave function as

$$\psi_{l,k}(r) = \frac{u_l(r)}{r} \tag{4.54}$$

and employ (4.6). Then we obtain an equation for the radial function,

$$u_l'' - \left[\frac{l(l+1)}{r^2} + \frac{2\xi k}{r} - k^2\right]u_l = 0 \quad . \tag{4.55}$$

This equation has two linearly independent solutions $F_l(r)$ and $G_l(r)$. It is convenient to choose them in the following manner. The regular solution $F_l(r)$, which vanishes at $r = 0$, must have the $r \to \infty$ asymptotic given by

$$F_l(r) \underset{r \to \infty}{\longrightarrow} \sin\left(kr - \frac{l\pi}{2} - \xi \ln 2kr + \eta_l\right) \quad . \tag{4.56}$$

The Coulomb phase shift η_l is given by

$$\exp(2i\eta_l) = \frac{\Gamma(l+1+i\xi)}{\Gamma(l+1-i\xi)} \quad . \tag{4.57}$$

According to (4.19), the asymptotic of $F_l(r)$ differs from (4.26) by an additional phase shift $\xi \ln 2kr$ which appears due to the long-range nature of the Coulomb interaction.

The irregular solution $G_l(r)$ for $r \to \infty$ must have asymptotics of the form

$$G_l(r) \to -\cos\left(kr - \frac{l\pi}{2} - \xi \ln 2kr + \eta_l\right) \quad . \tag{4.58}$$

Both functions $F_l(r)$ and $G_l(r)$ are real. Using their asymptotics, we find the Wronskian to be

$$F_l \frac{dG_l}{dr} - G_l \frac{dF_l}{dr} = k \quad . \tag{4.59}$$

For $r \to 0$, $F_l(r)$ and $G_l(r)$ are described by the approximate expressions

$$F_l(r) \simeq C_l(kr)^{l+1} \left\{1 + \frac{\xi}{l+1} kr + \ldots\right\} \quad , \tag{4.60}$$

$$G_l(r) \simeq -\frac{1}{(2l+1)C_l} \left\{\frac{1}{(kr)^l}\left(1 - \frac{\xi}{l} kr + \ldots\right)\right.$$
$$\left. + p(kr)^{l+1}\left(1 + \frac{\xi}{l+1} kr + \ldots\right)(\ln 2kr + \text{const})\right\} \quad , \tag{4.61}$$

where

$$C_l^2 + \frac{2^{2l}}{[(2l+1)!]^2} (l^2 + \xi^2)\left[(l-1)^2 + \xi^2\right] \ldots (1 + \xi^2) \frac{2\pi\xi}{e^{2\pi\xi} - 1} \quad , \tag{4.62}$$

$$p = \frac{2^{2l+1}}{(2l+1)(2l!)^2} (l^2 + \xi^2)\left[(l-1)^2 + \xi^2\right] \ldots (1 + \xi^2)\xi \quad . \tag{4.63}$$

In the most important case of $l = 0$, (4.60, 61) are simplified to

$$F_0(r) \simeq C_0 kr(1 + \xi kr + \ldots) \quad , \tag{4.64}$$

$$G_0(r) \simeq -\frac{1}{C_0} \left\{1 + 2\xi kr \left[\ln 2kr + 2\gamma - 1 + h(\xi) + \ln \xi\right] + \ldots\right\} \quad , \tag{4.65}$$

where

$$C_0^2 = \frac{2\pi\xi}{e^{2\pi\xi} - 1} \quad , \tag{4.66}$$

$$h(\xi) = \xi^2 \sum_{\nu=1}^{\infty} \frac{1}{\nu(\nu^2 + \xi^2)} - \ln \xi - \gamma \quad , \tag{4.67}$$

and $\gamma = 0.57721$ is the Euler constant.

The particle wave function of a state with definite momentum, normalized to unit incident wave amplitude, may be written in terms of the regular solution $F_l(r)$ as

$$\psi_k(r) = 4\pi \sum_{lm} i^l e^{i\eta_l} \frac{F_l(r)}{kr} Y_{lm}^*(n) Y_{lm}(n') \quad . \tag{4.68}$$

The asymptotic of the function (4.68) is

$$\psi_k(r) \underset{r \to \infty}{\longrightarrow} \exp\left[ikr + i\xi \ln(kr - kr)\right]$$

$$+ f(k, k') \frac{\exp(ikr - i\xi \ln 2kr)}{r} \quad , \tag{4.69}$$

where $f(k, k')$ is the scattering amplitude in the Coulomb field given by

$$f(k, k') = \frac{i}{2k} \sum_{l=0}^{\infty} (2l + 1)\left[1 - \exp(2i\eta_l)\right] P_l(\cos \vartheta)$$

$$= \frac{Ze^2}{2\mu v^2 \sin^2(\vartheta/2)} \exp\left(-i\xi \ln \sin^2 \frac{\vartheta}{2} + 2i\eta_0 + i\pi\right) \quad . \tag{4.70}$$

The logarithmic term enters the phase shift of the first term in (4.69) because the incident plane wave is distorted by the long-range Coulomb interaction. Making use of (4.70), we obtain the well known Rutherford formula for the scattering cross section in the Coulomb field,

$$d\sigma_e = \left(\frac{Ze^2}{2\mu v^2}\right)^2 \frac{do}{\sin^4(\vartheta/2)} \quad , \tag{4.71}$$

which reproduces the result obtained in classical mechanics.

4.4 Partial Green Functions and the Scattering Matrix

When considering particle scattering in a central field, it is convenient to expand the Green functions $G_0(E; r, r')$ and $G(E; r, r')$ in terms of partial components. The free motion Green function $G_0(E; r, r')$ may be written as a series

$$G_0(E; r, r') = \sum_l (2l + 1) G_l^0(k; r, r') P_l(\cos \theta)$$

$$= 4\pi \sum_l G_l^0(k; r, r') Y_{lm}^*(n') Y_{lm}(n) \quad , \tag{4.72}$$

where $Y_{lm}(n)$ and $Y_{lm}(n')$ are the spherical functions whose arguments are the angles associated with the directions of the vectors r and r'; θ is the angle between the vectors r and r'; $G_l^0(k; r, r')$ is the partial Green function. Substituting the explicit expressions (3.9, 10) for the Green functions $G_0^{(+)}(E; r, r')$ and $G_0^{(-)}(E; r, r')$, we obtain the partial retarded and advanced Green functions, i.e.,

$$G_l^0(k; r, r') \equiv \frac{\mu}{2\pi\hbar^2} g_{l,k}^0(r, r') \quad,$$

$$g_{l,k}^{0(\pm)}(r, r') = \mp ik \begin{cases} j_l(kr) h_l^{(\pm)}(kr') \quad, & r < r' \quad, \\ j_l(kr') h_l^{(\pm)}(kr) \quad, & r > r' \quad. \end{cases} \tag{4.73}$$

(The factor $\mu/2\pi\hbar^2$ is separated for the sake of convenience). The partial Green functions $g_{l,k}^0(r, r')$ satisfy the radial Schrödinger equation for the free motion with a point source in the right-hand part,

$$\left\{ \frac{1}{r^2} \frac{d}{dr} \left(r^2 \frac{d}{dr} \right) - \frac{l(l+1)}{r^2} + k^2 \right\} g_{l,k}^0(r, r') = \frac{1}{r^2} \delta(r - r') \quad. \tag{4.74}$$

The functions $g_{l,k}^0(r, r')$ are symmetric with respect to the variables r and r'. Making use of $g_{l,k}^{(+)}(r, r')$, we reduce the Schrödinger equation (4.9) with the boundary conditions (4.24) to an integral equation for the radial function $\psi_{l,k}(r)$,

$$\psi_{l,k}(r) = \psi_{l,k}^0(r) + \int_0^\infty dr' \, r'^2 \, g_{l,k}^{0(+)}(r, r') \, V_r(r') \, \psi_{l,k}(r') \quad, \tag{4.75}$$

where $\psi_{l,k}^0(r)$ is the incident wave radial function given by (4.22). The integral equation (4.75) is a partial component of the Lippmann–Schwinger equation (3.12).

Substituting the explicit expression for the Green function $g_{l,k}^{0(+)}(r, r')$, we can find the asymptotics of $\psi_{l,k}(r)$. The partial scattering amplitude is then given by

$$f_l(k) = -\frac{1}{4\pi} \int_0^\infty dr \, r^2 \, j_l(kr) \, V_r(r) \, \psi_{l,k}(r) \quad. \tag{4.76}$$

Since the Green function $g_{l,k}^{0(+)}(r, r')$ is k-independent for small k (and fixed r and r'), the k-dependence of the radial function $\psi_{l,k}(r)$ is similar to that of $\psi_{l,k}^0(r)$, i.e.,

$$\psi_{l,k}(r) \sim k^l \quad \text{as} \quad k \to 0 \quad.$$

Therefore, if the potential $V_r(r)$ is of finite range, then the functions $j_l(kr)$ and $\psi_{l,k}(r)$ for small k may be replaced by their expansions. Then, within the context of (4.76), we find that

$$f_l(k) \underset{k \to 0}{\longrightarrow} - a_l k^{2l} \quad, \tag{4.77}$$

where a_l is a constant. For $l = 0$, a_l has dimensions of length. We have already noted that it is referred to as the scattering length. Thus, as $k \to 0$, all the partial

amplitudes other than that for the S-wave, vanish by virtue of (4.77). The greater
the l, the faster the amplitude tends to zero.

Having expressed the partial scattering amplitudes in terms of the scattering
phase shift,

$$f_l(k) = \frac{1}{k} e^{i\delta_l} \sin \delta_l \quad ,$$

we find that for $k \to 0$ the scattering phase shift δ_l tends to a multiple of π, so
that

$$\delta_l(k) \underset{k \to 0}{\to} n\pi - a_l k^{2l+1} \quad . \tag{4.78}$$

Usually, the scattering phase shift decreases with growing energy and vanishes
for $k \to \infty$ if the potential is of finite range. Thus, for $l = 0$, the zero energy
scattering phase shift δ_0 can differ from zero (it may be equal to $n\pi$) and then
the partial scattering cross section is finite,

$$\sigma_e(0) = 4\pi a_0^2 \quad .$$

As the energy increases, the scattering phase shift δ_0 decreases and the cross
section $\sigma_l^{(0)}$ vanishes for $\delta_0(k) = (n-1)\pi$. Because all other partial cross sections ($l > 0$) are small, for such energies scattering does not occur (Ramsauer–
Townsend effect).

The total Green function $G(E; r, r')$ may be written in a series form similarly
to (4.72), i.e., as

$$G(E; r, r') = \sum_l (2l + 1) G_l(k; r, r') P_l(\cos \theta)$$

$$= 4\pi \sum_{lm} G_l(k; r, r') Y_{lm}^*(n') Y_{lm}(n) \quad , \tag{4.79}$$

where $G_l(k; r, r')$ is the partial Green function associated with particle scattering
by the central field $V(r)$. As in the previous case, we separate the factor $\mu/2\pi\hbar^2$
to obtain

$$G_l(k; r, r') \equiv \frac{\mu}{2\pi\hbar^2} g_{l,k}(r, r') \quad . \tag{4.80}$$

The partial Green function $g_{l,k}(r, r')$, by definition, is symmetric with respect to
all its arguments and satisfies the inhomogeneous equation

$$\left\{ \frac{1}{r^2} \frac{d}{dr} \left(r^2 \frac{d}{dr} \right) - \frac{l(l+1)}{r^2} - V_r(r) + k^2 \right\} g_{l,k}(r, r')$$

$$= \frac{1}{r^2} \delta(r - r') \tag{4.81}$$

with the following boundary conditions at large distances ($r \to \infty$): the retarded
Green function $g_{l,k}^{(+)}(r, r')$ must tend to a diverging spherical wave, the advanced
Green function $g_{l,k}^{(-)}(r, r')$ must become a converging spherical wave.

In order to construct the Green functions with the required asymptotics, we have to supplement the physical solution of (4.9), which has asymptotics (4.26) at infinity, with other solutions of (4.9) whose asymptotics are either converging or diverging spherical waves. We denote the relevant independent solutions of (4.9) by $\psi_{l,k}^{(+)}(r)$ and $\psi_{l,k}^{(-)}(r)$ and assume that their asymptotics at infinity are given by

$$\psi_{l,k}^{(\pm)}(r) \to \mp 4\pi i \, \frac{\exp\{\pm i[kr - (l\pi/2) + \delta_l]\}}{kr} \quad . \tag{4.82}$$

The physical solution $\psi_{l,k}(r)$ may be expressed in terms of this function, i.e.,

$$\psi_{l,k}(r) = \tfrac{1}{2} \left[\psi_{l,k}^{(-)}(r) + \psi_{l,k}^{(+)}(r) \right] \exp(i\delta_l) \quad . \tag{4.83}$$

Though the physical solution $\psi_{l,k}(r)$ is finite in the entire space and varies as r^l when $r \to 0$, both $\psi_{l,k}^{(+)}(r)$ and $\psi_{l,k}^{(-)}(r)$ for $r \to 0$ are singular and behave as $r^{-(l+1)}$.

It is clear from (4.81) that the Green function $g_{l,k}(r,r')$ satisfies the homogeneous equation (4.9) for $r \neq r'$. For $r = r'$, the Green function $g_{l,k}(r,r')$ is continuous, whereas its derivative is discontinuous, i.e.,

$$\frac{\partial}{\partial r} g_{l,k}(r,r')\Big|_{r=r'+\varepsilon} - \frac{\partial}{\partial r} g_{l,k}(r,r')\Big|_{r=r'-\varepsilon} = \frac{1}{r^2} \quad , \quad \varepsilon \to 0 \quad . \tag{4.84}$$

Therefore, it is not difficult to find the Green functions $g_{l,k}^{(\pm)}(r,r')$ by explicitly making use of $\psi_{l,k}^{(+)}(r)$, $\psi_{l,k}^{(-)}(r)$, and the regular solution $\psi_{l,k}(r)$. Indeed, bearing in mind that $g_{l,k}^{(+)}(r,r')$ is regular at the point $r = 0$ and assuming that for $r \to \infty$ it behaves as a diverging spherical wave, we can write it in the form

$$g_{l,k}^{(+)}(r,r') = \begin{cases} \psi_{l,k}(r) \, \Phi_1(r') & , \quad r < r' \quad , \\ \psi_{l,k}^{(+)}(r) \, \Phi_2(r') & , \quad r \gtrless r' \quad . \end{cases} \tag{4.85}$$

The functions $\Phi_1(r')$ and $\Phi_2(r')$ can easily be found from the continuity and symmetry conditions for the Green function. Thus, we obtain the final expression for the retarded partial Green functions:

$$g_{l,k}^{(+)}(r,r') = -\frac{ik}{16\pi^2} \exp(-i\delta_l) \begin{cases} \psi_{l,k}(r) \, \psi_{l,k}^{(+)}(r') & , \quad r < r' \quad , \\ \psi_{l,k}(r') \, \psi_{l,k}^{(+)}(r) & , \quad r > r' \quad . \end{cases} \tag{4.86}$$

Within the context of (1.36), the advanced and retarded partial Green functions $g_{l,k}^{(-)}(r,r')$ and $g_{l,k}^{(+)}(r,r')$ satisfy the relation

$$g_{l,k}^{(-)}(r,r') = \left[g_{l,k}^{(+)}(r,r') \right]^* \quad . \tag{4.87}$$

For sufficiently large r, the Green function (4.86) may be written as

$$g_{l,k}^{(+)}(r,r') \underset{r\to\infty}{\longrightarrow} -\frac{1}{4\pi i^l} \frac{e^{ikr}}{r} \psi_{l,k}(r') \quad . \tag{4.88}$$

If r' also is large, then

$$g_{l,k}^{(+)}(r,r') \underset{\substack{r \to \infty \\ r' \to \infty}}{\longrightarrow} (-1)^{l+1} \frac{i}{2k} \frac{e^{ikr}}{rr'} \left\{ (-1)^l e^{-ikr'} - S_l e^{ikr'} \right\} \quad . \tag{4.89}$$

Thus, the asymptotics of the partial Green function $g_{l,k}^{(+)}(r,r')$ are completely determined by the scattering matrix S_l.

4.5 Variable Phase Approach

According to (3.75), the scattering amplitude is completely determined by the given phase shifts. Since the latter are governed by the radial function asymptotics (4.26), the amplitude can be obtained without solving the complete scattering problem, i.e., there is no need to calculate the radial functions $\psi_{l,k}(r)$ in the whole space. This suggests the idea of the so-called variable phase approach, in which the phase function (variable phase) $\delta_l(r)$, with a very straightforward physical meaning, is introduced. If the potential $V(r)$ is given, then the phase function at the point r determines the scattering phase shift produced by the part of the potential contained within the sphere of radius r. The scattering phase shift for the total potential is equal to the asymptotic value of the phase function $\delta_l = \delta_l(\infty)$. Instead of considering the Schrödinger equation, the phase approach introduces an equation for the phase function which establishes direct relationship between the scattering phase shift and the potential.[1]

We introduce two functions $\delta_l(r)$ and $A_l(r)$, defined by their relation to the radial wave function $\psi_{l,k}(r)$,

$$\psi_{l,k}(r) = A_l(r) \left[\cos \delta_l(r) \, j_l(kr) - \sin \delta_l(r) \, n_l(kr) \right] \quad . \tag{4.90}$$

It is clear that the radial wave function $\psi_{l,k}(r)$ satisfies the Schrödinger equation (4.9). To ensure unambiguity of the functions $\delta_l(r)$ and $A_l(r)$, we have to supplement (4.90) with some condition; we take this to be the requirement that the radial wave function derivative must be given by

$$\frac{d\psi_{l,k}(r)}{dr} = A_l(r) \left[\cos \delta_l(r) \frac{dj_l(kr)}{dr} - \sin \delta_l(r) \frac{dn_l(kr)}{dr} \right] \quad , \tag{4.91}$$

which is equivalent to the additional condition

$$\frac{dA_l(r)}{dr} \left[\cos \delta_l(r) \, j_l(kr) - \sin \delta_l(r) \, n_l(kr) \right]$$

$$- \frac{d\delta_l(r)}{dr} A_l(r) \left[\sin \delta_l(r) \, j_l(kr) + \cos \delta_l(r) \, n_l(kr) \right] = 0 \quad . \tag{4.92}$$

---

[1] The equation for the phase function was derived by *Drukarev* [4.1]. A detailed account of the phase approach is given in the books [4.2, 3].

Thus, we have two equations, (4.9, 92), sufficient for an unambiguous calculation of the functions $\delta_l(r)$ and $A_l(r)$ which are referred to as the phase function and the amplitude function, respectively. Indeed, comparing (4.31) to (4.90), we see that if the potential $V(r)$ is truncated at some point $r = r_m$,

$$V(r, r_m) = V(r)\,\Theta(r_m - r) \quad , \tag{4.93}$$

[$\Theta(x)$ is the Heaviside function], then in the region $r > r_m$ the functions $\delta_l(r)$ and $A_l(r)$ take constant values equal to the phase shift and the amplitude of the asymptotic wave function which corresponds to the scattering by this truncated potential. For $r \to \infty$, the phase function $\delta_l(r)$ is equal to the scattering phase shift for the total potential, $\delta_l = \delta_l(\infty)$. Note that the phase function vanishes as $r \to 0$ since the condition $r_m = 0$ means that interaction does not occur.

Differentiating (4.91) and making use of the Schrödinger equation (4.9), we obtain

$$\frac{dA_l(r)}{dr}\left[\cos \delta_l(r)\,\frac{dj_l(kr)}{dr} - \sin \delta_l(r)\,\frac{dn_l(kr)}{dr}\right]$$
$$-\frac{d\delta_l(r)}{dr}\,A_l(r)\left[\sin \delta_l(r)\,\frac{dj_l(kr)}{dr} + \cos \delta_l(r)\,\frac{dn_l(kr)}{dr}\right]$$
$$-V_r(r)\,A_l(r)\left[\cos \delta_l(r)\,j_l(kr) - \sin \delta_l(r)\,n_l(kr)\right] = 0 \quad . \tag{4.94}$$

Equations (4.92, 94) form a set of first-order differential equations sufficient for the unambiguous description of the functions $\delta_l(r)$ and $A_l(r)$. (We remind the reader that the initial Schrödinger equation (4.9) for the radial wave function is a differential equation of the second order.)

Having eliminated from (4.92, 94) the derivative of the amplitude, we obtain an equation for the phase function $\delta_l(r)$,

$$\frac{d\delta_l(r)}{dr} = -kr^2 V_r(r)\left[\cos \delta_l(r)\,j_l(kr) - \sin \delta_l(r)\,n_l(kr)\right]^2 \quad , \tag{4.95}$$

$$\delta_l(0) = 0 \quad .$$

This equation does not depend on the amplitude, which is natural since the wave function normalization must not influence the scattering phase shift. According to (4.95), calculating the scattering phase shift for a given potential reduces to solving a first-order differential equation with an initial condition.

In a similar manner, having eliminated from (4.92, 94) the phase function derivatives, we obtain an equation for the amplitude function $A_l(r)$,

$$\frac{dA_l(r)}{dr} = -kr^2 V_r(r)\,A_l(r)\left[\cos \delta_l(r)\,j_l(kr) - \sin \delta_l(r)\,n_l(kr)\right]$$
$$\times \left[\sin \delta_l(r)\,j_l(kr) + \cos \delta_l(r)\,n_l(kr)\right] \quad , \tag{4.96}$$

which, in contrast to (4.95), is linear and contains the phase function. Integrating (4.96) yields

$$A_l(r) = A_l(r_0) \exp\left\{-k \int_{r_0}^r dr \, r^2 \, V_r(r)\right.$$

$$\times \left[\cos \delta_l(r) \, j_l(kr) - \sin \delta_l(r) \, n_l(kr)\right]$$

$$\left. \times \left[\sin \delta_l(r) \, j_l(kr) + \cos \delta_l(r) \, n_l(kr)\right]\right\} \quad , \tag{4.97}$$

where $A_l(r_0)$ is the amplitude function at the point r_0.

The phase given by the solution of (4.95) and the amplitude (4.97) completely determine the radial wave function (4.90). The phase function is described by the first-order (though nonlinear) equation and this simplifies approximate and numerical calculations in the scattering problem.

Equation (4.95) for the phase function immediately yields the relation between the scattering phase shift and the potential. In particular, since the expression

$$\left[\cos \delta_l(r) \, j_l(kr) - \sin \delta_l(r) \, n_l(kr)\right]^2$$

is always positive, the sign of the phase function derivative is uniquely determined by the sign of the potential. In the case of attraction $(V_r(r) < 0)$, the phase function derivative, and hence the scattering phase shift, are positive; in the case of repulsion $(V_r(r) > 0)$, both the phase function derivative and the scattering phase shift are negative.

Let us study how the behavior of the solution of (4.95) depends on the properties of the potential as $r \to 0$. We write (4.95) in integral form,

$$\delta_l(r) = -k \int_0^r dr \, r^2 \, V_r(r) \left[\cos \delta_l(r) \, j_l(kr) - \sin \delta_l(r) \, n_l(kr)\right]^2 . \tag{4.98}$$

If the potential is nonsingular (or weakly singular) and satisfies the condition $r^2 V_r(r) \to 0$ as $r \to 0$, then substituting $\cos \delta_l(r) \simeq 1$ and $\sin \delta_l(r) \simeq \delta_l(r)$ into the integrand, we have

$$\delta_l(r) \simeq -k \int_0^r dr \, r^2 \, V_r(r) \, j_l^2(kr)$$

$$+ 2k \int_0^r dr \, r^2 \, V_r(r) \, \delta_l(r) \, j_l(kr) \, n_l(kr) \quad .$$

For sufficiently small r, the first term is much smaller then the second one and can be disregarded, so that

$$\delta_l(r) \simeq -\frac{k^{2l+1}}{\left[(2l+1)!!\right]^2} \int_0^r dr \, V_r(r) \, r^{2l+2} \quad , \qquad r \to 0 \quad . \tag{4.99}$$

In particular, for the potential of the form $V_r(r) = C r^p$ $(p > -2)$, we have

$$\delta_l(r) \simeq -\frac{C (kr)^{2l+1} r^{2+p}}{(2l+3+p)\left[(2l+1)!!\right]^2} \quad . \tag{4.100}$$

One more example is a strongly singular repulsive potential which satisfies the condition $r^2 V_r(r) \to \infty$ as $r \to 0$. Since both the phase function and its derivative are always finite, the expression within the square brackets in (4.95) vanishes for such potentials. Assuming that for small r $\cos \delta_l(r) \simeq 1$ and $\sin \delta_l(r) \simeq \delta_l(r)$, we find that

$$\delta_l(r) \simeq -\frac{(kr)^{2l+1}}{(2l+1)!! \, (2l-1)!!} \quad , \qquad r \to 0 \quad . \tag{4.101}$$

The same expression for the scattering phase shift occurs in the case of a hard repulsive core of small radius.

Problems

4.1 Find the wave number dependence of the scattering phase shift $\delta_l(k)$ for small k.

Suppose the particle wavelength is much greater than the potential range R ($kR \ll 1$), and the particle energy is much lower than the potential in the interaction region.

The term with k^2 in (4.9) may be disregarded for $r \ll k^{-1}$. If $R \ll r \ll k^{-1}$, then the potential may be also disregarded and we have

$$\psi_l'' + \frac{2}{r} \psi_l' - \frac{l(l+1)}{r^2} \psi_l = 0 \quad . \tag{4.102}$$

The general solution of this equation may be written in the form

$$\psi_l = c_1 r^l + c_2 r^{-(l+1)} \quad , \tag{4.103}$$

where c_1 and c_2 are k-independent constants determined by the solution in the inner region $r \leq R$. If $r \sim k^{-1}$, then one may neglect the potential in (4.9) but must retain the terms with k^2. In this case, (4.9) coincides with the free motion equation (4.10). The general solution of the latter may be written as

$$\psi_{l,k} = c_1 \frac{(2l+1)!!}{k^l} j_l(kr) - c_2 \frac{k^{l+1}}{(2l-1)!!} n_l(kr) \quad . \tag{4.104}$$

The coefficient in (4.104) is fitted in such a way that for $r \ll k^{-1}$ this solution reduces to (4.103).

Substituting the asymptotic (4.13), we can write the solution (4.104) for $kr \gg 1$ as

$$\psi_{l,k} \to c_1 \frac{(2l+1)!!}{k^{l+1}} \left(1 + \delta_l^2\right)^{1/2} \frac{1}{r} \sin\left(kr - \frac{l\pi}{2} + \delta_l\right) \quad , \tag{4.105}$$

where the phase shift δ_l is given by

The scattering matrix $S_l(k)$ can be expressed in terms of the scattering amplitude $f(k, z)$:

$$S_l(k) = 1 + ik \int_{-1}^{1} dz \, f(k, z) \, P_l(z) \quad . \tag{9.73}$$

Making use of the relation

$$\int_{-1}^{1} dz \, P_l(z) \, P_{\alpha_r}(-z) = \frac{2}{\pi} \frac{\sin \alpha_r \pi}{(l + \alpha_r + 1)(l - \alpha_r)} \quad ,$$

which holds for integer l, and substituting (9.55) for $f(k, z)$, we find the contribution of the r-th pole of the amplitude (9.55) to $S_l(k)$ to be

$$S_l^{(r)}(k) \simeq \frac{i}{k} \frac{b_r(k)}{l - \alpha_r(k)} \quad . \tag{9.74}$$

If α_r is close to an integer l, then the energy dependence of S_l is determined by (9.74). Expanding $\alpha_r(k)$ in a series about the value k_0 for which $\mathrm{Re}\,\{\alpha_r(k_0)\} = l$, we obtain

$$S_l(k) \simeq -i \frac{\hbar^2}{\mu} \left[\frac{d\alpha(k_0)}{dk_0} \right]^{-1} \left[\frac{b_r(k_0)}{E - E_0 + \frac{i}{2} \Gamma_0} \right] \quad , \tag{9.75}$$

where E_0 and Γ_0 are given by (9.34). Thus, the energy dependence of the scattering matrix $S_l(k)$ near the Regge pole is of resonance type. Evidently, this resonance-like dependence can be manifested only if Γ_0 is small compared to the characteristic energies on which the scattering amplitude depends.

9.3 Consider the behavior of the Regge trajectories in the vicinity of the threshold $E = 0$ [9.3].

If the potential $V(r)$ satisfies the condition (7.31), then the Jost function $f(\lambda, k)$ is an analytic function of k in the region $\mathrm{Im}\,\{k\} < m/2$ except for the point $k = 0$ which is a branch point for arbitrary λ and a multiple pole of the order $|\lambda| - 1/2$ for real half-integer λ. Thus, to define $f(\lambda, k)$ unambiguously, we have to cut the complex k plane along the imaginary axis from $k = 0$ to $k = im/2$. In the region $\mathrm{Im}\,\{k\} < m/2$ and $k \neq i\kappa$ ($0 \leq \kappa \leq m/2$), the function $f(\lambda, k)$ is uniquely determined by the iteration expansion; its value is usually referred to as the principal value of $f(\lambda, k)$. The Jost function values on other sheets of the Riemann surface are defined by analytic continuation. Suppose $f(\lambda, k \exp(in\pi))$ is the value of the Jost function that is obtained by analytic continuation along the path that lies in the region $\mathrm{Im}\,\{k\} < m/2$ and encircles the point $k = 0$ anticlockwise as the argument continuously varies from φ to $\varphi + n\pi$. By using (9.2, 6), we obtain the relation

$$f(\lambda, k e^{i\pi}) = f(\lambda, k e^{-i\pi}) + 2i \cos \lambda\pi \, f(\lambda, k) \quad , \tag{9.76}$$

4.3 Show that if the potential range R is finite, then the S-state scattering phase shift δ_l, the interaction range R, and the wave number k satisfy the inequality

$$\frac{d\delta_0}{dk} + R - \frac{1}{2k}\sin(2kR + 2\delta_0) > 0 \quad . \tag{4.113}$$

The Schrödinger equation for two close wave numbers k and $\tilde{k}$ may be written as

$$\frac{1}{r}\frac{d^2}{dr^2}(r\psi) - \left[V_r(r) - k^2\right]\psi = 0 \quad ,$$

$$\frac{1}{r}\frac{d^2}{dr^2}(r\tilde{\psi}^*) - \left[V_r(r) - \tilde{k}^2\right]\tilde{\psi}^* = 0 \quad .$$

We multiply the first equation by $\tilde{\psi}^*$, the second one by ψ, and subtract the second result from the first one. Then we multiply the equality obtained by r^2 and integrate over r from zero to R. Then, for $\tilde{k} \to k$, we have

$$\frac{1}{2k}\left(\psi'\frac{\partial\psi^*}{\partial k} - \psi\frac{\partial\psi'^*}{\partial k}\right)r^2\bigg|_R = \int_0^R dr\, r^2|\psi|^2 \quad . \tag{4.114}$$

The potential vanishes at $r = R$, hence we may substitute the asymptotic

$$\psi = 4\pi\, e^{i\delta_0}\,\frac{\sin(kr + \delta_0)}{kr}$$

into the left-hand part of (4.114). The result is

$$\frac{8\pi^2}{k^2}\left\{\frac{d\delta_0}{dk} + R - \frac{1}{2k}\sin(2kR + 2\delta_0)\right\} = \int_0^R dr\, r^2\,|\psi|^2 \quad . \tag{4.115}$$

Because the right-hand part is positive, the inequality (4.113) follows from (4.115) immediately.

4.4 Verify that the wavefunctions of stationary particle states in the central field,

$$\psi_k^{(+)}(r) = \sum_{lm} i^l\,\psi_{l,k}(r)\,Y_{lm}^*(n_k)\,Y_{lm}(n_r) \quad , \tag{4.116}$$

which contain plane and diverging spherical waves at infinity, are mutually orthogonal and normalized by the condition

$$\int dr\,\psi_{k'}^*(r)\,\psi_k(r) = (2\pi)^3\,\delta(k - k') \quad . \tag{4.117}$$

Substituting (4.116) in the right-hand part of (4.117), making use of the orthogonality condition for the spherical functions, and integrating over the angles, we find that

$$\int d\boldsymbol{r}\, \psi_{\boldsymbol{k}'}^*(\boldsymbol{r})\, \psi_{\boldsymbol{k}}(\boldsymbol{r}) = \sum_{lm} Y_{l,m}(\boldsymbol{n}_{k'})\, Y_{lm}(\boldsymbol{n}_k)$$

$$\times \int_0^\infty dr\, r^2\, \psi_{l,k}^*(r)\, \psi_{l,k}(r) \quad . \tag{4.118}$$

The functions $\psi_{l,k}$ and $\psi_{l,k'}$ are orthogonal, i.e.,

$$\int_0^\infty dr\, r^2\, \psi_{l,k'}^*(r)\, \psi_{l,k}(r) = C\, \delta(k - k') \quad , \tag{4.119}$$

where C is a constant. If the functions $\psi_k^{(+)}(r)$ are normalized to the unit amplitude in the plane wave, then

$$C = \frac{8\pi^3}{k^2} \quad . \tag{4.120}$$

Within the context of the completeness condition for the spherical functions,

$$\sum_{lm} Y_{lm}^*(\boldsymbol{n}')\, Y_{lm}(\boldsymbol{n}) = \delta(\boldsymbol{n} - \boldsymbol{n}') \quad , \tag{4.121}$$

we thus obtain

$$\int d\boldsymbol{r}\, \psi_{\boldsymbol{k}'}^*(\boldsymbol{r})\, \psi_{\boldsymbol{k}}(\boldsymbol{r}) = \frac{8\pi^3}{k^2}\, \delta(k - k')\, \delta(\boldsymbol{n} - \boldsymbol{n}') \quad . \tag{4.122}$$

Multiplying the expression in the right-hand part by $d\boldsymbol{k}$ and integrating over the whole wave vector space yield $(2\pi)^3$, which is the $\boldsymbol{k}$-space delta function multiplied by $(2\pi)^3$. This proves the relation (4.117).

4.5   Find the asymptotics of the total Green function $G(E; \boldsymbol{r}, \boldsymbol{r}')$ for a particle in the central field $V(r)$.

We substitute the asymptotic representation (4.88) for the partial Green function $g_{l,k}^{(+)}(r, r')$ in (4.79, 80) to obtain

$$G(E; \boldsymbol{r}, \boldsymbol{r}') \xrightarrow[r\to\infty]{} -\frac{\mu}{2\pi\hbar^2}\, \frac{\exp(ikr)}{r}\, \psi_{\boldsymbol{k}'}^{(-)*}(\boldsymbol{r}') \quad , \tag{4.123}$$

where $\boldsymbol{k}' = (\boldsymbol{r}/r)k$ and

$$\psi_{\boldsymbol{k}}^{(-)}(\boldsymbol{r}) = \sum_{lm} i^l\, \psi_{l,k}^*(r)\, Y_{lm}^*(\boldsymbol{n}_k)\, Y_{lm}(\boldsymbol{n}_r) \quad . \tag{4.124}$$

The function $\psi_{\boldsymbol{k}}^{(-)}(\boldsymbol{r})$ describes a stationary particle state in the central field $V(r)$. As distinct from (4.8), $\psi_{\boldsymbol{k}}^{(-)}(\boldsymbol{r})$ at infinity contains a plane wave along with the diverging spherical wave, i.e.,

$$\psi_{\boldsymbol{k}}^{(-)}(\boldsymbol{r}) \xrightarrow[r\to\infty]{} e^{i\boldsymbol{k}\boldsymbol{r}} + f(\boldsymbol{k}, \boldsymbol{k}')\, \frac{e^{-ikr}}{r} \quad . \tag{4.125}$$

The function $\psi_k^{(-)}(r)$ can be immediately derived from $\psi_k^{(+)}(r)$. To do this, we have to perform complex conjugation and substitute $-k$ for k, i.e.,

$$\psi_k^{(-)}(r) = \left[\psi_{-k}^{(+)}(r)\right]^* \quad . \tag{4.126}$$

It is not difficult to show that if by virtue of some perturbation the particle undergoes transition into a continuum state, then its final state wave function is $\psi_k^{(-)}(r)$. Indeed, suppose the particle is in the field of the potential $V(r)$ and its transition is caused by the excitation $U(r)$. With the initial wave function being denoted as $\psi_k^{(+)}(r)$, the wave function distortion due to the perturbation $U(r)$ is given by

$$\Delta\psi(r) = \int dr' \, G(E; r, r') \, U(r') \, \psi_k^{(+)}(r') \quad .$$

Making use of the Green function asymptotics (4.123), we find the transition amplitude to be

$$\mathcal{F}(k, k') = -\frac{\mu}{2\pi\hbar^2} \int dr \, \psi_{k'}^{(-)*}(r) \, U(r) \, \psi_k^{(+)}(r) \quad . \tag{4.127}$$

This formula determines the amplitude of particle transition between the continuum states in the field of the potential $V(r)$, induced by the external perturbation $U(r)$.

5. Optical Theorem

5.1 The Total Cross Section and the Elastic Scattering Amplitude

In Chap. 2, the scattering matrix was shown to be unitary by virtue of the conservation of probability in time. Unitarity of the scattering matrix leads to important corollaries; in particular, it is responsible for the relationship between the total cross section of all processes occuring in the system, and the imaginary part of the elastic scattering amplitude.

Unitarity of the scattering matrix S implies that

$$S^+ S = 1 \quad \text{and} \quad SS^+ = 1 \quad . \tag{5.1}$$

We introduce the transition operator $\mathfrak{T} = S - 1$. The $\mathfrak{T}$ matrix must satisfy the condition

$$\mathfrak{T} + \mathfrak{T}^+ = -\mathfrak{T}^+ \mathfrak{T} \quad , \tag{5.2}$$

which follows from the unitarity of the S matrix. We adopt the energy representation and introduce the transition matrix t on the energy shell. Then, in view of (2.33), (5.2) may be rewritten in the matrix form

$$\mathrm{i} \left\{ t_{\beta\alpha} - t_{\alpha\beta}^* \right\} = 2\pi \sum_{\gamma} t_{\gamma\beta}^* t_{\gamma\alpha} \, \delta(E_\alpha - E_\gamma) \quad , \tag{5.3}$$

where $E_\alpha = E_\beta$; summation in the right-hand part extends over all intermediate states γ.

In the special case $\alpha = \beta$, (5.3) reduces to

$$-2 \operatorname{Im} \left\{ t_{\alpha\alpha} \right\} = 2\pi \sum_{\gamma} \left| t_{\gamma\alpha} \right|^2 \delta(E_\alpha - E_\gamma) \quad . \tag{5.4}$$

The right-hand part of (5.4) may be expressed in terms of the sum of cross sections of different transitions $\alpha \rightarrow \gamma$. Indeed, according to (2.34, 38), the cross section of the transition $\alpha \rightarrow \gamma$ is

$$\sigma_{\alpha \rightarrow \gamma} = \frac{2\pi\mu}{\hbar^2 k} \left| t_{\gamma\alpha} \right|^2 \delta(E_\alpha - E_\gamma) \quad ,$$

where $\hbar k$ is the momentum of relative motion of the particles involved in the collision. Therefore, (5.4) may be rewritten as

$$-\frac{2\mu}{\hbar^2 k}\,\text{Im}\{t_{\alpha\alpha}\} = \sum_{\gamma} \sigma_{\alpha\to\gamma} \quad . \tag{5.5}$$

Since the elastic scattering amplitude and the transition matrix satisfy the relation

$$f(k,k') = -\frac{\mu}{2\pi\hbar^2}\,\langle k'|t|k\rangle \quad , \tag{5.6}$$

the left-hand part of (5.5) can be immediately expressed in terms of the forward direction elastic scattering amplitude. Having denoted the sum of cross sections by σ_t,

$$\sigma_t = \sum_{\gamma} \sigma_{\alpha\to\gamma} \quad , \tag{5.7}$$

we obtain from (5.5) the final relation

$$\sigma_t = \frac{4\pi}{k}\,\text{Im}\{f(0)\} \quad , \tag{5.8}$$

which is referred to as the optical theorem. It determines the general relationship between the total cross section of all processes, including elastic scattering and every probable inelastic transition, and the imaginary part of the forward direction elastic scattering amplitude.

5.2 Unitarity Relation for the Elastic Scattering Amplitude

Let us consider (5.3) for the general case of $\alpha \neq \beta$, but assume that no processes other than elastic scattering occur. The states of the system are then completely determined by the momenta of relative motion

$$\alpha \to k \quad , \qquad \beta \to k' \quad , \quad \text{and} \quad \gamma \to k'' \quad ,$$

and we go from summation to integration,

$$\sum_{\gamma} \cdots \to \int \frac{dk''}{(2\pi)^3} \cdots \quad .$$

Having expressed the transition matrix elements in terms of the scattering amplitude by means of (5.6), and carried out integration over intermediate energies by virtue of delta function, we obtain from (5.3)

$$f(k,k') - f^*(k',k) = \frac{ik}{2\pi} \int do'' \, f^*(k',k'') \, f(k,k'') \quad , \tag{5.9}$$

with the integration in the right-hand part extending over all directions of the vector k''. Relation (5.9) is usually referred to as the unitarity condition for the elastic scattering amplitude. If the particles are spinless, the scattering amplitude $f(k, k')$ is symmetric with respect to the vectors k and k', hence the unitarity condition (5.9) may be rewritten as

$$\text{Im}\left\{f(k, k')\right\} = \frac{k}{4\pi} \int do'' \, f^*(k', k'') \, f(k, k'') \quad . \tag{5.10}$$

As it should be anticipated, when $k = k'$, (5.9) yields the optical theorem (5.8).

If the interaction is central, the elastic scattering amplitude depends only on the angle between the vectors k and k'. Making use of the expansion (3.49) for the amplitude and the addition rule for the spherical functions, we obtain a relation for the partial amplitudes

$$\text{Im}\left\{f_l\right\} = k\left|f_l\right|^2 \quad . \tag{5.11}$$

This formula may be rewritten as $\text{Im}\left\{f_l^{-1}\right\} = -k$, then the amplitude f_l must be of the form

$$f_l = \frac{1}{g_l - ik} \quad , \tag{5.12}$$

where $g_l = g_l(k)$ is a real function that is related to the scattering phase shift δ_l by

$$g_l = k \cot \delta_l \quad , \tag{5.13}$$

and therefore,

$$f_l = \frac{1}{k} e^{i\delta_l} \sin \delta_l \quad . \tag{5.14}$$

The unitarity condition (5.9) for the scattering amplitude in principle enables one to find the amplitude if the cross section is known for all angles. Indeed, if we write the elastic scattering amplitude as

$$f(\vartheta) = \sqrt{\sigma(\vartheta)} \, e^{i\alpha(\vartheta)} \quad , \tag{5.15}$$

where $\alpha(\vartheta)$ is the phase of the scattering amplitude, then, with the scattering cross section $\sigma(\vartheta)$ known for all angles, (5.9) may be regarded as an integral equation for the unknown phase $\alpha(\vartheta)$, i.e.,

$$\sin \alpha(\vartheta) = \frac{k}{4\pi} \int do'' \, \sqrt{\frac{\sigma(\theta) \, \sigma(\theta')}{\sigma(\vartheta)}} \, \cos\left[\alpha(\theta) - \alpha(\theta')\right] \quad . \tag{5.16}$$

Here θ is the angle between the vectors k and k'', θ' is the angle between k' and k''. Integration extends over all directions of the vector k''.

Equation (5.16) is invariant with respect to the substitution $\alpha \rightarrow \pi - \alpha$, hence the phase of the scattering amplitude can be calculated only within the uncertainty of the transformation

$$f(\vartheta) \to -f^*(\vartheta) \quad . \tag{5.17}$$

Thus, measuring the elastic scattering cross section for all angles at fixed energy makes it possible to find the elastic scattering amplitude within the uncertainty of the transformation (5.17).

Problems

5.1 Rewrite the unitarity condition (5.9) by assuming the elastic scattering amplitude $f(\boldsymbol{k}, \boldsymbol{k}')$ to be a function of independent variables k and $z = \cos \vartheta$.

We introduce the notation

$$f(\boldsymbol{k}, \boldsymbol{k}') \equiv f(k, z) \quad . \tag{5.18}$$

Let z_1 be the cosine of the angle between the vectors $\boldsymbol{k}$ and $\boldsymbol{k}''$, z_2 the cosine of the angle between $\boldsymbol{k}'$ and $\boldsymbol{k}''$, and φ the angle between the planes $(\boldsymbol{k}, \boldsymbol{k}'')$ and $(\boldsymbol{k}', \boldsymbol{k}'')$. The quantities z, z_1, z_2, and φ satisfy the relation

$$z_2 = z z_1 + \sqrt{1 - z^2} \sqrt{1 - z_1^2} \cos \varphi \quad . \tag{5.19}$$

In the right-hand part of (5.9), we transform from integration over the directions of the vector $\boldsymbol{k}''$, i.e., z_1 and the angle φ, to integration over z_1 and z_2, so that

$$\int do'' \, f(k, z_1) \, f^*(k, z_2)$$

$$= \int_{-1}^{1} dz_1 \int_{0}^{\pi} d\varphi \int_{-1}^{1} dz_2 \, \delta\left(z_2 - z z_1 - \sqrt{1 - z^2} \sqrt{1 - z_1^2} \cos \varphi\right)$$
$$\times f(k, z_1) \, f^*(k, z_2)$$

$$= 2 \int_{-1}^{1} dz_1 \int_{-1}^{1} dz_2 \, \frac{\theta\left(1 - z^2 - z_1^2 - z_2^2 + 2 z z_1 z_2\right)}{\left(1 - z^2 - z_1^2 - z_2^2 + 2 z z_1 z_2\right)^{1/2}}$$
$$\times f(k, z_1) \, f^*(k, z_2) \quad .$$

As a result, the unitarity condition (5.9) reduces to

$$\text{Im}\left\{f(k, z)\right\} = \frac{k}{2\pi} \int_{-1}^{1} dz_1 \int_{-1}^{1} dz_2$$
$$\times \frac{\theta\left(1 - z^2 - z_1^2 - z_2^2 + 2 z z_1 z_2\right)}{\left(1 - z^2 - z_1^2 - z_2^2 + 2 z z_1 z_2\right)^{1/2}}$$
$$\times f(k, z_1) \, f^*(k, z_2) \quad . \tag{5.20}$$

The integration limits in (5.20) are actually determined by the θ-function. By using the Regge expansion

$$\frac{\theta(1 - z^2 - z_1^2 - z_2^2 + 2zz_1z_2)}{(1 - z^2 - z_1^2 - z_2^2 + 2zz_1z_2)^{1/2}}$$

$$= \frac{\pi}{2} \sum_{l=0}^{\infty} (2l + 1) \, P_l(z) \, P_l(z_1) \, P_l(z_2) \quad , \tag{5.21}$$

(5.20) immediately yields the unitary condition (5.11) for the partial amplitudes.

5.2   Find the mean momentum transfer in a scattering process per unit time [4.4].

Let us consider an operator Q which does not depend on time explicitly and commutes with H_0. The time derivative of the operator Q is given by

$$\dot{Q} = \frac{1}{i\hbar} [Q, H] \quad , \tag{5.22}$$

where H is the total Hamiltonian of the system, $H = H_0 + V$. We denote the eigenfunctions of the operators Q and H_0 by φ_α; they correspond to the total Hamiltonian eigenfunctions $\psi_\alpha^{(+)}$. Since Q commutes with H_0, the average value of (5.22) in the state ψ_α is

$$\left(\psi_\alpha, \dot{Q}\psi_\alpha\right) = \left(\psi_\alpha, \frac{1}{i\hbar} [Q, V]\psi_\alpha\right) \quad . \tag{5.23}$$

Within the context of the completeness condition for the set of functions ψ_α, the right-hand part of (5.23) can be rewritten as

$$\left(\psi_\alpha, \frac{1}{i\hbar} [Q, V]\psi_\alpha\right) = \frac{1}{i\hbar} \sum_\gamma Q_\gamma \Big\{ (\psi_\alpha, \varphi) (\varphi_\gamma, V\psi_\alpha)$$
$$- (\psi_\alpha, V\varphi_\gamma) (\varphi_\gamma, \psi_\alpha) \Big\} \quad ,$$

or, since V is Hermitian, as

$$\left(\psi_\alpha, \frac{1}{i\hbar} [Q, V]\psi_\alpha\right) = \frac{2}{\hbar} \sum_\gamma Q_\gamma \, \mathrm{Im} \left\{ (\psi_\alpha, \varphi_\gamma) (\varphi_\gamma, V\psi_\alpha) \right\} \quad . \tag{5.24}$$

Substituting $V\psi_\alpha = t\varphi_\alpha$, where t is the transition operator on the energy shell, and making use of (3.22), we find that

$$\left(\psi_\alpha, \frac{1}{i\hbar} [Q, V]\psi_\alpha\right) = \frac{2}{\hbar} Q_\alpha \, \mathrm{Im} \{t_{\alpha\alpha}\}$$
$$+ \frac{2\pi}{\hbar} \sum_\gamma Q_\gamma |t_{\alpha\alpha}|^2 \, \delta(E_\alpha - E_\gamma) \quad . \tag{5.25}$$

If Q is the unit operator, then (5.25) yields the optical theorem (5.4).

We employ the optical theorem (5.4) and the definition (2.34) of the transition probability per unit time. Then (5.25) becomes

$$\left(\psi_\alpha, \frac{1}{i\hbar}[Q,V]\psi_\alpha\right) = \sum_\gamma (Q_\gamma - Q_\alpha)\, w_{\alpha\to\gamma} \qquad (5.26)$$

With (5.23), we have

$$\left(\psi_\alpha, \dot{Q}\psi_\alpha\right) = \sum_\gamma (Q_\gamma - Q_\alpha)\, w_{\alpha\to\gamma} \quad, \qquad (5.27)$$

i.e., the mean change per unit time of the quantity Q in the state ψ_α is directly expressed in terms of the sum of products of the changes of Q produced by individual transitions, by the relevant transition probabilities.

Substituting the momentum p for Q, we obtain from (5.26)

$$\sum_{p'} (p' - p)\, w_{p\to p'} = \left(\psi_p, (-\nabla V)\psi_p\right) \quad. \qquad (5.28)$$

Thus, the mean change in the momentum per unit time in a scattering event that is described by the wave function ψ_p, is directly expressed in terms of the mean value of the force in the state ψ_p.

6. Time Inversion and Reciprocity Theorem

6.1 Transformation of Wave Functions and Operators Under Inversion of Time

In classical mechanics, equations of motion for a conservative system are invariant with respect to the change of the sign of time, i.e., when t is replaced by $-t$. This implies that in such systems time is reversible: if $r(t)$ is an allowed trajectory of motion, then $r(-t)$ is also allowed. In quantum mechanics, time reversibility occurs if the Hamiltonian of the system is invariant as the sign of time is changed. Hamiltonian invariance with respect to time inversion gives rise to certain relations between direct and inverse transition probabilities.

Let us denote time inversion to be the transformation

$$t \rightarrow t' = -t \quad . \tag{6.1}$$

Suppose the system is in the state $\psi(t)$ and $\psi'(t')$ is the function into which $\psi(t)$ is transformed by time inversion. It is clear that

$$\psi'(t) \equiv \psi(-t) \quad . \tag{6.2}$$

We assume the transformation (6.1) to be associated with the time reflection operator R. By definition, R transforms the wave function $\psi(t)$ into the one $\psi'(t)$, i.e.,

$$\psi'(t) = R\,\psi(t) \quad . \tag{6.3}$$

It is not difficult to show that the time inversion transformation R is equivalent to successive operations of complex conjugation and some unitary transformation. To do this, we consider the time-dependent Schrödinger equation

$$i\hbar\,\frac{\partial \psi}{\partial t} = H\,\psi \quad . \tag{6.4}$$

If the Hamiltonian of the system is invariant under time inversion, substituting $-t$ for t in (6.4) yields

$$-i\hbar\,\frac{\partial \psi'}{\partial t} = H\,\psi' \quad . \tag{6.5}$$

In order to find the relation between $\psi'(t)$ and $\psi(t)$, we consider the complex conjugate to (6.4),

$$-i\hbar \frac{\partial \psi^*}{\partial t} = H^* \psi^* \quad . \tag{6.6}$$

Since H is a Hermitian operator, H and H^* have similar eigenvalues though their eigenfunctions can be different. This implies that H and H^* are related by some unitary transformation W, i.e.,

$$H = WH^*W^{-1} \qquad (W^+W = WW^+ = 1) \quad . \tag{6.7}$$

Applying the operator W from the left to both parts of (6.6), we obtain

$$-i\hbar \frac{\partial(W \psi^*)}{\partial t} = HW\psi^* \quad . \tag{6.8}$$

Comparing this equation to (6.5), we find that

$$\psi' = WK\psi \quad , \tag{6.9}$$

where K is the complex conjugation operator, W is a unitary operator which satisfies the condition (6.7). Thus, the time-inversed wave function $\psi'(t)$ is obtained from the initial function $\psi(t)$ by means of complex conjugation and some unitary tranformation [4.5], i.e.,

$$R = WK \quad . \tag{6.10}$$

The complex conjugation operator K is not linear since

$$K \sum a\psi = \sum a^* K\psi \quad . \tag{6.11}$$

However, complex conjugation conserves the absolute value of the scalar product of two arbitrary functions and hence does not change the wave function normalization, so that

$$\left|(K\psi, K\varphi)\right| = \left|(\psi^*, \varphi^*)\right| = \left|(\psi, \varphi)\right| \quad . \tag{6.12}$$

An operator that satisfies conditions (6.11, 12) is referred to as an anti-unitary operator. The product of a unitary operator and an anti-unitary one is an anti-unitary operator. Therefore, the time reflection operator R is anti-unitary.

Reflection of time is accompanied not only by transformation of functions, but also by transformation of operators. To define the time-reflected operator Q' corresponding to Q, we impose the requirement that the matrix element of Q over the states φ and ψ must be equal to the matrix element of the time-inversed operator Q' over the time inversed states ψ' and φ', i.e.,

$$(\psi, Q\varphi) = (\varphi', Q'\psi') \quad . \tag{6.13}$$

As follows from this condition,

$$Q^+ = R^{-1}Q'R \quad , \tag{6.14}$$

or

$$Q' = RQ^+R^{-1} \quad . \tag{6.15}$$

This equation determines the relationship between the operator Q and the time-inversed operator Q'.

By their behavior under time reflection, all physical quantities can be divided into two groups. The first group includes quantities which are not changed under the reflection of time,

$$Q' = Q \quad . \tag{6.16}$$

Such quantities are the coordinate of a point, the energy, the square of the angular momentum, etc. The second group consists of physical quantities which change their signs under time reflection,

$$Q' = -Q \quad . \tag{6.17}$$

Examples of such quantities are the velocity, momentum, angular momentum, and others. Obviously, in the state described by the time-inversed wave function ψ', all physical quantities of the first group maintain the same values as in the state ψ, and physical quantities of the second group acquire opposite signs.

6.2 Time Inversion Operators for Particular Systems

The explicit form of the time inversion operator R depends on the properties of the system and the representation in which the system is considered.

A) A System of Spinless Particles in Zero Electromagnetic Field. It is not difficult to show that in the coordinate representation the time inversion operator of this system reduces to the complex conjugation operator, i.e.,

$$R = K \quad . \tag{6.18}$$

Indeed, in the coordinate representation $H = H^*$, hence $W = 1$ satisfies the condition (6.7). The operators of coordinate r and momentum $p = -i\hbar \nabla$ are given by

$$r' = KrK^{-1} = r \quad , \qquad p' = K(-i\hbar \nabla) K^{-1} = i\hbar \nabla = -p \quad .$$

In the momentum representation, $r = i\hbar \nabla_p$, $\hat{p} = p$, and $H^* \neq H$. Then we have to put

$$R = W_p K \quad , \tag{6.19}$$

where W_p is the operator which interchanges p and $-p$, so that

$$W_p p W_p^{-1} = -p \quad .$$

It is not difficult to show that condition (6.7) is satisfied; we have $r' = r$ and $p' = -p$.

B) A System of Charged Particles in an External Electromagnetic Field Described by the Vector Potential A. In this case the Hamiltonian contains the term

$$\frac{1}{2m}\left(p - \frac{e}{c}A\right)^2 \quad .$$

Condition (6.7) is satisfied in the coordinate representation if

$$R = W_A K \quad , \tag{6.20}$$

where W_A is the operator whose action on the vector potential A produces $-A$.

C) A System of Spin-Possessing Particles in an Electromagnetic Field A. In this case the Hamiltonian contains the term

$$-\mu\sigma \text{ curl } A \quad ,$$

which describes magnetic interaction (μ is the magnetic moment, σ is the Pauli spin matrix). To satisfy relation (6.7) in the coordinate representation, the condition

$$W = W_A W_\sigma$$

must hold, where W_A is the operator introduced in the previous subsection, W_σ is the spin operator defined by

$$W_\sigma \sigma^* W_\sigma^{-1} = -\sigma \quad .$$

It may be directly verified that the operator

$$W_\sigma = i\sigma_y \tag{6.21}$$

satisfies (6.7). The phase i is introduced in (6.21) because for particles with spin 1/2 the time reflection operator R must satisfy the condition $R^2 = -1$.

6.3 Time-Inversed Wave Function

Let us consider the complete set of eigenfunctions ψ_α of the Hamiltonian H:

$$H\psi_\alpha = E_\alpha \psi_\alpha \quad . \tag{6.22}$$

In the states described by the function ψ_α, some physical quantities can have definite values along with energy. Suppose the set of physical quantities A unambiguously determines the state ψ_α. Then

$$A \psi_\alpha = \alpha \psi_\alpha \quad . \tag{6.23}$$

The time inversion procedure transforms this equation into

$$A' \psi'_\alpha = \alpha \psi'_\alpha \quad , \tag{6.24}$$

where

$$\psi'_\alpha = R \psi_\alpha \quad , \qquad A' = RAR^{-1} \qquad (A^+ = A) \quad .$$

As follows from (6.16, 17), the time-inversed operators A' can differ from the initial operators A only in the sign. Therefore,

$$A' = \pm A \quad , \tag{6.25}$$

where the upper and the lower signs correspond to the quantities of the first and the second groups, respectively. Substituting (6.25) into (6.24), we have

$$A \psi'_\alpha = \pm \alpha \psi'_\alpha \quad . \tag{6.26}$$

Comparing (6.26) to (6.23), we see that if A is not changed by time reflection, then

$$\psi'_\alpha = \psi_\alpha \quad , \tag{6.27}$$

and if it does change the sign, then

$$\psi'_\alpha = \beta_{-\alpha} \psi_{-\alpha} \quad , \tag{6.28}$$

where $\beta_{-\alpha}$ is the phase factor, $|\beta_{-\alpha}| = 1$. Thus we have shown that the time-inversed wave function ψ'_α reproduces, with an uncertainty of the phase factor, the function ψ'_α describing the state that differs from ψ_α by the signs of the quantities from the second group contained in the set α.

As an example we consider the time inversion of the wave function of the state with well defined square of the momentum and projection of the momentum, $\psi_\alpha \equiv \psi_{jm}$. We assume all other quantum numbers to be invariant with respect to time inversion and omit them. The functions ψ_{jm} satisfy the equation

$$j_z \psi_{jm} = m \psi_{jm} \quad . \tag{6.29}$$

We perform time inversion, bearing in mind that $j'_z = -j_z$. Then we have

$$j_z \psi_{jm} = -m \psi'_{jm} \quad , \tag{6.30}$$

from which we find that

$$\psi'_{jm} = \beta^j_{-m} \psi_{j-m} \quad . \tag{6.31}$$

In view of further applications, we calculate the phase factor β^j_{-m} . We introduce the operators $j_\mp = j_x \mp j_y$. Applying $j_\mp$ to the function ψ_{jm} , we obtain

$$j_\mp \psi_{jm} = \left[(j \pm m)(j \mp m + 1)\right]^{1/2} \psi_{jm\mp1} \quad . \tag{6.32}$$

Making use of the definition of a time-inversed operator (6.15), since $j_\mp$ changes its sign under time inversion, we find that

$$R j_\mp R^{-1} = -j_\pm \quad ,$$

or

$$R j_\mp = -j_\pm R \quad . \tag{6.33}$$

Applying the time reflection operator R to (6.32), we obtain

$$R j_- \psi_{jm} = \beta^j_{-m+1} \left[(j + m)(j - m + 1)\right]^{1/2} \psi_{j-m+1} \quad . \tag{6.34}$$

Within the context of the relation (6.33),

$$R j_- \psi_{jm} = -j_+ R \psi_{jm} = -\beta^j_{-m} j_+ \psi_{j-m} \quad ,$$

and then

$$R j_- \psi_{jm} = -\beta^j_{-m} \left[(j + m)(j - m + 1)\right]^{1/2} \psi_{j-m+1} \quad . \tag{6.35}$$

Comparing (6.34) to (6.35) yields

$$\beta^j_{-m+1} = -\beta^j_{-m} \quad ;$$

this equality is satisfied if

$$\beta^j_{-m} = (-1)^m \beta^j \quad . \tag{6.36}$$

Thus, the explicit m-dependence of the phase factor is determined by the properties of the angular momentum.

To find the phase factor β^j, an additional condition must be imposed. One of the ways is to require that time reflection must not modify the rules for addition of angular momenta. We consider the wave function of a system with a well defined angular momentum consisting of two subsystems whose momenta are also well defined, i.e.,

$$\psi_{IM} = \sum_{m_1 m_2} (j_1 m_1 j_2 m_2 \,|\, IM) \, \psi_{j_1 m_1} \psi_{j_2 m_2} \quad . \tag{6.37}$$

The time reflection procedure transforms this equality into

$$\psi'_{IM} = \beta^{j_1} \beta^{j_2} \sum_{m_1 m_2} (j_1 m_1 j_2 m_2 \,|\, IM)(-1)^{m_1+m_2} \psi_{j_1 - m_1} \psi_{j_2 - m_2} \quad . \tag{6.38}$$

In deriving (6.38), use was made of the relations

$$\psi'_{j_1 m_1} = (-1)^{m_1} \beta^{j_1} \psi_{j_1 - m_1} \quad ,$$

$$\psi'_{j_2 m_2} = (-1)^{m_2} \beta^{j_2} \psi_{j_2 - m_2} \quad .$$

After renotation of the summation indices $m_1 \to -m_1$, $m_2 \to -m_2$, and noting the symmetry property of the Clebsch–Gordon coefficients,

$$(j_1 - m_1 j_2 - m_2 \,|\, IM) = (-1)^{j_1 + j_2 - I} \left(j_1 m_1 j_2 m_2 \,|\, I - M\right) \quad , \tag{6.39}$$

(6.38) reduces to

$$\psi'_{IM} = (-1)^{-I + M + j_1 + j_2} \, \beta^{j_1} \, \beta^{j_2} \, \psi_{I-M} \quad , \tag{6.40}$$

whereas we had to obtain

$$\psi'_{IM} = (-1)^M \, \beta^I \, \psi_{I-M} \quad . \tag{6.41}$$

Comparing (6.40) to (6.41) yields

$$\beta^j = (-1)^{-j} \quad . \tag{6.42}$$

Thus, the wave function of a state with well defined angular momentum is transformed by time inversion according to the law

$$\psi'_{jm} = (-1)^{-j+m} \, \psi_{j-m} \quad . \tag{6.43}$$

In Chap. 1 we noted that angular dependences of the eigenfunctions of the square of the angular momentum and its projection in the coordinate representation are similar. They are described by the spherical functions $Y_{lm}(n_r)$ and $\psi_{lm}(n_k)$. Time inversion transforms the spherical function in the momentum representation, $Y_{lm}(n_k)$, according to (6.43) so that

$$Y'_{lm}(n_k) = Y^*_{lm}(-n_k) = (-1)^{-l+m} \, Y_{l-m}(n_k) \quad ,$$

and therefore it may be regarded as the wave function of the state with definite angular momentum in the momentum representation,

$$\psi_{lm}(n_k) = Y_{lm}(n_k) \quad . \tag{6.44}$$

In the coordinate representation, the state with this angular momentum is described by the wave function

$$\psi_{lm}(n_r) = i^l \, Y_{lm}(n_r) \quad , \tag{6.45}$$

which contains an additional phase factor i^l that is introduced in order to ensure that time inversion transforms the function (6.45) according to (6.43). Note that only separating the factor i^l in (4.21, 22) enables one to treat them as expansions in terms of states with well defined angular momenta of the system.

6.4 The Reciprocity Theorem and Detailed Balance

We have already mentioned that time reversibility occurs if the Hamiltonian of the system is invariant under time inversion, $H' = H$. It is not difficult to verify that this property of the Hamiltonian is responsible for the invariance of the scattering matrix, so that

$$S' = S \quad . \tag{6.46}$$

The property (6.46) of the scattering matrix enables one to derive simple relations between the direct and inverse transition probabilities. According to (6.14),

$$S = \left(R^{-1} S' R \right)^{+} \quad . \tag{6.47}$$

Having calculated the matrix element of (6.47) corresponding to the transition $\alpha \to \beta$, according to the definition of a time-inversed state (6.3) we obtain

$$\left(\varphi_\beta , S \varphi_\alpha \right) = \left(\varphi'_\alpha , S' \varphi_\beta \right) \quad . \tag{6.48}$$

Substituting S for S' in the right-hand part and using (6.28, 46) we come to the relation that is usually called the reciprocity theorem or condition:

$$S_{\beta \alpha} = \beta^{*}_{-\alpha} \beta_{-\beta} S_{-\alpha -\beta} \quad . \tag{6.49}$$

The reciprocity theorem determines the general relation between the scattering matrix elements for the direct ($\alpha \to \beta$) and time-inversed ($-\beta \to -\alpha$) transitions. The reciprocity theorem is a direct corollary of the scattering matrix invariance with respect to the inversion of time.

Of course, a relation analogous to (6.49) also holds for the matrix elements of the transition operator t, i.e.,

$$t_{\beta \alpha} = \beta^{*}_{-\alpha} \beta_{-\beta} t_{-\alpha -\beta} \quad . \tag{6.50}$$

According to (2.37), the probability of the transition from the state α into the state β per unit time is proportional to the squared modulus of the matrix element $t_{\beta \alpha}$ and the final state density $\varrho(E)$. Therefore, we may employ the reciprocity theorem (6.50) to obtain the relation between the probabilities of the direct ($\alpha \to \beta$) and time-inversed ($-\beta \to -\alpha$) transitions. The result is

$$\frac{w_{\alpha \to \beta}}{\varrho(E_\beta)} = \frac{w_{-\beta \to -\alpha}}{\varrho(E_\alpha)} \quad . \tag{6.51}$$

Note that the time-inversed transition ($-\beta \to -\alpha$) is the transition from the state $-\beta$ into the state $-\alpha$ which differ from the final β and initial α states of the direct transition ($\alpha \to \beta$) by the signs of all the quantities from the second group. If the final state densities of the two transitions are equal, then the direct and time-inversed transition probabilities are equal, too.

For example, we consider a transition in the system with the initial and final states determined by given relative momenta, particle spins, and spin projections, i.e.,

$$\alpha \equiv k \; ; \; j_1, m_1 \; ; \; j_2, m_2 \qquad \text{and} \qquad \beta \equiv k' \; ; \; j_1', m_1' \; ; \; j_2', m_2' \; .$$

(For elastic scattering, $k = k'$, $j_1 = j_1'$, $j_2 = j_2'$; in the general case of some reaction, $k \neq k'$, $j_1 \neq j_1'$, $j_2 \neq j_2'$). If the Hamiltonian is invariant under time inversion, then the reciprocity condition for the transition under consideration and the time-inversed one may be written as

$$\langle k' j_1' m_1' \, j_2' m_2' \, | \, S \, | \, k \, j_1 m_1 \, j_2 m_2 \rangle$$

$$= (-1)^{-j_1 - j_2 - j_1' - j_2' + m_1 + m_2 + m_1' + m_2'}$$

$$\times \langle -k j_1 - m_1 j_2 - m_2 \, | \, S \, | \, -k' j_1' - m_1' j_2' - m_2' \rangle \; . \qquad (6.52)$$

When deriving (6.52), we employed the phase factor definition (6.43).

If the Hamiltonian is also invariant under spatial reflections, then $S = PSP^{-1}$, where P is the spatial reflection operator, and hence

$$\langle k' j_1' m_1' j_2' m_2' \, | \, S \, | \, k j_1 m_1 j_2 m_2 \rangle$$

$$= (-1)^{-j_1 - j_2 - j_1' - j_2' + m_1 + m_2 + m_1' + m_2'}$$

$$\times \langle k j_1 - m_1 j_2 - m_2 \, | \, S \, | \, k' j_1' - m_1' j_2' - m_2' \rangle \; . \qquad (6.53)$$

If the states $-\alpha$ and $-\beta$ differ from α and β only by the signs of spin projections, then the condition (6.53) may be written in the form (6.49). With the states being defined in this manner, the probabilities of the transitions $-\beta \to -\alpha$ and $\beta \to \alpha$, averaged over initial and final spin projections, coincide, i.e.,

$$\sum w_{-\beta \to -\alpha} = \sum w_{\beta \to \alpha} \; . \qquad (6.54)$$

Thus, for the systems invariant with respect to both time inversion and spatial coordinate reflections, semi-detailed balance occurs: the direct and inverse transition probabilities, averaged over initial and final spin projections and referenced to the same final state, are equal, i.e.,

$$\frac{\sum w_{\alpha \to \beta}}{\varrho(E_\beta)} = \frac{\sum w_{\beta \to \alpha}}{\varrho(E_\alpha)} \; . \qquad (6.55)$$

If the Hamiltonian is invariant under spatial rotations, too, then, as was shown in Chap. 2, the scattering matrix does not depend on spin projections. Thus, the reciprocity theorem immediately yields the relation between the scattering operator matrix elements for the direct and inverse transitions,

$$S_{\beta\alpha} = S_{\alpha\beta} \; . \qquad (6.56)$$

This relation implies that detailed balance occurs in the system: the direct and inverse transition probabilities referenced to the same final state, are equal, so that

$$\frac{w_{\alpha \to \beta}}{\varrho(E_\beta)} = \frac{w_{\beta \to \alpha}}{\varrho(E_\alpha)} \quad . \tag{6.57}$$

The detailed balance occurs in the systems whose properties are invariant with respect to rotations and reflections of spatial coordinates and to time inversion.

Problems

6.1   Show that if N channels are open, then the scattering matrix is completely determined by $N(N+1)/2$ real parameters.

The reciprocity and unitarity conditions decrease the number of independent parameters which determine the scattering matrix. If N channels are open, then the scattering matrix is complex and in the general case contains $2N^2$ real parameters. Among these, only $N(N+1)/2$ are independent by virtue of the unitarity and reciprocity conditions. The scattering matrix may be written as

$$S = \frac{1 - (i/2) K}{1 + (i/2) K} \quad , \tag{6.58}$$

where $K^+ = K$. As follows from this representation, $S^+ = S^{-1}$. Reversing the equality (6.58) yields

$$\frac{i}{2} K = \frac{1 - S}{1 + S} \quad . \tag{6.59}$$

The matrix K possesses the same symmetry properties as S. In the total moment representation, the K matrix is Hermitian and symmetric, i.e., depends on $N(N+1)/2$ real parameters which are sufficient for the complete description of scattering and reactions.

6.2   Show that if two channels are open, then the scattering matrix is described by the expression

$$S = \begin{pmatrix} \sqrt{1 - r^2} \, e^{2i\delta} & ir \, e^{i(\delta + \eta)} \\ ir \, e^{i(\delta + \eta)} & \sqrt{1 + r^2} \, e^{2i\eta} \end{pmatrix} \quad , \tag{6.60}$$

where r, δ, and η are real parameters.

7. Analytic Properties
of the Scattering Matrix

7.1 Analytic Properties of Radial Wave Functions

If the angular momentum quantum number l is fixed, then the scattering matrix is a function either of energy E or of the wave number k. Being defined for real wave numbers, the scattering matrix $S_l(k)$ can be analytically continued to complex k. Analyticity of the scattering matrix is a corollary of the physical causality principle which infers that the cause must preceed the consequence. However, peculiar analytic properties of the scattering matrix, e.g., the nature and dislocation of singularities, are determined by the properties of the physical system. Therefore, investigation of analytic properties of the scattering matrix can provide information concerning the properties of the physical system itself. The remarkable success of the theory is the unified description of scattering and bound states of the system.

Let us discuss analytic properties of the scattering matrix in the case of nonrelativstic particle scattering in a central field. First of all we study the wave number dependence of the radial wave functions. For the sake of simplicity, we consider only S-scattering.

The radial wave function in the S-state is governed by the equation

$$\frac{d^2u}{dr^2} + \left[k^2 - V_r(r)\right] u = 0 \quad . \tag{7.1}$$

This is a linear homogeneous equation of the second order; its general solution may be written as a linear combination of two independent solutions. We remind the reader that the solutions u_1 and u_2 are independent if their Wronskian is not equal to zero,

$$W(u_1, u_2) \equiv u_1 u_2' - u_1' u_2 \neq 0 \quad .$$

Let us study various solutions of (7.1) as functions of k. We assume that the singularity of the potential at the point $r = 0$ is weaker than r^{-2}, and that the potential decreases at infinity faster than r^{-3}, i.e., that the potential $V_r(r)$ satisfies the conditions

$$\int_0^\infty dr \, r \, V_r(r) = M < \infty \quad , \tag{7.2}$$

$$\int_0^\infty dr \, r^2 \, V_r(r) = N < \infty \quad .$$ (7.3)

The various solutions of (7.1) are determined by the relevant boundary conditions, which usually are imposed either on the solution and its derivative at some point r, or on the asymptotics of the solution at infinity.

The Regular Solution.   We introduce the function $\Phi(k, r)$ which satisfies (7.1) and the following boundary conditions at $r = 0$:

$$\Phi(k, 0) = 0 \quad , \qquad \Phi'(k, 0) = 1 \quad .$$ (7.4)

Since both (7.1) and the boundary conditions (7.4) are real for real k, the function $\Phi(k, r)$ is real and depends on k^2, i.e., is an even function of k.

If k is complex, then the expression within the square brackets in (7.1) is an entire (regular) function of k, i.e., it has no singularities if k is integer. According to the Poincaré theorem, if (i) the coefficient entering the second-order differential equation is a function of the coordinate and an entire function of some parameter, and (ii) the boundary conditions do not depend on this parameter, then the solution is an entire function of this parameter for any coordinate. Therefore, the function $\Phi(k, r)$, which is a solution of (7.1) and satisfies the boundary conditions (7.4), is an entire (regular) function of k, i.e., an analytic nonsingular function in the open plane of the complex variable k for any r.

To prove this property of the function $\Phi(k, r)$, we reduce the differential equation (7.1) and the boundary conditions (7.1) at the point $r = 0$ to an integral equation. We rewrite (7.1) as

$$\Phi'' + k^2 \Phi = V_r \Phi$$ (7.5)

and regard (7.5) as an inhomogeneous equation with a given right-hand part. We assume the two independent solutions of the relevant homogeneous equation to be

$$\Phi_1 = \frac{1}{k} \sin kr \qquad \text{and} \qquad \Phi_2 = \cos kr \quad ,$$ (7.6)

and impose the condition

$$\Phi_1' \Phi_2 - \Phi_1 \Phi_2' = 1$$ (7.7)

which determines the coefficients. Then the general solution of the inhomogeneous equation may be written as

$$\Phi = \Phi_1 \int_a^r dr' \, \Phi_2 \, V_r \Phi + \Phi_2 \int_r^b dr' \, \Phi_1 V_r \Phi \quad ,$$ (7.8)

where a and b are arbitrary constants. To satisfy the boundary condition (7.4), we must take a and b such that

$$\int_0^a dr' \; \Phi_2 V_r \Phi = -1 \qquad \text{and} \qquad b = 0 \quad,$$

then (7.8) reduces to an integral equation for $\Phi(k, r)$, i.e.,

$$\Phi(k, r) = \frac{1}{k} \sin kr + \int_0^r dr' \; \frac{\sin k(r - r')}{k} V_r(r') \Phi(k, r') \quad, \qquad (7.9)$$

which also may be written as

$$\Phi(k, r) = \frac{1}{k} \sin kr + \int_0^\infty dr' \; G_k(r - r') V_r(r') \Phi(k, r') \quad, \qquad (7.10)$$

where $G_k(r - r')$ is the Green function determined by the equation (7.5) and the boundary conditions (7.4) imposed on the function $\Phi(k, r)$. It is given by

$$G_k(r - r') = \begin{cases} \frac{\sin k(r - r')}{k} \quad, & r' < r \quad, \\ 0 \quad, & r' > r \quad. \end{cases} \qquad (7.11)$$

Note that the integration range in (7.9) is limited from above by $r' = r$, therefore, (7.9) is a Volterra integral equation that can be solved by the iteration method.

The solution of (7.9) may be written as a series

$$\Phi(k, r) = \sum_{n=0}^\infty \Phi^{(n)}(k, r) \quad, \qquad (7.12)$$

where

$$\Phi^{(0)}(k, r) = \frac{1}{k} \sin kr \qquad (7.13)$$

and

$$\Phi^{(n)}(k, r) = \int_0^r dr' \; \frac{\sin k(r - r')}{k} V_r(r') \Phi^{(n-1)}(k, r') \quad. \qquad (7.14)$$

To verify whether the integrals appearing in successive iterations converge, we impose the following restriction on the sine functions:

$$\left| \frac{\sin k(r - r')}{k} \right| \leq r \quad. \qquad (7.15)$$

It is clear that for the integrals to be convergent, the potential must satisfy the condition (7.2). To prove that the series (7.12) is convergent, it is sufficient to substantiate the validity of the inequalities

$$\frac{1}{r} \left| \Phi^{(1)}(k, r) \right| \leq q(r) \quad, \qquad q(r) \equiv \int_0^r dr' \; r' V_r(r') \quad,$$

$$\frac{1}{r} \left| \Phi^{(2)}(k, r) \right| \leq \int_0^r dr' \; r' \; V_r(r') q(r') = \int_0^{q(r)} dq \; q = \frac{q(r)^2}{2!} \quad,$$

$$\frac{1}{r} \left| \Phi^{(3)}(k,r) \right| \le \int_0^r dr' \, r' \, V_r(r') \frac{q(r')^2}{2!} = \frac{q(r)^3}{3!} \quad , \ldots \quad .$$

Thus, for all k and r we have

$$\frac{1}{r} \left| \Phi^{(n)}(k,r) \right| \le \frac{q(r)^n}{n!} \le \frac{M^n}{n!} \quad , \tag{7.16}$$

with the quantity M being defined by (7.2). Substituting the inequalities obtained in the expression (7.12), we find that

$$\left| \Phi(k,r) \right| \le r \sum_{n=0}^{\infty} \frac{M^n}{n!} = r \, e^M \quad , \tag{7.17}$$

i.e., each term of the series (7.12) is majorized by the relevant term of the exponential expansion and hence the series (7.12) converges uniformly if the potential $V_r(r)$ satisfies the condition (7.2) at the point $r = 0$. It is not difficult to show by direct substitution that the solution in the series form (by virtue of uniform convergence of the letter) indeed satisfies (7.9).

The restriction (7.15) is suitable for estimates when the argument of the sine function is small; for large arguments, it is better to employ the inequality

$$\left| \sin x \right| \le 1 \quad .$$

Combining the two estimates, we find the restriction that holds for all values of x to be

$$\left| \sin x \right| \le \beta \frac{x}{1+x} \qquad (x \ge 0) \quad , \tag{7.18}$$

where β is a constant whose numerical value is irrelevant. The right-hand part of (7.18) is a monotonic function of x; the inequality remains valid when any number greater than x is substituted for x in the right-hand side.

When considering complex k, we have to bear in mind that the function $\sin z$ grows with increasing $|\mathrm{Im}\{z\}|$ as $\exp[|\mathrm{Im}\{z\}|]$. Hence, in the general case instead of (7.18) we must consider the inequality

$$\left| \sin z \right| \le \beta \frac{|z|}{1+|z|} \exp\left[|\mathrm{Im}\{z\}|\right] \quad ,$$

and the function (7.14) can be estimated by the same technique; the result is

$$\left| \Phi^{(n)}(k,r) \right| \le \beta \frac{r}{1+|kr|} \exp\left[|\mathrm{Im}\{kr\}|\right] \frac{M^n}{n!} \quad . \tag{7.19}$$

Similarly to the case of real k ($k \ge 0$), this infers that the series converges and satisfies the initial integral equation. And since the inequality (7.19) ensures uniform convergence of (7.12) for arbitrary k in any limited region, the function $\Phi(k,r)$ is analytic for any k provided the individual terms $\Phi^{(n)}(k,r)$ are analytic. The zero iteration $\Phi^{(0)}(k,r) = (\sin kr)/k$ is an analytic (regular) function for complex k; analyticity (regularity) of subsequent iterations can be proved by induction.

Thus, the physical solution $\Phi(k, r)$, initially defined for real k $(k \geq 0)$, can be analytically continued to complex k; the resultant function satisfies, as before, the radial Schrödinger equation (which is complex in this case).

The Jost Solutions. We introduce another solution of (7.1), $f(k, r)$, which satisfies the boundary condition

$$\lim_{r \to \infty} e^{ikr} f(k, r) = 1 \qquad (7.20)$$

and usually is referred to as the Jost solution. From (7.20) we find the asymptotics of the function $f(k, r)$ at infinity to be

$$f(k, r) \to e^{-ikr} \quad , \qquad r \to \infty \quad . \qquad (7.21)$$

The boundary condition (7.20) enables one to find $f(k, r)$ in the lower half-plane of complex k, Im $\{k\} \leq 0$. It is not difficult to show that $f(k, r)$ is an analytic function of complex k for Im $\{k\} < 0$ and is continuous along the real axis Im $\{k\} = 0$.

To prove this assertion, we reduce the differential equation (7.1) and the boundary condition (7.20) to an integral equation. The equation (7.1) may be treated as an inhomogeneous equation

$$f'' + k^2 f = V_r f \quad , \qquad (7.22)$$

the coefficients being determined by the requirement $f_1' f_2 - f_1 f_2' = 1$. We take the two independent solutions of the homogeneous equation to be

$$f_1 = -e^{-ikr} \qquad \text{and} \qquad f_2 = -\frac{1}{2ik} e^{ikr} \quad . \qquad (7.23)$$

The general solution of the inhomogeneous equation (7.22) may be written as

$$f = f_1 \int_a^r dr' \, f_2 V_r f + f_2 \int_r^b dr' \, f_1 V_r f \quad ,$$

where a and b are constants. By virtue of the boundary condition (7.20), we have

$$\int_a^\infty dr' \, e^{ikr'} \, V_r f = -2ik \qquad \text{and} \qquad b = \infty \quad ,$$

then $f(k, r)$ is governed by the integral equation

$$f(k, r) = e^{-ikr} + \int_r^\infty dr' \, \frac{\sin k(r' - r)}{k} \, V_r(r') f(k, r') \quad , \qquad (7.24)$$

because the Green function for (7.22) with the boundary condition (7.20) is

$$G_k(r - r') = \begin{cases} 0 & , \quad r' < r \\ \frac{\sin k(r'-r)}{k} & , \quad r' > r \end{cases} . \tag{7.25}$$

Having introduced the function

$$g(k, r) \equiv e^{ikr} f(k, r) \quad , \tag{7.26}$$

we rewrite (7.24) in the form

$$g(k, r) = 1 + \int_r^\infty dr' \, \tilde{G}_k(r' - r) \, V_r(r') \, g(k, r') \quad , \tag{7.27}$$

where

$$\tilde{G}_k(x) \equiv \frac{1 - e^{-2ikx}}{2ik} = \int_0^x dy \, e^{-2iky} \quad .$$

The solution of (7.27) may be written in the series form, i.e.,

$$g(k, r) = \sum_{n=0}^\infty g^{(n)}(k, r) \quad , \tag{7.28}$$

where

$$g^{(0)} = 1 \quad ,$$

$$g^{(n)} = \int_r^\infty dr' \, \tilde{G}_k(r' - r) \, V_r(r') \, g^{(n-1)} \quad . \tag{7.29}$$

The series is convergent for $\mathrm{Im}\,\{k\} < 0$. Indeed, in this case

$$\left| G_k(r' - r) \right| \leq r' - r \leq r' \quad ,$$

and it is a simple matter to prove the inequalities

$$\left| g^{(1)} \right| \leq p(r) \quad , \qquad p(r) \equiv \int_r^\infty dr' \, r' \, \left| V_r(r') \right| \quad ,$$

$$\left| g^{(2)} \right| \leq \int_r^\infty dr' \, r' \, \left| V_r(r') \right| p(r') = \int_0^{p(r)} dp \, p = \frac{p(r)^2}{2!} \quad ,$$

$$\left| g^{(3)} \right| \leq \int_r^\infty dr' \, r' \, \left| V_r(r') \right| \frac{p(r')^2}{2!} = \frac{p(r)^3}{3!} \quad , \, \ldots \quad ,$$

from which we find that

$$\left| g^{(n)} \right| \leq \frac{p(r)^n}{n!} \leq \frac{M^n}{n!} \quad . \tag{7.30}$$

As follows from (7.30), each term of the series (7.28) is majorized by the relevant term of the exponential expansion and hence the series converges uniformly if the potential $V_r(r)$ satisfies condition (7.2) at $r = 0$. To prove that $g(k, r)$ is an analytic function of k, it is sufficient to show that the sequence of k-derivatives uniformly converges and is continuous with respect to k. Having differentiated the series (7.28) with respect to k, one obtains a series that uniformly converges if the potential $V_r(r)$ satisfies condition (7.3) at infinity, for $r \to \infty$. Thus we have shown that the function $f(k, r)$, which is related to $g(k, r)$ by (7.26), is analytic, has no singularities in the lower half-plane of the complex variable k, and is continuous along the real k axis.

Extension of the analyticity region of $f(k, r)$ to the upper half-plane of complex k is determined by the decrease of the potential at infinity. For example, if the potential satisfies the condition

$$\int_0^\infty dr\, \mathrm{e}^{mr} \left|V_r(r)\right| < \infty \quad , \tag{7.31}$$

where m is real and positive, then the function $f(k, r)$ is analytic for $\mathrm{Im}\,\{k\} < m/2$, with the exception of the point $k = 0$. The proof is analogous to the previous cases, the only difference is that $|G_k(x)| < |x \exp(imx)|$.

As follows from the boundary condition (7.20) and (7.1), in the analyticity region including the real axis, the Jost solution satisfies the condition

$$f^*(-k^*, r) = f(k, r) \quad . \tag{7.32}$$

The Jost solutions $f(k, r)$ and $f(-k, r)$ of (7.1) are independent. This can be easily verified because the Wronskian of $f(k, r)$ and $f(-k, r)$ is not equal to zero. Since the Wronskian does not depend on r, we can calculate it by substituting the $r \to \infty$ asymptotics for $f(k, r)$ and $f(-k, r)$ and we have

$$W\big(f(k, r)\, f(-k, r)\big) = 2ik \quad . \tag{7.33}$$

The function $\Phi(k, r)$ may be written as a linear combination of $f(k, r)$ and $f(-k, r)$, i.e.,

$$\Phi(k, r) = a\, f(k, r) + b\, f(-k, r) \quad .$$

The coefficients a and b are determined by condition (7.4) imposed on the function $\Phi(k, r)$ at the point $r = 0$. Thus we obtain

$$\Phi(k, r) = -\frac{1}{2ik} \left\{ f(-k)\, f(k, r) - f(k)\, f(-k, r) \right\} \quad , \tag{7.34}$$

where

$$f(k) = f(k, 0) \qquad \text{and} \qquad f(-k) = f(-k, 0) \quad . \tag{7.35}$$

Equation (7.33) immediately yields the relation

$$f(k) = W\big[f(k, r)\,, \Phi(k, r)\big] \quad , \tag{7.36}$$

which will be employed in our future analysis.

Since the scattering matrix can be expressed in terms of the Jost functions $f(k)$ and $f(-k)$, they play an important role in scattering theory [4.6]. In order to find the relation between the Jost functions and the scattering matrix $S(k)$, we consider the asymptotic of the function (7.34) for large r that is given by

$$\Phi(k,r) \underset{r\to\infty}{\to} -\frac{f(-k)}{2ik}\left\{e^{-ikr} - \frac{f(k)}{f(-k)}\, e^{ikr}\right\} \ . \tag{7.37}$$

Comparing this expression to (4.24), we find that

$$S(k) = \frac{f(k)}{f(-k)} \ . \tag{7.38}$$

We have already considered the analytic properties of the Jost functions, therefore, (7.38) can aid in the analysis of the analytic properties of the scattering matrix.

Within the context of the integral equation (7.9) for the regular solution, we can write the asymptotics of the function $\Phi(k,r)$ as

$$\begin{aligned}\Phi(k,r) \underset{r\to\infty}{\to} -\frac{1}{2ik}&\left\{\left[1 + \int_0^\infty dr\, e^{ikr}\, V_r(r)\, \Phi(k,r)\right] e^{-ikr}\right. \\ &\left. - \left[1 + \int_0^\infty dr\, e^{-ikr}\, V_r(r)\, \Phi(k,r)\right] e^{ikr}\right\} \ .\end{aligned}$$

Comparing this relation to (7.37), we obtain the general relation between the Jost function $f(k)$ and the regular solution $\Phi(k,r)$:

$$f(k) = 1 + \int_0^\infty dr\, e^{-ikr}\, V_r(r)\, \Phi(k,r) \ . \tag{7.39}$$

For physical (real) values of k ($k \geq 0$), (7.17) yields the estimate

$$\left|f(k) - 1\right| \leq \beta\, e^M \int_0^\infty dr\, \frac{r}{1 + kr}\, \left|V_r(r)\right| \ , \tag{7.40}$$

from which we find that in the limiting cases $V_r(r) \to 0$ or $k \to \infty$ the Jost function tends to the value that occurs for free motion, $f(k) \to 1$.

For complex k, the following estimate holds:

$$\begin{aligned}\left|f(k) - 1\right| \leq \beta\, e^M &\int_0^\infty dr\, \frac{r}{1 + |k|r}\, \left|V_r(r)\right| \\ &\times \exp\left[\,\left|\mathrm{Im}\,\{k\}\right| + \mathrm{Im}\,\{k\}\right]r \ .\end{aligned} \tag{7.41}$$

If $\mathrm{Im}\,\{k\} < 0$, then the exponent vanishes and the restriction reduces to that for physical values of k. Thus, the integral in (7.39) is convergent for all k in the lower half-plane, and hence the Jost function $f(k)$ is regular in this region. If $\mathrm{Im}\,\{k\} > 0$, then the integral in (7.39) is divergent at its upper limit unless conditions more stringent than (7.3) are imposed on the decrease of the potential. The region $\mathrm{Im}\,\{k\} < 0$, in which the Jost function $f(k)$ is regular, does not contain the physical region (real positive semi-axis). However, the integral (7.39) is continuous for $\mathrm{Im}\,\{k\} \leq 0$ and hence the physically meaningful Jost function (for real positive k) is continuously joint to the function $f(k)$ that is analytic for $\text{Im}\{k\} < 0$.

7.2 Generalization to Include Nonzero Angular Momenta

The results of the previous section can be easily extended to the case of arbitrary angular momentum quantum number l. The radial function of the state with quantum number l must satisfy the equation

$$\frac{d^2 u_l}{dr^2} + \left[k^2 - \frac{l(l+1)}{r^2} - V_r(r) \right] u_l = 0 \quad . \tag{7.42}$$

The various solutions of this equation are determined by the boundary conditions either at the point $r = 0$ or $r \to \infty$. We assume, as before, that the potential $V_r(r)$ is real and satisfies conditions (7.2, 3).

First of all we consider the solutions determined by the boundary conditions at $r = 0$. If r is small, the centrifugal energy in (7.42) is much greater than $k^2 - V_r(r)$ and these terms may be disregarded. Thus we obtain a truncated equation

$$\frac{d^2 \tilde{u}_l}{dr^2} - \frac{l(l+1)}{r^2} \tilde{u}_l = 0 \quad , \tag{7.43}$$

the general solution of which is of the form

$$\tilde{u}_l = \alpha\, r^{l+1} + \beta\, r^{-l} \quad . \tag{7.44}$$

As follows from (7.44), the appropriate two independent solutions of (7.42), $\Phi_l(k,r)$ and $\Phi_l^{(1)}(k,r)$, are the ones which satisfy the boundary conditions

$$\Phi_l(k,r) \underset{r\to 0}{\longrightarrow} r^{l+1} \quad , \tag{7.45}$$

$$\Phi_l^{(1)}(k,r) \underset{r\to 0}{\longrightarrow} r^{-l} \quad . \tag{7.46}$$

When considering the solution of (7.42) with boundary conditions at infinity, it is convenient to proceed from the truncated equation

$$\frac{d^2 \tilde{u}_l}{dr^2} + k^2 \tilde{u}_l = 0 \quad , \tag{7.47}$$

which follows from (7.42) if both the potential $V_r(r)$ and the centrifugal energy are neglected. The general solution of (7.47) is given by

$$\tilde{u}_l = \alpha\, e^{-ikr} + \beta\, e^{ikr} \quad , \tag{7.48}$$

so we consider the solutions $f_l(k,r)$ and $f_l(-k,r)$ with the asymptotic properties given by

$$\lim_{r\to\infty} e^{ikr} f_l(k,r) = 1 \quad . \tag{7.49}$$

It is not difficult to verify that the solutions $f_l(k,r)$ and $f_l(-k,r)$ satisfy the condition

$$W\big(f_l(k,r),\, f_l(-k,r)\big) = 2ik \tag{7.50}$$

Similarly to the S-scattering case, we shall call $f_l(k,r)$ and $f_l(-k,r)$ the Jost solutions.

We define the Jost function $f_l(k)$ by means of the equality[1]

$$f_l(k) \equiv W\big(f_l(k,r),\, \Phi_l(k,r)\big) \quad , \tag{7.51}$$

then the regular solution $\Phi_l(k,r)$ is expressed in terms of the Jost solutions $f_l(k,r)$ and $f_l(-k,r)$ as given by

$$\Phi_l(k,r) = -\frac{1}{2ik}\left\{ f_l(-k)\, f_l(k,r) - f_l(k)\, f_l(-k,r) \right\} \quad . \tag{7.52}$$

The asymptotic of (7.52) at infinity is

$$\Phi_l(k,r) \rightarrow -\frac{1}{2ik}\left\{ f_l(-k)\, e^{-ikr} - f_l(k)\, e^{ikr} \right\} \quad .$$

Comparing it to (4.24), we find that

$$S_l(k) = (-1)^l\, \frac{f_l(k)}{f_l(-k)} \quad . \tag{7.53}$$

Formula (7.53) describes the general relationship between the scattering matrix $S_l(k)$ and the Jost functions $f_l(k)$ and $f_l(-k)$.

The general relation between the Jost function $f_l(k)$ and the regular solution $\Phi_l(k,r)$ is

$$f_l(k) = 1 - i \int_0^\infty dr\, r h_l^{(-)}(kr)\, V_r(r)\, \Phi_l(k,r) \quad . \tag{7.54}$$

Similarly to the S-scattering case, the regular solution $\Phi_l(k,r)$ for $l \neq 0$ is an entire function in the complex k plane, and the Jost solution $f_l(k,r)$ is analytic in the lower half-plane of complex k. By virtue of the boundary conditions (7.45, 49), Φ_l and f_l for complex k satisfy the conditions

$$\Phi_l^*(k^*,r) = \Phi_l(k,r) \quad , \tag{7.55}$$

$$f_l^*(-k^*,r) = f_l(k,r) \quad . \tag{7.56}$$

The regular solution $\Phi_l(k,r)$ differs only by a factor from the radial wave function $\psi_{l,k}(r)$ normalized as in (4.26). It is not difficult to show that

---

[1] Note that the Jost function may also be defined as $f_l(k) \equiv \lim_{r\to 0}(2l+1)\, r^l\, f_l(k,r)$.

$$\psi_{l,k}(r) = \frac{4\pi i^l}{f_l(-k)} \frac{\Phi_l(k,r)}{r} \quad . \tag{7.57}$$

We must indicate, however, an essential difference between the functions $\psi_{l,k}(r)$ and $\Phi_l(k,r)$. The normalized radial wave function $\psi_{l,k}(r)$ is a solution of the radial Schrödinger equation (4.9) and satisfies the boundary conditions imposed both at the point $r = 0$ and at infinity, namely, it tends to zero as $j_l(kr)$ at the point $r = 0$,

$$\psi_{l,k}(r) \underset{r \to 0}{\to} 4\pi \, j_l(kr) \quad ,$$

and behaves as a sum of incident and converging spherical waves as $r \to \infty$,

$$\psi_{l,k}(r) \underset{r \to \infty}{\to} 4\pi \left\{ j_l(kr) + ik \, f_l(k) \, h_l^{(+)}(kr) \right\} \quad .$$

At the same time, the regular solution $\Phi_l(k,r)$, which is also a solution of the radial Schrödinger equation, is determined only by the boundary conditions at the point $r = 0$, i.e.,

$$\Phi_l(k,0) = 0 \quad \text{and} \quad \Phi_l'(k,0) = 1 \quad .$$

In contrast to $\psi_{l,k}(r)$, the regular solution $\Phi_l(k,r)$ is real because both the boundary conditions and the radial equation are real.

The solutions $\psi_{l,k}^{(+)}(r)$ and $\psi_{l,k}^{(-)}(r)$, normalized according to (4.82), may be expressed in terms of the Jost solutions $f_l(-k,r)$ and $f_l(k,r)$:

$$\psi_{l,k}^{(+)}(r) = -4\pi i \frac{f_l(-k,r)}{kr} \exp\left[i\left(\delta_l - \frac{l\pi}{2} \right) \right] \quad ,$$

$$\tag{7.58}$$

$$\psi_{l,k}^{(-)}(r) = -4\pi i \frac{f_l(k,r)}{kr} \exp\left[-i\left(\delta_l - \frac{l\pi}{2} \right) \right] \quad .$$

These formulas together with (4.86, 87) enable one to express the retarded and advanced Green functions $g_{l,k}^{(+)}(r,r')$ and $g_{l,k}^{(-)}(r,r')$ in terms of the regular solution $\Phi_l(k,r)$ and the Jost solutions $f_l(k,r)$ and $f_l(-k,r)$ and thus reveal their analytic properties.

7.3 Jost Function Zeros and Bound States

The stationary states of the system which are associated with discrete energy levels and describe finite motion, are referred to as bound states. The bound states are described by the square integrable solutions of the Schrödinger equation.

Let us show that the zeros of the function $f_l(k)$ in the lower half-plane of complex k are associated with the bound states of the system. According to (7.52), if $f_l(k)$ vanishes for some $k = k_n$, i.e.,

$$f_l(k_n) = 0 \quad , \tag{7.59}$$

then the relation

$$\Phi_l(k_n, r) = \frac{1}{\gamma_n} f_l(k_n, r) \tag{7.60}$$

holds, where γ_n is a constant. Since $\Phi_l(k_n, r)$ vanishes for $r = 0$, and $f_l(k_n, r)$, by virtue of $\mathrm{Im}\{k_n\} < 0$, exponentially decreases as $r \to \infty$, the function $\Phi_l(k_n, r)$ is square integrable and describes the finite motion of the system.

Let us show that k_n^2 is real. Making use of (7.42) for Φ_l and the analogous equation for Φ_l^*, we find that

$$\frac{d}{dr} W\Big(\Phi_l(k_n, r), \Phi_l^*(k_n, r)\Big) = 2i\,\mathrm{Im}\{k_n^2\}\,|\Phi_l(k_n, r)|^2 \quad . \tag{7.61}$$

According to (7.55), we have

$$\Phi_l^*(k_n, r) = \frac{1}{\gamma_n^*} f_l(-k_n^*, r) \quad ,$$

and hence, if $\mathrm{Im}\{k_n\} < 0$, then the function $\Phi_l^*(k_n, r)$ exponentially decreases as $r \to \infty$. We integrate (7.61) with regard to this property of Φ_l^* and the analogous property of Φ_l. The result is

$$\mathrm{Im}\{k_n^2\} \int_0^\infty dr\,|\Phi_l(k_n, r)|^2 = 0 \quad .$$

This relation holds only provided that k_n^2 is real and Φ_l is a square integrable function, i.e.,

$$\mathrm{Im}\{k_n^2\} = 0 \tag{7.62}$$

and

$$\int_0^\infty dr\,|\Phi_l(k_n, r)|^2 < \infty \quad . \tag{7.63}$$

Thus we have proved that k_n^2 is real.

Let us show that $k_n^2 < 0$. If we put $k_n^2 > 0$, then k_n is real. However, $f_l(k)$ cannot be equal to zero for real k since otherwise $f_l(-k)$ must also vanish by virtue of (7.56). This would lead to $\Phi_l(k_n, r) \equiv 0$ which is impossible due to the boundary condition (7.45). Therefore, the function $f_l(k)$ can be equal to zero only for the wave numbers k_n, for which $k_n^2 < 0$. This implies that the zeros of the function $f_l(k)$ can lie only in the lower half-plane on the imaginary axis, i.e.,

$$k_n = -i\kappa_n \quad , \qquad \kappa_n > 0 \quad . \tag{7.64}$$

(Here n is the number of the zero.)

The zeros of the function $f_l(k)$ are associated with a discrete sequence of energy levels

$$E_n = -\frac{\hbar^2 \kappa_n^2}{2\mu} \quad . \tag{7.65}$$

Each level, in turn, is associated with the wave function $\Phi_l(-i\,\kappa_n\,,r)$ which is a solution of (7.42) and satisfies the square integrability condition (7.43). Therefore, it describes a bound state of the system. The functions $\Phi_l(-i\,\kappa_n\,,r)$ vanish at $r = 0$ and have asymptotics

$$\Phi_l(-i\,\kappa_n\,,r) \underset{r\to\infty}{\to} \frac{1}{\gamma_n} e^{-\kappa_n r} \quad . \tag{7.66}$$

By using (7.52), the coefficient γ_n can easily be found to be

$$\gamma_n = -\frac{2\kappa_n}{f_l(i\,\kappa_n)} \quad . \tag{7.67}$$

Note that $\Phi_l(-i\,\kappa_n\,,r)$ are real functions.

Let us show that the zeros of the function $f_l(k)$ are simple. We differentiate (7.51) with respect to k^2 and then put $k^2 = k_n^2$. Making use of (7.60), we have

$$\frac{df_l(k_n)}{dk_n^2} = \frac{1}{\gamma_n} W\left(\frac{d}{dk_n^2} f_l(k_n\,,r),\, f_l(k_n\,,r)\right)$$

$$+ \gamma_n W\left(\Phi_l(k_n\,,r),\, \frac{d}{dk_n^2} \Phi_l(k_n\,,r)\right) \quad . \tag{7.68}$$

Because both f_l and Φ_l satisfy (7.42), Φ_l vanishes at $r = 0$, and $f_l(k_n\,,r)$ has the asymptotics (7.66), we obtain

$$W\left(\Phi_l(k_n\,,r),\, \Phi_l(k,r)\right) = (k_n^2 - k^2) \int_0^r dr'\, \Phi_l(k_n\,,r')\,\Phi_l(k,r') \quad ;$$

$$W\left(f_l(k_n\,,r),\, f_l(k,r)\right) = (k^2 - k_n^2) \int_r^\infty dr'\, f_l(k_n\,,r')\, f_l(k,r') \quad .$$

We differentiate these expressions with respect to k^2 and put $k^2 = k_n^2$, then

$$W\left(\Phi_l(k_n\,,r),\, \frac{d}{dk_n^2} \Phi_l(k_n\,,r)\right) = -\int_0^r dr'\, \Phi_l^2(k_n\,,r') \quad ;$$

$$W\left(\frac{d}{dk_n^2} f_l(k_n\,,r),\, f_l(k_n\,,r)\right) = -\int_r^\infty dr'\, f_l^2(k_n\,,r') \quad .$$

Substituting this result into (7.68) yields

$$\frac{d}{dk_n} f_l(k_n) = -2k_n\gamma_n \int_0^\infty dr\, \Phi_l^2(k_n\,,r) \quad . \tag{7.69}$$

Since γ_n is not equal to zero and $\Phi_l(k_n, r)$ is real, the derivative $df_l(k_n)/dk_n$ does not vanish. Therefore, the zeros of the function $f_l(k)$ in the lower half-plane are simple.

It is clear that $f_l(-k)$ is analytic in the upper half-plane of complex k and possesses simple zeros located on the positive imaginary semi-axis. These zeros correspond to the bound states of the system. According to the definition (7.53), the zeros of the function $f_l(-k)$ correspond to the poles of the scattering matrix $S_l(k)$. Thus, each bound state of the system is associated with a zero of the scattering matrix located on the negative imaginary semi-axis, and a symmetric (with respect to the real axis) pole of the scattering matrix on the positive imaginary semi-axis.

We denote the normalized wave function of the bound state by $u_l(r)$,

$$u_l(r) = \alpha_n \Phi_l(-\mathrm{i}\kappa_n, r) \quad . \tag{7.70}$$

The normalization constant α_n is determined by the condition

$$\int_0^\infty dr \, u_l^2(r) = 1 \quad . \tag{7.71}$$

Applying (7.69), we have

$$\alpha_n^2 = \frac{2\kappa_n \gamma_n}{df_l(-\mathrm{i}\kappa_n)/d\kappa_n} \quad . \tag{7.72}$$

It is sometimes convenient to express the bound state wave function in terms of the Jost function $f_l(-\mathrm{i}\kappa_n, r)$, so that

$$u_l(r) = \beta_n f_l(-\mathrm{i}\kappa_n, r) \quad , \tag{7.73}$$

where

$$\beta_n^2 = \frac{\alpha_n^2}{\gamma_n^2} = -f_l(\mathrm{i}\kappa_n) \left[\frac{df_l(-\mathrm{i}\kappa_n)}{d\kappa_n} \right]^{-1} \quad . \tag{7.74}$$

Making use of the scattering matrix definition (7.53), we obtain the normalization constant β_n in the integral form,

$$\beta_n^2 = \frac{(-1)^l}{2\pi} \oint dk \, S_l(k) \quad , \tag{7.75}$$

the integration contour being a small circle that goes clockwise around the pole $k = \mathrm{i}\kappa_n$. The scattering matrix in the vicinity of the pole may be written as

$$S_l(k) \simeq \frac{c_n}{k - \mathrm{i}\kappa_n} \qquad (k \to \mathrm{i}\kappa_n) \quad .$$

According to (7.75), the residue c_n is determined by the normalization constant of the bound state wave function, i.e.,

$$c_n = (-1)^{l+1} \mathrm{i}\,\beta_n^2 \quad . \tag{7.76}$$

Note that β_n is expressed in terms of the Jost function $f(k)$ in the upper half-plane for which the Jost function itself is indefinite in the general case. For this reason, in the inverse scattering problem β_n is regarded as an independent parameter.

7.4 Symmetry and Dislocation of Scattering Matrix Singularities in the Complex k Plane

Let us study analytic properties of the scattering matrix $S_l(k)$ in the complex k plane. Off the interaction region, scattering in the state with a given l is described by the time-dependent wave function

$$\psi_l(r,t) \sim \frac{1}{r} \left\{ (-1)^l e^{-ikr} - S_l(k) e^{ikr} \right\} \exp\left(-\frac{i}{\hbar} Et\right) , \tag{7.77}$$

$$E = \frac{\hbar^2 k^2}{2\mu} ,$$

and the relation between the scattering matrix and the Jost functions is given by (7.53), so that

$$S_l(k) = (-1)^l \frac{f_l(k)}{f_l(-k)} .$$

Since both functions $f_l(k)$ and $f_l(-k)$ are defined for complex k, this relation defines the scattering matrix $S_l(k)$ in the whole complex k plane.

We denote the real and imaginary parts of the complex wave number k as k' and k'', i.e.,

$$k = k' + ik'' . \tag{7.78}$$

The complex wave numbers are associated with comlex energies W:

$$W = E - \frac{i}{2} \Gamma , \quad E = \frac{\hbar^2}{2} \left(k'^2 - k''^2 \right) , \quad \Gamma = -\frac{2\hbar^2}{\mu} k'k'' . \tag{7.79}$$

Here E is the real part of the complex energy W. Complex energies describe nonstationary states of the system for which the probability densities vary exponentially with time,

$$|\psi_l|^2 \sim \exp\left(-\frac{\Gamma t}{\hbar}\right) .$$

The quantity Γ entering the imaginary part of the energy is called the disintegration constant. If Γ is positive, the system undergoes disintegration which leads to an exponentially decreasing (with time) probability density at any point of the space. If Γ is negative, capture occurs, resulting in probability density increase. According to (7.79), the first and the third quadrants of the complex k plane are associated with captures, the second and the fourth quadrants correspond to disintegration.

As follows from (7.53), the scattering matrix possesses the symmetry property

$$S_l(-k) = S_l^{-1}(k) \quad . \tag{7.80}$$

Substituting the definition $S_l(k) = \exp[2i\,\delta_l(k)]$ for the scattering phase shift, we find $\delta_l(k)$ to be an odd function of k for real k, so that

$$\delta_l(-k) = -\delta_l(k) \quad . \tag{7.81}$$

Since the definition of the scattering phase shift is ambiguous, (7.81) is valid within the uncertainty of the term $n\pi$, where n is an arbitrary integer.

The symmetry property (7.80) occurs because the Schrödinger equation contains the square of the wave number and hence is invariant with respect to the changes of its sign. Therefore, when $-k$ is substituted for k in (7.77), the resultant function must again satisfy the Schrödinger equation. Equation (7.80) just follows from this requirement.

Note that (7.80) is valid only for integer l. In the general case of arbitrary l, we have to employ the relation

$$S_l(-k) = e^{2il\pi}\, S_l^{(-1)}(k) \quad , \tag{7.80'}$$

rather than (7.80) and thus the scattering phase shift is transformed, under the interchange $k \to -k$, according to

$$\delta_l(-k) = -\delta_l(k) + l\pi \quad .$$

When arbitrary l are considered, it is convenient to introduce the phase shift $\Delta_l(k) \equiv \delta_l(k) - (l\pi/2)$ which is an odd function of k, $\Delta_l(-k) = -\Delta_l(k)$.

Making use of (7.56) for the Jost function, we obtain the relation

$$S_l^*(k^*) = S_l^{-1}(k) \quad . \tag{7.82}$$

For real k, it reproduces the unitarity condition (5.1) for the scattering matrix.

Equations (7.80, 82) establish a one-to-one correspondence between the values of the scattering matrix in different quadrants of the complex k plane. If the scattering matrix $S_l(\tilde{k})$ takes the value $\tilde{S}_l$ at some point $\tilde{k}$,

$$S_l(\tilde{k}) = \tilde{S}_l \quad , \tag{7.83}$$

then at the symmetric points it is given by

$$S_l(\tilde{k}^*) = \tilde{S}_l^{*-1} \quad , \qquad S_l(-\tilde{k}^*) = \tilde{S}_l^* \quad , \qquad S_l(-\tilde{k}) = \tilde{S}_l^{-1} \quad . \tag{7.84}$$

Therefore, the scattering matrix in the whole k plane can be found by the known form of the $S_l(k)$ in some quadrant.

According to (7.83, 84), the values of the scattering matrix $S_l(k)$ at the points symmetric with respect to the imaginary axis are complex conjugate. Therefore, on the imaginary axis $S_l(k)$ is real and the phase shift $\delta_l(k)$ is imaginary,

$$\delta_l(i\kappa) = -\delta_l^*(i\kappa) \quad , \qquad \kappa \text{ is real.}$$

For the points symmetric with respect to the real axis, equality (7.82) holds. Then $|S_l(k)| = 1$ on the real axis and therefore the phase shift $\delta_l(k)$ is real. Note that on the real axis $S_l(k) \rightarrow 1$ as $k \rightarrow \infty$.

Relations (7.80, 82) lead to symmetric dislocation of the scattering matrix zeros and poles in the k plane. As follows from (7.80), if $S_l(k)$ has a zero at the point k, then it must necessarily have a pole at the point $-k$, symmetric with respect to the center of the plane. Condition (7.82) implies that existence of a scattering matrix zero at the point k causes a pole at the point k^* symmetric with respect to the real axis. Thus we see that both zeros and poles of the scattering matrix are dislocated by pairs, symmetric with respect to the imaginary axis.

It is not difficult to show that in the upper half-plane the scattering matrix poles can be located anywhere, the only condition is that if a pole in the lower half-plane does not lie on the imaginary axis, then another pole must occur at the point symmetric with respect to this axis. Since each pole of $S_l(k)$ is associated with a zero at the point symmetric with respect to the real axis, in the lower half-plane the zeros of the scattering matrix are located on the imaginary axis; in the upper half-plane, zeros can lie anywhere, in correspondence with the dislocation of poles in the lower half-plane.

In order to prove the above assertion, we employ the time-dependent Schrödinger equation (1.12), which immediately yields the probability conservation law

$$\frac{\partial}{\partial t} \int_V d\boldsymbol{r} \, |\psi|^2 = -\int_s d\boldsymbol{s} \, \boldsymbol{j} \quad , \tag{7.85}$$

where $\boldsymbol{j}$ is the probability flux density,

$$\boldsymbol{j} = \frac{i\hbar}{2\mu} \left(\psi \nabla \psi^* - \psi^* \nabla \psi \right) \quad .$$

Integration in the left-hand part extends over an arbitrary volume V enveloped by the surface s, and in the right-hand part over the surface s. If the scattering matrix $S_l(k)$ has a pole at the point $\tilde{k}$, then the first term in the wave function (7.77) may be ignored in the vicinity of this point. Therefore,

$$\psi_{l,k}(r,t) = \frac{u_l(r)}{r} \exp\left(-\frac{i}{\hbar} Wt\right)$$

$$\underset{r\rightarrow\infty}{\longrightarrow} -\frac{1}{r} S_l(k) \exp\left(ikr - \frac{i}{\hbar} Wt\right) \qquad (k \rightarrow \tilde{k}) \quad . \tag{7.86}$$

Suppose V is the volume of a sphere of radius R. We assume that R is sufficiently large, so that we can use the asymptotic expression (7.86). Then the probability conservation condition takes the form

$$\tilde{k}' \, \tilde{k}'' \int_0^R dr \, u_l^2 = -\frac{\tilde{k}'}{2} \left| S_l(k) \right|^2 e^{-2\tilde{k}'' R} \quad . \tag{7.87}$$

Since the right-hand part is negative, this equality holds either if $\bar{k}' = 0$, i.e., the pole of $S_l(k)$ lies on the imaginary axis, or if $\bar{k}' \neq 0$ and $\bar{k}'' < 0$, i.e., the pole of $S_l(k)$ lies in the lower half-plane. Thus we have proved that in the upper half-plane of complex k, the poles of the scattering matrix $S_l(k)$ can occur only on the imaginary axis.

The scattering matrix $S_l(k)$, regarded as a function of the complex variable k, describes various physical processes. If k is real, the scattering matrix describes realistic scattering and is expressed in terms of real phase shifts which determine the cross section. The poles of the scattering matrix associated with imaginary and complex wave numbers are responsible for the bound, virtual, and quasistationary states of the system.

7.5 Bound States and Extra Zeros

Let us consider imaginary wave numbers $k = -i\kappa$ (κ is real) which correspond to negative energies $E = -\hbar^2 \kappa^2/2\mu$. If the wave function is square integrable, then negative energy is associated with a bound state. The square integrability condition within the context of (7.77) may be written as

$$\int_0^\infty dr \left|(-1)^l e^{-\kappa r} - S_l(-i\kappa) e^{\kappa r}\right|^2 < \infty \quad ;$$

it is satisfied if

$$\kappa > 0 \quad \text{and} \quad S_l(-i\kappa) = 0 \quad .$$

Thus, the bound state of the system corresponds to a zero of $S_l(k)$ located on the negative imaginary semi-axis. By virtue of the above mentioned symmetry properties of $S_l(k)$, this zero is associated with a pole of $S_l(k)$ at the point $i\kappa$ on the positive imaginary semi-axis. If k is close to $\pm i\kappa$, then the scattering matrix reduces to

$$S_l(k) = \exp\left(2i \arg C_l\right) \frac{k + i\kappa}{k - i\kappa} \quad , \quad \kappa > 0 \quad , \tag{7.88}$$

where C_l is a constant.

Note that though each bound state is associated with a zero of the function $S_l(k)$ on the negative imaginary semi-axis, the inverse assertion is not always true. In some cases the scattering matrix has so-called extra zeros which are not related with bound states [7.1]. To verify this, we express the scattering matrix in terms of the Jost functions according to (7.53). We have already shown that the zeros of the function $f_l(k)$ in the lower half-plane correspond to the bound states. If $f_l(k)$ vanishes, the scattering matrix must vanish, too. However, the scattering matrix also vanishes at the points for which the function $f_l(-k)$ tends to infinity. These extra zeros of the scattering matrix $S_l(k)$ obviously are not associated with bound states. Inasmuch as extra zeros of $S_l(k)$ are given rise to by the poles of $f_l(-k)$, they do not appear if the Jost function $f_l(k)$ is regular in the whole k-plane. This imposes the following condition on the potential:

$$\int_0^\infty dr\, e^{mr}\, |V_r(r)| < \infty \qquad (7.89)$$

for any positive m. Therefore, the scattering matrix possesses no extra zeros if the potential is finite-range. Thus we come to the following formal procedure separating extra zeros of the scattering matrix $S_l(k)$ from physical ones. One has to truncate the potential at some sufficiently large distance R, find the zeros of $S_l(k)$ in the lower half-plane, and then allow R to extend to infinity. The limiting values of $\kappa(R)$ for $R \to \infty$ determine the bound state energies of the system.

An example of the potential that gives rise to extra zeros of the scattering matrix is the exponential well

$$V_r(r) = -V_{r0}\, \exp\left(-\frac{r}{a}\right) \quad . \qquad (7.90)$$

We introduce instead of r a new variable $y = \alpha \exp(-r/2a)$, where $\alpha = 2a\, V_{r0}^{1/2}$. Then the equation for the radial function (7.1) (we consider only the S-case) reduces to the Bessel equation

$$\frac{d^2u}{dy^2} + \frac{1}{y}\frac{du}{dy} + \left(1 + \frac{\nu^2}{y^2}\right) u = 0 \quad , \qquad \nu = 2ak \quad , \qquad (7.91)$$

the general solution of which may be written as

$$u = c_1\, J_{i\nu}(y) + c_2\, J_{-i\nu}(y) \quad , \qquad (7.92)$$

c_1 and c_2 being arbitrary constants.

For $r = 0$, we have $y = \alpha$. If $r \to \infty$, then $y \to 0$ and

$$J_{i\nu}(y) \to \frac{y^{i\nu}}{2^{i\nu}\, \Gamma(i\nu + 1)} = \left(\frac{\alpha}{2}\right)^{i\nu} \frac{e^{-ikr}}{\Gamma(i\nu + 1)} \quad .$$

Then the Jost solution is given by

$$f(k, r) = \left(\frac{\alpha}{2}\right)^{-i\nu} \Gamma(i\nu + 1)\, J_{i\nu}(y) \quad ,$$

and for the scattering matrix we obtain

$$S(k) = \left(\frac{\alpha}{2}\right)^{-2i\nu} \frac{\Gamma(i\nu + 1)\, J_{i\nu}(\alpha)}{\Gamma(-i\nu + 1)\, J_{-i\nu}(\alpha)} \quad , \qquad \nu = 2ak \quad . \qquad (7.93)$$

The regular solution of (7.91), which vanishes at $r = 0$, is

$$\Phi(k, r) = -\frac{1}{2ik}\, \Gamma(i\nu + 1)\, \Gamma(-i\nu + 1) \left\{ J_{-i\nu}(\alpha)\, J_{i\nu}\left(\alpha\, e^{-r/2a}\right) \right.$$

$$\left. - J_{i\nu}(\alpha)\, J_{-i\nu}\left(\alpha\, e^{-r/2a}\right) \right\} \quad . \qquad (7.94)$$

On the imaginary axis $k = -i\kappa$ ($\kappa > 0$), the scattering matrix (7.93) vanishes at the points κ_n for which

$$J_{2a\,\kappa_n}(\alpha) = 0 \quad . \tag{7.95}$$

These zeros satisfy the condition $f(k) = 0$ and are associated with the bound states because the wave function

$$\Phi(-i\,\kappa_n\,,r) = -\frac{1}{2\,\kappa_n}\,\Gamma(2a\,\kappa_n + 1)\,\Gamma(-2a\,\kappa_n + 1)$$

$$\times\,J_{-2a\,\kappa_n}(\alpha)\,J_{2a\,\kappa_n}\left(\alpha\,e^{-r/2a}\right)$$

exponentially decreases as $r \to \infty$.

Along with the above mentioned zeros, the scattering matrix vanishes at the points for which

$$\Gamma(-i\nu + 1) = \infty \tag{7.96}$$

(this corresponds to the condition $f(-k) = \infty$). Equation (7.96) holds for $k = -i\kappa_m$ and $2a\,\kappa_m = m$ (i.e., $i\nu = m$), where m is a positive integer. It is not difficult to show that the sequence of zeros (7.96) does not give rise to bound states. The expression for the wave function is

$$\Phi(-i\,\kappa_m\,,r) = -\frac{1}{2\,\kappa_m}\,\Gamma(m+1)\,\Gamma(-m+1)\left\{J_{-m}(\alpha)\,J_m\left(\alpha\,e^{-r/2a}\right)\right.$$

$$\left. -\,J_m(\alpha)\,J_{-m}\left(\alpha\,e^{-r/2a}\right)\right\} \quad .$$

The factor $\Gamma(-m+1)$ tends to infinity, but the expression within the curly braces vanishes by virtue of the Bessel function symmetry,

$$J_{-m}(\alpha) = (-1)^m\,J_m(\alpha) \quad ,$$

where m is an integer. In order to remove the indeterminate form, we express the negative-index Bessel function in terms of the Neumann function,

$$J_{-m}(\alpha) = \cos m\pi\,J_m(\alpha) - \sin m\pi\,N_m(\alpha) \quad .$$

Since $\sin m\pi\,\Gamma(-m+1) = \pi/m!$, we find that

$$\Phi(-i\,\kappa_m\,,r) = \frac{1}{2\,\kappa_m}\,\frac{\pi}{m!}\,\Gamma(m+1)\left\{N_m(\alpha)\,J_m\left(\alpha\,e^{-r/2a}\right)\right.$$

$$\left. -\,J_m(\alpha)\,N_m\left(\alpha\,e^{-r/2a}\right)\right\} \quad .$$

This solution is regular at the point $r = 0$ but grows exponentially as $r \to \infty$. Hence, it is not associated with a bound state.

We employ this example of an exponential well to show how to separate extra zeros from physical ones by truncating the potential. We substitute the modified potential,

$$V_{rR}(r) = \begin{cases} -V_{r0}\,e^{-r/a}\,, & r < R \quad, \\ 0 & , \quad r > R \quad, \end{cases} \tag{7.97}$$

for the potential (7.90). In the region $r > R$, the solution of (7.1) with the potential (7.97) is given by

$$\Phi(k, r) \sim \left(e^{-ikr} - S(k)\,e^{ikr}\right) \quad. \tag{7.98}$$

In the region $r > R$, it is given by (7.94) as before. To find $S(k)$, we make use of the requirement that the logarithmic derivatives of the functions (7.94, 98) must be equal at the point $r = R$ to obtain

$$S(k) = e^{-2ikR}\,\frac{J_{-i\nu}(\alpha)\,J_{i\nu+1}\left(\alpha\,e^{-R/2a}\right) + J_{i\nu}(\alpha)\,J_{-i\nu-1}\left(\alpha\,e^{-R/2a}\right)}{J_{-i\nu}(\alpha)\,J_{i\nu-1}\left(\alpha\,e^{-R/2a}\right) + J_{i\nu}(\alpha)\,J_{-i\nu+1}\left(\alpha\,e^{-R/2a}\right)} \quad.$$

$$\tag{7.99}$$

The requirement that $S(k)$ must vanish yields the condition

$$J_{-i\nu}(\alpha)\,J_{i\nu+1}\left(\alpha\,e^{-R/2a}\right) + J_{i\nu}(\alpha)\,J_{-i\nu-1}\left(\alpha\,e^{-R/2a}\right) = 0 \quad, \tag{7.100}$$

which is equivalent to (7.95) for $R \to \infty$. Therefore, to exclude extra zeros when calculating the spectrum of bound states, one has to substitute the modified potential (7.97) for the one (7.90) and allow $R \to \infty$ in the result obtained.

7.6 Quasistationary States and Resonances

The probability conservation condition (7.85) implies that the scattering matrix singularities can lie either on the imaginary axis in the upper half-plane, or in the lower half-plane of complex k. The poles on the imaginary axis in the upper half-plane are associated with bound states whose energies can be found from the condition that $S_l(k)$ tends to infinity. This condition is equivalent to the requirement that the regular solution (7.52) of the Schrödinger equation, associated with a bound state, must contain only the second term which exponentially decreases at infinity. The scattering matrix poles located in the lower half-plane of the imaginary axis correspond to quasistationary states which describe either decays or captures. The complex energies of quasistationary states, similarly to the bound state energies, can be found from the condition that $S_l(k)$ tends to infinity. The wave function of the quasistationary state, as in the case of bound states, is given by the second term of (7.52) which at infinity reduces to a diverging wave. On the other hand, the quasistationary state energies also can be found from the condition that the coefficient of the converging wave in (7.52) must vanish.

Note that if the energies are positive real, then the probability conservation requires retaining both terms in the regular solution (7.52). Indeed, if $f_l(-k) = 0$

and $f_l(k) \neq 0$, then a probability flux must exist in the radial direction, $j_r > 0$, but this contradicts the probability conservation law since for real E the solution is stationary. If the energy is complex, then, because the probability density exponentially decreases with time, the probability conservation can be provided only taking into account the diverging wave in the asymptotics of the solution (7.52).

Let us consider a pole of $S_l(k)$ in the fourth quadrant associated with a decay quasistationary state, i.e.,

$$k_r = k'_r + i k''_r \quad , \qquad k'_r > 0 \quad , \qquad k''_r < 0 \quad .$$

Its energy is

$$W_r = E_r - \frac{i}{2}\, \Gamma_r \quad ,$$

$$E_r = \frac{\hbar^2}{2\mu}\left(k'^2_r - k''^2_r\right) \quad , \tag{7.101}$$

$$\Gamma_r = -\frac{2\hbar^2}{\mu}\, k'_r\, k''_r > 0 \quad .$$

The probability density of the decay state is determined by the square of the wave function modulus and decreases exponentially as

$$|\psi(t)|^2 \sim \exp\left(-\frac{\Gamma_r t}{\hbar}\right) \quad . \tag{7.102}$$

In particular, such is the law of time evolution of the probability that a particle will be found in some limited spatial region. According to (7.102), the decay probability per unit time is

$$w_r = \frac{\Gamma_r}{\hbar} \quad . \tag{7.103}$$

Therefore, the lifetime τ_r of the decay quasistationary state may be defined by the equality

$$\tau_r = \frac{\hbar}{\Gamma_r} \quad . \tag{7.104}$$

Note that a system which decays with time has no discrete energy spectrum. The escaping particle goes to infinity, therefore, the motion of the system is infinite and hence its energy spectrum is continuous. However, if for some reason the decay probability is low, then the quasistationary state concept can be introduced. Being in a quasistationary state, the particle moves within the system for a long time and escapes only after a considerable period $\tau_r = 1/w_r$. The energy spectrum of such quasistationary states is quasidiscrete and contains a sequence of broadened levels E_r whose widths Γ_r are determined by their lifetimes, $\Gamma_r = \hbar/\tau_r$. It is clear that introducing quasistationary states is meaningful only if the quasidiscrete level width is much narrower than the interlevel distance, $\Gamma_r \ll D$, where D is the mean distance between the quasidiscrete levels.

The radial wave function of the quasistationary decay state grows exponentially at large distances,

$$\Phi(k,r) \sim \exp\left(ik'_r r - k''_r r\right) \qquad (k''_r < 0) \quad , \qquad r \to \infty \quad , \tag{7.105}$$

because the probability flux is directed along the radius ($j_r > 0$) and, at large distances, waves occur which have abandoned the system long before the time instant t. And since the probability density in the system is proportional to $\exp\left(-\Gamma_r t/\hbar\right)$, it must have been very large at $t = -\infty$. Obviously, the radial function of the quasistationary state cannot be normalized because the normalization integral $\int dr \, |\Phi|^2$ is divergent.

Note that the quasistationary decay states associated with diverging waves actually do not exist. Indeed, the decay must occur after the formation of the decay state which must involve a converging wave, whereas the condition $f_l(-k_r) = 0$ excludes converging waves for all times. A decay state can be realized approximately by superposing stationary waves with energies E close to the resonance energy E_r. Let $\psi_E(r,t)$ be the stationary state wave function. Off the interaction region, this function is given by

$$\psi_E(r,t) = \frac{1}{r} \left\{ e^{-ikr} - S(k) e^{ikr} \right\} e^{(-i/\hbar)Et} \quad , \qquad r > R \quad . \tag{7.106}$$

We take a superposition of the states $\psi_E(r,t)$ with energies within the interval Δ about the resonance value E_r ($\Gamma_r \ll \Delta \ll D$) and form a wave packet localized in the region $r < R$ at $t = 0$, so that

$$\psi(r,t) = \int_{E_r-\Delta/2}^{E_r+\Delta/2} dE \, a(E) \, \psi_E(r,t) \quad , \tag{7.107}$$

where $a(E)$ is a slowly varying function of E. Let us find the shape of the wave packet for $t > 0$ in the region $r > R$. Since $\Delta \ll D$, for $r > R$ we can disregard the converging wave contribution in (7.107) due to the pole nature of the scattering matrix

$$S(k) \simeq \frac{c(E)}{E - E_r + (i/2)\Gamma_r} \quad .$$

We make use of the relation $k - k'_r \simeq (E - E_r)/\hbar v_r$, where $v_r = \hbar k'_r/\mu$ is the particle velocity corresponding to the resonance energy E_r, and put the slowly varying functions before the integration symbol. Thus we find that

$$\psi(r,t) \simeq - a(E_r) \frac{1}{r} \exp\left(ik'_r - \frac{i}{\hbar} E_r t\right)$$

$$\times \int_{-\Delta/2\Gamma_r}^{\Delta/2\Gamma_r} dx \, \frac{\exp\left\{(-i/\hbar) \Gamma_r \left[t - (r/v_r)\right] x\right\}}{x + (i/2)} \quad , \tag{7.108}$$

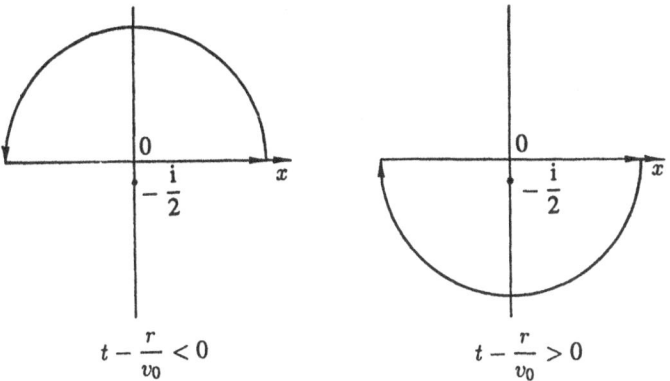

Fig. 7.1. Integration contours for (7.108) for the two signs of the quantity $t - r/v_0$

$$x = \frac{E - E_r}{\Gamma_r} \quad , \qquad r > R \quad .$$

Since $\Delta \gg \Gamma$, integration may be extended from $-\infty$ to ∞. We regard x as a complex variable and perform integration in (7.108) over the contour shown in Fig. 7.1. The integrand has a pole at $x = -i/2$. If $t - (r/v_r) < 0$, then the contour must be closed in the upper half-plane and the integral vanishes. If $t - (r/v_r) > 0$, then the contour must be closed in the lower half-plane and the integral is determined by the residue at the point $x = -i/2$. Thus we have

$$\psi(r,t) = \begin{cases} 0 \quad , & t - \frac{r}{v_r} < 0 \\[2mm] 2\pi i\, a(E_r)\, c(E_r)\, \frac{1}{r} & \\[1mm] \times \exp\left[ik'_r r - \frac{i}{\hbar} E_r t - \frac{\Gamma_r}{2\hbar}\left(t - \frac{r}{v_r}\right)\right] , & t - \frac{r}{v_r} > 0 \end{cases}$$

(7.109)

$$r > R \quad .$$

The wave function (7.109) approximately describes the quasistationary decay state off the interaction region for $r > R$. At the initial time $t = 0$, the wave function is not equal to zero only for $r < R$. At time t, the wave function is of the form shown in Fig. 7.2. The maximum in the region $r > R$ corresponds to the decay of the quasistationary state immediately after its formation; this maximum moves with the velocity v_r. The continuous decay is responsible for the exponential decrease, $\sim \exp\left(-\Gamma_r t/\hbar\right)$, of the particle density at the point with fixed r [7.2].

The poles of the scattering matrix in the third quadrant of the complex k plane correspond to the quasistationary states associated with capture. For the capture states, the radial probability flux is directed inwards, $j_r < 0$, and the probability density at a fixed point grows by the exponential law, $\sim \exp\left(-i\Gamma t/\hbar\right)$, $\Gamma < 0$. Within the context of (7.80, 82), the existing capture states give rise to decay states and vice versa. We remind the reader that the poles in the lower half-plane

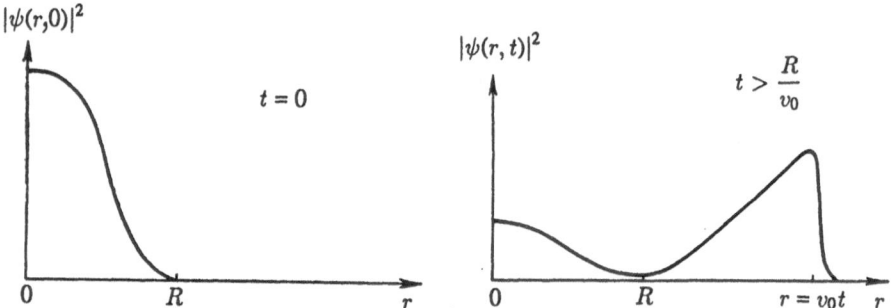

Fig. 7.2. Probability density distribution for a quasistationary decay state at the initial time $t = 0$ and for $t > R/v_0$ (R is the interaction range)

associated with quasistationary states correspond to zeros in the upper half-plane symmetric with respect to the real axis.

In the energy space, the poles associated with the quasistationary states lie on the second physical sheet of the Riemann energy surface.

The quasistationary states of the system manifest themselves under scattering for energies of the system close to the quasidiscrete levels. In this case, the energy dependence of the scattering cross section is of resonance type. Let us find the scattering amplitude for the case when incident particle energy is close to one of the quasidiscrete levels of the system. To do this, we calculate the scattering matrix $S_l(k)$ in the vicinity of the pole $k_r = k'_r + ik''_r$ ($k''_r < 0$) which is given rise to by the quasidiscrete level. We expand the functions $f_l(k)$ and $f_l(-k)$ in a power series of $k - k^*_r$ and $k - k_r$ and retain only the first-order terms. Bearing in mind that the Jost function is Hermitian by virtue of (7.56), we have

$$f_l(k) \simeq b_l^{(r)}\left(k - k'_r + ik''_r\right) ,$$

$$f_l(-k) \simeq b_l^{(r)*}\left(k - k'_r - ik''_r\right) ,$$

where $b_l^{(r)}$ is a constant. Therefore, the scattering matrix in the vicinity of the pole is given by

$$S_l(k) \simeq \exp\left(2i\,\delta_l^{(r)}\right) \frac{k - k'_r + ik''_r}{k - k'_r - ik''_r} , \tag{7.110}$$

where

$$\exp\left(2i\,\delta_l^{(r)}\right) = (-1)^l \frac{b_l^{(r)}}{b_l^{(r)*}} . \tag{7.111}$$

After both the numerator and the denominator of (7.110) are multiplied by $\hbar^2 k/\mu$ and the relation $\hbar^2 k\left(k - k'_r\right)/\mu \simeq E - E_r$ is applied, we obtain

$$S_l(E) \simeq \exp\left(2i\,\delta_l^{(r)}\right)\frac{E - E_r - (i/2)\,\gamma_r}{E - E_r + \frac{i}{2}\,\gamma_r}$$

$$= \exp\left(2i\,\delta_l^{(r)}\right)\left\{1 - \frac{i\gamma_r}{E - E_r + (i/2)\,\gamma_r}\right\} \quad , \tag{7.112}$$

where $\gamma_r(k) = -(2\hbar^2/\mu)\,k k_r''$ is the resonance width. Note that the resonance width $\gamma_r(k)$ differs from the quasidiscrete level width Γ_r by the factor $k/k_r' = (E/E_r)^{1/2}$.

Making use of (7.112), we find the partial scattering cross section to be

$$\sigma_e^{(l)} = \frac{\pi}{k^2}(2l+1)\left\{4\sin^2\delta_l^{(r)} - 4\,\mathrm{Re}\left(e^{i\delta_l^{(r)}}\right)\sin\delta_l^{(r)}\right.$$

$$\left. \times \frac{\gamma_r}{E - E_r + (i/2)\gamma_r} + \frac{\gamma_r^2}{\left(E - E_r\right)^2 + \gamma_r^2/4}\right\} \quad . \tag{7.113}$$

The first term in (7.113) describes potential scattering, the second term is responsible for interference of potential and resonant scattering, the last term is associated with resonant scattering by a quasidiscrete level. As follows from (7.113), the resonant scattering cross section has a sharp maximum at $E = E_r$; the maximum value is

$$\sigma_{e,\mathrm{res}}^{(l)} = \frac{4\pi}{k^2}(2l+1) \quad . \tag{7.114}$$

If the potential scattering phase shift is small $\left(\delta_l^{(r)} \ll 1\right)$, then the resonance term makes the dominant contribution to the scattering cross section.

Since $S_l(E) = \exp\left[2i\,\delta_l(E)\right]$, (7.112) can be rewritten as

$$\delta_l(E) = \delta_l^{(r)} - \arctan\frac{\gamma_r}{2(E - E_r)} \quad . \tag{7.115}$$

For $|E - E_r| \gg \gamma_r$, the scattering phase shift δ_l coincides with the potential scattering phase shift $\delta_l^{(r)}$. The scattering phase shift (7.115) increases by π as the energy varies in the resonance region from $E \ll E_r$ to $E \gg E_r$. If the phase shift δ_l is close to zero for energies lower than the resonance value, then its value attains π for energies higher than the resonance energy at which the scattering phase shift is equal to $\pi/2$.

The scattering matrix (7.112) may be employed to find the elastic scattering amplitude for a particle with energy E close to the quasidiscrete level E_r. To do this, we substitute (7.112) into (3.75) in the term which corresponds to l associated with the level E_r. Then the elastic scattering amplitude is given by the sum

$$f(\vartheta) = f(\vartheta)_{\mathrm{pot}} - \frac{2l+1}{2k}\frac{\gamma_r}{E - E_r + (i/2)\gamma_r}\exp\left(2i\,\delta_l^{(r)}\right)P_l(\cos\vartheta) \quad , \tag{7.116}$$

where $f(\vartheta)_{\mathrm{pot}}$ is the potential scattering amplitude which coincides with the total amplitude far from resonance; the second term describes the resonant scattering amplitude. The applicability region for (7.116) is determined by the condition $|E - E_r| \ll D$.

7.7 Virtual States

The poles of the scattering matrix dislocated on the imaginary axis in the lower half-plane of complex k are associated with the so-called antibound, or virtual states. In the upper half-plane, the virtual states correspond to the scattering matrix zeros which lie on the imaginary axis. Suppose the pole $k_m = -i\,\kappa_m$ ($\kappa_m > 0$) is associated with a virtual state; then the scattering matrix has a zero at the point $k_m^* = i\,\kappa_m$.

Making use of (7.52), it is not difficult to show the virtual state wave function asymptotic to be

$$\Phi(-i\,\kappa_m, r) \sim e^{\kappa_m r} \qquad (\kappa_m > 0) \quad, \qquad r \to \infty \quad, \tag{7.117}$$

so that the virtual state wave function grows exponentially at large distances. It is evident that the wave functions of virtual states cannot be normalized.

The virtual states correspond to negative energy levels

$$E_m = -\frac{\hbar^2 \kappa_m^2}{2\mu} \quad, \tag{7.118}$$

which are associated with the scattering matrix poles on the second nonphysical sheet of the energy shell.

Similarly to the quasistationary states, the virtual states are manifested under elastic scattering. If the virtual state has sufficiently low energy, then the virtual level appreciably modifies the energy dependence of the low energy scattering cross section. Indeed, for sufficiently small κ_m and k, we can employ the expansions

$$f_0(k) \simeq b_0(k - i\,\kappa_m) \quad,$$

$$f_0(-k) \simeq b_0^*(k + i\,\kappa_m) \quad, \tag{7.119}$$

where b_0 is imaginary since $f_0(k)$ must be equal to $f_0(-k)$ for $k \to 0$ (if k is small, the consideration can be restricted to S-scattering). By virtue of (7.119), the scattering matrix reduces to the form

$$S_0(k) \simeq \frac{\kappa_m + ik}{\kappa_m - ik} \quad. \tag{7.120}$$

Therefore, the scattering cross section in the S-state is given by

$$\sigma_e = \frac{4\pi}{\kappa_m^2 + k^2} \quad. \tag{7.121}$$

In the zero energy limiting case ($k \to 0$) we have

$$\sigma_e = \frac{4\pi}{\kappa_m^2} \; .$$ (7.122)

It is clear that if the virtual level energy is very low (κ_m is small), then the low energy scattering cross section is anomalously large.

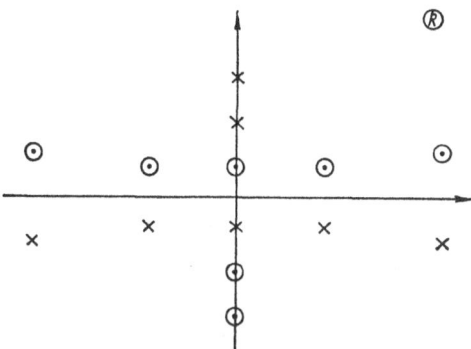

Fig. 7.3. Dislocation of poles ($\times$) and zeros ($\circ$) of the scattering matrix $S_l(k)$ in the complex k plane

Note that (7.121, 122) are even with respect to κ_m, therefore analogous energy dependences of the scattering cross sections can also be associated with bound states with rather low binding energies. Hence, one can be sure that virtual states really exist in the system only if bound states, which might be responsible for such low energy behavior of the scattering cross section, do not occur.

Figure 7.3 shows the dislocation of poles and zeros of the scattering matrix $S_l(k)$ in the complex k plane. If inelastic processes can occur, then $|S_l| < 1$ and poles and zeros in the right and left half-planes move downwards and upwards, respectively.

7.8 The Scattering Matrix in the Case of a Square Well Potential

To illustrate the obtained results, we consider the scattering matrix for the square well potential

$$V_r(r) = \begin{cases} -V_{r0} & , \quad r < R \; , \\ 0 & , \quad r > R \; . \end{cases}$$ (7.123)

For the sake of simplicity we consider only spherically symmetric states ($l = 0$). The Jost solution that satisfies (7.20) is given by

$$f(k, r) = \begin{cases} C_1 e^{ik_0 r} + C_2 e^{-ik_0 r} & , \quad r < R \; , \\ e^{-ikr} & , \quad r > R \; , \end{cases}$$ (7.124)

where $k_0^2 = V_{r0} + k^2$. The coefficients C_1 and C_2 are governed by the joining condition at $r = R$, so that

$$C_1(k) = \frac{1}{2}\left(1 - \frac{k}{k_0}\right) e^{-i(k+k_0)R} \quad,$$

$$C_2(k) = \frac{1}{2}\left(1 + \frac{k}{k_0}\right) e^{-i(k-k_0)R} \quad.$$

$$(7.125)$$

It can be easily verified that C_1 and C_2 satisfy the relation

$$C_2(k) = C_1^*(-k) \quad.$$

Making use of (7.124, 125), we find the Jost function to be

$$f(k) \equiv C_1(k) + C_2(k)$$

$$= e^{-ikR}\left(\cos k_0 R + i\,\frac{k}{k_0}\,\sin k_0 R\right) \quad.$$

$$(7.126)$$

In accordance with the previous general consideration, $f(k)$ is an entire function.

The solution of the Schrödinger equation with the potential (7.123) which is regular at $r = 0$ is given by

$$\Phi(k,r) = \begin{cases} C_0 \sin k_0 r \quad, & r < R \quad, \\ C\left(e^{-ikr} - S(k)\,e^{ikr}\right) \quad, & r > R \quad, \end{cases}$$

$$(7.127)$$

where

$$C_0 = \frac{|C_2|^2 - |C_1|^2}{k} \quad, \qquad C = -\frac{f(-k)}{2ik} \quad,$$

$$(7.128)$$

and the scattering matrix is then described by the expression

$$S(k) = e^{-2ikR}\,\frac{k_0 \cot k_0 R + ik}{k_0 \cot k_0 R - ik} \quad.$$

$$(7.129)$$

Being treated as a function of the complex variable k, the scattering matrix (7.129) has simple zeros at the points where

$$k_0 \cot k_0 R = -ik \quad,$$

simple poles at the points for which

$$k_0 \cot k_0 R = ik \quad,$$

$$(7.130)$$

and an essential singularity at the infinite point $k = \infty$.

The bound state energies are determined either by the scattering matrix zeros dislocated on the imaginary axis in the lower half-plane of k, or by its poles on the imaginary axis in the upper half-plane. We put the bound states in correspondence to the scattering matrix poles and introduce the notation

$$k = i\kappa \ , \qquad \kappa > 0 \ .$$

Then we obtain from (7.130) relations which enable us to find the energy levels associated with the bound states, i.e.,

$$k_0 \cot k_0 R = -\kappa \ , \qquad k_0^2 = V_{r0} - \kappa^2 \ . \tag{7.131}$$

For the sake of convenience we introduce the dimensionless variables

$$k_0 R = x \qquad \text{and} \qquad \kappa R = y$$

and rewrite (7.131) as

$$y = -x \cot x \ , \qquad x^2 + y^2 = V_{r0} R^2 \ . \tag{7.132}$$

One can find numerical or graphic solutions of these equations. The graphic solution of (7.132) determines the relevant values x_n and y_n as intersection points of the curve $y = -x \cot x$ and the circle of radius $V_{r0}^{1/2} R$ with the center in the zero on the plane x, y. Because bound states correspond to $x > 0$ and hence $y > 0$, the solutions exist only for x which lie in the intervals $(\pi/2, \pi)$, $(3\pi/2, 2\pi)$, $(5\pi/2, 3\pi)$, and so on.

If the circle radius $V_{r0}^{1/2} R$ is smaller than $\pi/2$, then the circle is not intersected by the curve $y = -x \cot x > 0$ and no bound states occur in the system. Making use of the definition of the reduced depth of the potential well $V_{r0} = (2\mu/\hbar^2) V_0$, where V_0 is the well depth in the energy units, we can write the condition that the bound states do not exist as

$$\frac{2\mu}{\hbar^2} V_0 R^2 < \frac{\pi^2}{4} \ . \tag{7.133}$$

If the circle radius $V_{r0}^{1/2} R$ is greater than $\pi/2$ but smaller than $3\pi/2$, the circle is intersected by the curve $y = -x \cot x > 0$ at one point (x_1, y_1) and the system possesses one bound state with the energy

$$E_1 = -\frac{\hbar^2}{2\mu R^2} y_1^2 \ . \tag{7.134}$$

In Fig. 7.4 this situation is shown by the circle 1. The system possesses a single bound state if

$$\frac{\pi^2}{4} < \frac{2\mu}{\hbar^2} V_0 R^2 < \frac{9\pi^2}{4} \ . \tag{7.135}$$

In the bound state, the particle motion is restricted to the inner region of the potential well. The relevant discrete negative value of energy may be interpreted in the following way. To be unable to escape from the potential well, the particle must undergo total reflection at $r = R$, which is possible only for energies such that the waves are in phase before and after the reflection. If the phase condition

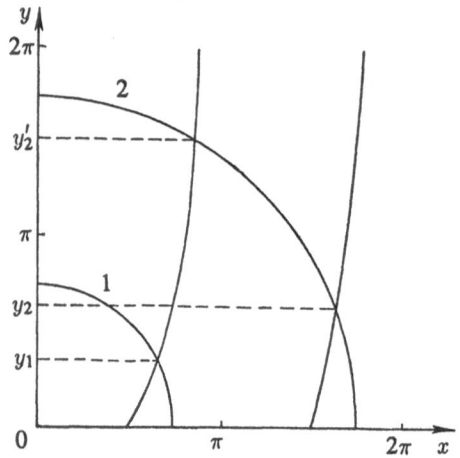

Fig. 7.4. Graphical aid to search for bound state energies in the case of a square well potential

is not satisfied, then reflection leads to interference which violates stationarity of the state.

If the potential satisfies the requirement

$$\frac{9\pi^2}{4} < \frac{2\mu}{\hbar^2} V_0 R^2 \quad , \tag{7.136}$$

then at least two bound states exist in the system. Circle 2 in Fig. 7.4 shows the case when the system possesses two bound states with energies

$$E_2 = -\frac{\hbar^2}{2\mu R^2} y_2^2 \quad , \qquad E_2' = -\frac{\hbar^2}{2\mu R^2} y_2'^2 \quad . \tag{7.137}$$

It is clear that the number of bound states increases as the quantity $(2\mu/\hbar^2) V_0 R^2$ grows. Thus, according to (7.133, 135, etc.), the presence or absence of some definite number of bound S-states in the square potential well is determined by the product of the particle mass, the well depth, and the radius squared.

The scattering matrix poles on the imaginary axis in the lower half-plane of complex k are associated with virtual states. We introduce the notation

$$k = -i\tilde{\kappa} \quad , \qquad \tilde{\kappa} > 0$$

and find from (7.130) that the virtual levels are determined by the equation

$$\tilde{y} = x \cot x \quad , \qquad x^2 + \tilde{y}^2 = V_{r0} R^2 \quad , \tag{7.138}$$

where $x = k_0 R$ and $\tilde{y} = \tilde{\kappa} R$. The virtual levels are determined by the intersections of the curve $\tilde{y} = x \cot x > 0$ with the circle of radius $V_{r0}^{1/2} R$ in the plane $x, \tilde{y}$ (Fig. 7.5). Let $V_{r0}^{1/2} R_1$ and $V_{r0}^{1/2} R_2$ be the radii of the circles 1 and 2 tangent to the first and second branches of the curve $\tilde{y} = x \cot x > 0$, respectively. Obviously, no virtual states occur for $R < R_1$ and $(\pi/2) V_{r0}^{-1/2} < R < R_2$.

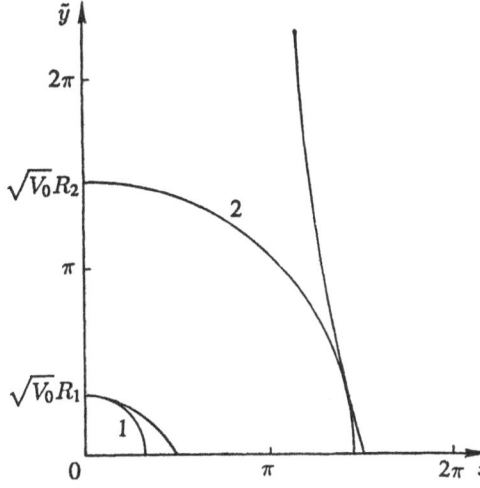

Fig. 7.5. Graphical aid to look for virtual energy levels in the case of a square well potential

The scattering matrix poles in the lower half-plane off the imaginary axis correspond to quasistationary states which are associated with the decays and captures in the system. To be precise, we consider a pole of $S(k)$ in the fourth quadrant, associated with a decay state. Suppose $(2\mu/\hbar^2)\, V_0 R^2 \gg 1$. In this case we can anticipate that the system possesses long-lived quasistationary states. Indeed, if the energy is positive, then the particle motion is infinite in spite of the well. However, particle reflection from the well edges can give rise to quasistationary states. The coefficient of wave penetration through the potential drop is equal to $4kk_0/(k_0 + k)^2$. Therefore, if $0 < E \ll V_0$ (which is possible if the well is sufficiently deep), the wave associated with the particle motion can undergo considerable reflection from the well edge and penetrates outside the well with very low probability. This can occur only for energies at which the incident and reflected waves satisfy some phase relation, as in the case of bound states.

We now consider a quasistationary decay state corresponding to a pole of $S(k)$ in the fourth quadrant of the complex k plane. The energy of this state can be found from the equation

$$k_0 \cot k_0 R = ik \quad , \qquad k = k' + ik'' \qquad (k'' < 0) \quad . \tag{7.139}$$

According to (7.127), the wave function is

$$\Phi(k, r) = \begin{cases} C_0 \sin k_0 r \quad , & r < R \quad , \\ C' e^{ikr} \quad , & r > R \quad , \end{cases} \tag{7.140}$$

where C_0 is given by (7.128), $C' = f(k)/2ik$. Note that the condition (7.139) is equivalent to the requirement that the solution of the Schrödinger equation associated with a quasistationary state must contain only a diverging wave for $r > R$. The condition (7.139) itself implies that the logarithmic derivatives of the internal and external wave functions are equal at the point $r = R$.

Equation (7.139) is complex, therefore, its roots, and hence the energies of quasistationary states, are also complex. The equation can be solved numerically, however, the approximate solution is more suitable for interpreting its physical meaning. For the sake of convenience, we introduce the notation

$$F(k) \equiv k_0 R \cot k_0 R \quad , \tag{7.141}$$

then (7.139) may be rewritten as

$$F(k) = ikR \quad . \tag{7.142}$$

Disregarding the right-hand part, we obtain the equation

$$F(k) = 0 \quad , \tag{7.143}$$

which does not contain imaginary quantities. Let k'_r be the real positive roots of (7.143); the relevant resonance energies are

$$E_r = \frac{\hbar^2 k_r'^2}{2\mu} \quad . \tag{7.144}$$

In order to find the solution of (7.142), we assume that k is close to some resonance value k'_r and expand the function $F(k)$ in a power series about the point k'_r, so that

$$F(k) = (k - k'_r) \left. \frac{dF}{dk} \right|_{k'_r} + \dots \quad . \tag{7.145}$$

Substituting this expansion into (7.142), we have

$$k \simeq k_r = k'_r + ik''_r \quad , \tag{7.146}$$

where

$$k''_r \simeq \frac{k'_r R}{\left. \frac{dF}{dk} \right|_{k'_r}} \quad . \tag{7.147}$$

Making use of (7.141, 143), we find dF/dk in the resonance to be

$$\left. \frac{dF}{dk} \right|_{k'_r} = -k_0 R \left. \frac{d}{dk} k_0 R \right|_{k'_r} < 0 \quad . \tag{7.148}$$

Therefore, the pole (7.146) indeed lies in the fourth quadrant of the complex k plane. The relevant complex energy of the quasistationary level is

$$W_r = E_r - \frac{i}{2} \Gamma_r \quad , \tag{7.149}$$

and the level width Γ_r is described by

$$\Gamma_r = -\frac{2k'_r R}{(dF/dE)\big|_{E_r}} \quad .$$

(7.150)

The level width (7.150) is positive by virtue of (7.148). It is not difficult to estimate Γ_r by the order of magnitude. When passing from some level E_r to another one, the cotangent argument $k_0 R$ in (7.143) acquires π. Therefore, the approximate relation

$$\frac{d}{dE} k_0 R \simeq \frac{\pi}{D}$$

occurs, where D is the distance between the resonances. Making use of (7.148, 150), we find that

$$\Gamma_r \simeq \frac{4k'_r}{k_0} \frac{D}{\pi} \quad .$$

(7.151)

As long as $k'_r \ll k_0$, the width is much smaller than D,

$$\Gamma_r \ll D \quad .$$

(7.152)

Let us consider scattering in the case when energy E is close to some resonance value E_r. After the notation (7.141) is introduced, the scattering matrix (7.129) reduces to

$$S(k) = e^{-2ikR} \frac{F(k) + ikR}{F(k) - ikR} \quad .$$

(7.153)

Expanding $F(k)$ in a power series about the resonance energy E_r and making use of (7.143), we obtain

$$S(k) = e^{-2ikR} \left\{ 1 - \frac{i\gamma_r}{E - E_r + (i/2)\gamma_r} \right\} \quad ,$$

(7.154)

where γ_r is the resonance width,

$$\gamma_r(k) = -\frac{2kR}{(dF/dE)\big|_{E_r}} \quad ,$$

(7.155)

which differs from the level width Γ_r by the additional factor k/k'_r.

The elastic scattering cross section can be calculated by using (7.154); the result is

$$\begin{aligned}
\sigma_e &= \frac{\pi}{k^2} \left| e^{2ikR} - 1 + \frac{i\gamma_r}{E - E_r + (i/2)\gamma_r} \right|^2 \\
&= \frac{\pi}{k^2} \left| e^{2ikR} - 1 \right|^2 + \frac{2\pi}{k^2} \operatorname{Re}\left(e^{2ikR} - 1 \right) \frac{i\gamma_r}{E - E_r + (i/2)\gamma_r} \\
&\quad + \frac{\pi}{k^2} \frac{\gamma_r^2}{(E - E_r)^2 + \frac{1}{4}\gamma_r^2} \quad .
\end{aligned}$$

(7.156)

The first term in (7.156) describes potential scattering. If $kR \ll 1$, the potential scattering cross section is

$$\sigma_{e,pot} = 4\pi R^2 \quad . \tag{7.157}$$

The second term in (7.156) is given rise to by the interference of potential and resonant scattering. The third term is associated with resonant scattering. If the energy is equal to the resonance value, then the resonant scattering cross section attains its maximum value

$$\sigma_{e,res}^{max} = \frac{4\pi}{k^2} \quad . \tag{7.158}$$

If $kR \ll 1$, the resonant scattering cross section in the vicinity of the resonance is much greater than the potential scattering cross section. For energies far from resonances, resonant scattering can be ignored because the contribution of potential scattering is dominant. Figure 7.6 shows the energy dependence of the elastic scattering cross section in the vicinity of a resonance.

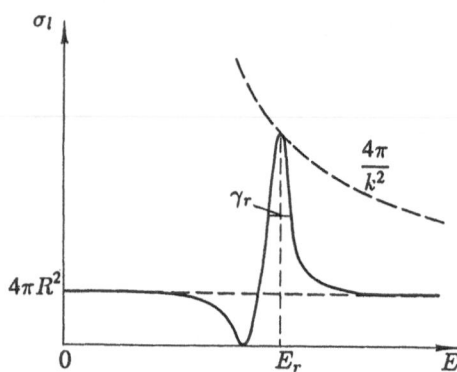

Fig. 7.6. Elastic scattering cross section as a function of energy in the vicinity of a resonance

In the case of scattering, the wave function of the system is given by (7.127). Let us find the ratio of the squared amplitude moduli of the wave functions corresponding to the inner and outer regions of the well. Making use of (7.128, 141), we obtain

$$\frac{|C_0|^2}{|C|^2} = \frac{1}{\sin^2 k_0 R} \frac{k^2 R^2}{F^2(k) + k^2 R^2} \quad . \tag{7.159}$$

In the vicinity of the resonance energy E_r expansion (7.145) can be employed, then

$$\frac{|C_0|^2}{|C|^2} = \frac{\frac{1}{4}\gamma_r^2}{(E - E_r)^2 + \frac{1}{4}\gamma_r^2} \quad . \tag{7.160}$$

We see that the incident particle penetrates into the potential well with appreciable probability only if its energy is close to some resonance value. According to (7.160), if $E = E_r$, then

$$\frac{|C_0|^2}{|C|^2} = 1 \quad .$$

Far from resonances, the particle undergoes almost total reflection from the well edge and hence the wave function in the inner region is very small.

Problems

7.1 Find the upper limit for the normalization constant of the bound state wave function in the case of a finite-range potential.

If the potential range R is finite, then the Jost solution $f_l(k, r)$ for the state with angular momentum quantum number l in the outer region $r > R$ can be expressed in terms of the Hankel function $h_l^{(-)}(kR)$, i.e.,

$$f_l(k, r) = i^{-l+3} \, kr \, h_l^{(-)}(kr) \quad . \tag{7.161}$$

The bound state wave function for $r > R$,

$$u_l(r) = \beta_l \, f_l(-i\kappa, r) \quad , \tag{7.162}$$

satisfies the normalization condition

$$\int_0^\infty dr \, u_l^2(r) = 1 \quad . \tag{7.163}$$

Separating the inner and outer integration ranges in the integral, we obtain

$$\int_0^R dr \, u_l^2(r) + (-1)^l \, \kappa^2 \beta_l^2 \int_R^\infty dr \, r^2 \left[h_l^{(-)}(-i\kappa r) \right]^2 = 1 \quad . \tag{7.164}$$

Since the first term in this relation is positive and smaller than one, the inequality

$$\beta_l^2 \leq \frac{1}{(-1)^l \kappa^2 \int_r^\infty dr \, r^2 \left[h_l^{(-)}(-i\kappa r) \right]^2} \tag{7.165}$$

holds. For the particular cases of $l = 0$ and $l = 1$ we have

$$\beta_0^2 \leq 2\kappa \, e^{2\kappa R} \quad , \qquad \beta_1^2 \leq 2\kappa \, e^{2\kappa R} \, \frac{1}{1 + (2/\kappa R)} \quad . \tag{7.166}$$

Thus, the upper limit of β_l^2 is determined by the potential range R and the radius of the bound state of the system, i.e., by the quantity κ^{-1}.

If the potential range R is not equal to zero, then β_0^2 is always smaller than $2\kappa \exp(2\kappa R)$. This relation can be used to estimate the potential range from the scattering data. Indeed, if the phase shift δ_0 is a known function of k, then $S_0(k)$ can be extrapolated to imaginary k and thus we can find the residue c_0 and hence β_0^2. If $\beta_0^2 \leq 2\kappa$, then the range R can be arbitrarily small; if $\beta_0^2 > 2\kappa$, then the potential range R is finite.

7.2   Show that the Bargmann potentials

$$V_{r1}(r) = \frac{(\varrho^2 - \sigma^2)\left\{\varrho^2 - \sigma^2 + \sigma^2 \cosh(\varrho r - 2\theta) - \varrho^2 \cosh \sigma r\right\}}{\left\{(\varrho - \sigma)\sinh\left[\tfrac{1}{2}(\varrho + \sigma)r - \theta\right] - (\varrho + \sigma)\sinh\left[\tfrac{1}{2}(\varrho - \sigma)r - \theta\right]\right\}^2}$$

and   (7.167)

$$V_{r2}(r) = \frac{(\varrho^2 - \sigma^2)\left\{\varrho^2 - \sigma^2 + \sigma^2 \cosh \varrho r - \varrho^2 \cosh(\sigma r + 2\theta)\right\}}{\left\{(\varrho - \sigma)\sinh\left[\tfrac{1}{2}(\varrho + \sigma)r + \theta\right] - (\varrho + \sigma)\sinh\left[\tfrac{1}{2}(\varrho - \sigma)r + \theta\right]\right\}^2}$$

(7.168)

(where $\varrho > \sigma > 0$, $\theta > 0$) lead to equal scattering phase shifts but different bound state energy levels [7.3].

Let $f_1(k)$ and $f_2(k)$ be the Jost functions which correspond to the potentials $V_{r1}(r)$ and $V_{r2}(r)$, respectively. The solution of (7.1) with the potential $V_{r1}(r)$ is

$$f_1(k) = \frac{2k + i\varrho}{2k - i\sigma} \quad ,$$ (7.169)

and for the bound state we have

$$k_1 = -\frac{i}{2}\varrho \quad , \qquad E_1 = -\frac{\hbar^2}{8\mu}\varrho^2 \quad .$$ (7.170)

In the same manner, substituting $V_{r2}(r)$ for the potential we find that $f_2(k)$ is given by

$$f_2(k) = \frac{2k + i\sigma}{2k - i\varrho}$$ (7.171)

and the bound state parameters are

$$k_2 = -\frac{i}{2}\sigma \quad , \qquad E_2 = -\frac{\hbar^2}{8\mu}\sigma^2 \quad .$$ (7.172)

However, both functions $f_1(k)$ and $f_2(k)$ yield the same scattering matrix

$$S(k) = \frac{(2k + i\sigma)(2k + i\varrho)}{(2k - i\sigma)(2k - i\varrho)} \quad ,$$ (7.173)

and hence equal scattering phase shifts for the potentials $V_{r1}(r)$ and $V_{r2}(r)$.

Since both potentials $V_{r1}(r)$ and $V_{r2}(r)$ depend on the parameter θ, whereas the Jost functions $f_1(k)$ and $f_2(k)$ (and hence the scattering matrix $S(k)$) are θ-independent, many continuous θ-dependent potentials can lead to equal phase shifts and bound state energies. The above example shows that phase shift $\delta(k)$ and bound state energies are insufficient for the calculation of the potential. To find the potential unambiguously, the normalization constants of the bound state wave functions are required.

7.3 Prove that the number N_l of bound states associated with given angular momentum quantum number l satisfies the Bargmann inequality

$$N_l < \frac{1}{2l+1} \int_0^\infty dr\, r\, |V_r(r)| \quad . \tag{7.174}$$

Suppose $V_r(r)$ is an attractive potential. We replace $V_r(r)$ by $\zeta V_r(r)$, where $0 \le \zeta \le 1$, and denote the number of bound states by $N_l(\zeta)$. It is clear that $N_l(\zeta)$ is an increasing function of ζ; $N_l(0) = 0$, and $N_l(1) = N_l$. The bound state energy is a decreasing function of ζ. To verify this assertion we employ the equality

$$\left(\Phi' \frac{\partial \Phi}{\partial \zeta} - \Phi \frac{\partial \Phi'}{\partial \zeta} \right)' = -\frac{\partial k^2}{\partial \zeta} \Phi^2 - |V_r| \Phi^2 \quad ,$$

integrating which yields

$$\frac{\partial k^2}{\partial \zeta} = -\frac{\int_0^\infty dr\, |V_r| \Phi^2}{\int_0^\infty dr\, \Phi^2} < 0 \quad . \tag{7.175}$$

As ζ decreases, the bound state energy grows up to zero; the further decrease of ζ causes disappearance of the bound state under consideration and $N_l(\zeta)$ becomes smaller by one.

Making use of the Schrödinger equation, the wave function of the bound state with energy $k^2 = 0$ can be shown to satisfy the integral equation

$$\Phi_l(0, r) = \zeta \int_0^\infty dr'\, G_l(r, r') |V_r(r')| \Phi_l(0, r') \quad , \tag{7.176}$$

where $G_l(r, r')$ is the Green function of the equation with no bound states,

$$G_l(r, r') = \frac{1}{2l+1} \begin{cases} r^{l+1} r'^{-l} & , \quad r < r' \\ r^{-l} r'^{l+1} & , \quad r > r' \end{cases} \quad .$$

The number of bound states N_l coincides with the total number of zero-energy bound states which appear as ζ varies from 0 to 1, i.e., N_l is equal to the number of eigenvalues ζ_n in the interval 0 to 1.

Since $|V_r(r)| > 0$, we can introduce

$$\tilde{G}_l(r,r') = \sqrt{|V_r(r)\,V_r(r')|}\;G_l(r,r') \quad,$$

$$\tilde{\Phi}_l(0,r) = \sqrt{|V_r(r)|}\;\Phi_l(0,r) \quad,$$

and then (7.176) reduces to

$$\zeta_n^{-1}\,\tilde{\Phi}_n(0,r) = \int_0^\infty dr'\;\tilde{G}_l(r,r')\,\tilde{\Phi}_n(0,r') \quad. \tag{7.177}$$

Because $\tilde{\Phi}_n(0,r)$ form a complete set of orthogonal functions, we find that

$$\sum_{n=1}^\infty \zeta_n^{-1} = \text{Tr}\,\{\tilde{G}\} = \frac{1}{2l+1}\int_0^\infty dr\,r\,|V_r(r)| \quad. \tag{7.178}$$

Since $\zeta_n \le 1$, we have

$$N_l \le \sum_{n=1}^{N_l} \zeta_n^{-1} \le \sum_{n=1}^\infty \zeta_n^{-1} = \frac{1}{2l+1}\int_0^\infty dr\,r\,|V_r(r)| \quad. \tag{7.179}$$

An important corollary follows from the Bargmann inequality (7.174): if the integral $\int_0^\infty dr\,r\,|V_r(r)|$ is finite, then the system possesses a finite number of bound states.

7.4 Express the particle time delay in the interaction region Q in terms of the scattering matrix S.

The particle wave function in the state with definite energy off the interaction region is described by (7.106). Suppose such functions form the wave packet

$$\psi(r,t) = \int dE\,a(E)\,\psi_E(r,t) \tag{7.180}$$

and assume that its amplitude $a(E)$ has a sharp maximum in the vicinity of some mean energy value E_0. The wave function of the packet may be written as

$$\psi(r,t) = \frac{1}{r}\left\{a_-(r,t)\,e^{-ik_0 r} - a_+(r,t)\,e^{ik_0 r}\right\}e^{(-i/\hbar)E_0 t} \quad, \tag{7.181}$$

where

$$a_-(r,t) = \int dE\,a(E)\,\exp\left[-i(k-k_0)r - \frac{i}{\hbar}(E-E_0)t\right] \quad, \tag{7.182}$$

$$a_+(r,t) = \int dE\,a(E)\,S(E)\,\exp\left[i(k-k_0)r - \frac{i}{\hbar}(E-E_0)t\right] \quad. \tag{7.183}$$

The first term in (7.181) describes the converging part of the packet associated with the incident wave; the second term describes the diverging part of the packet and is determined by particle interaction with the scattering field. If the wave packet is sufficiently narrow, then the logarithm of the scattering matrix $S(E)$

can be expanded in a power series of energy deviation from the mean value E_0, i.e.,

$$\ln S(E) = \ln S(E_0) + (E - E_0) \frac{d}{dE_0} \ln S(E_0) + \cdots \quad ,$$

and hence

$$S(E) \simeq S(E_0) \exp\left[(E - E_0) \frac{d}{dE_0} \ln S(E_0)\right] \quad . \tag{7.184}$$

Substituting (7.184) into (7.183) yields

$$a_+(r, t) = S(E_0) \int dE \, a(E) \tag{7.185}$$

$$\times \exp\left[i(k - k_0)r - \frac{i}{\hbar}(E - E_0)\left(t + i\hbar \frac{d}{dE_0} \ln S(E_0)\right)\right] \quad .$$

Comparing (7.185) to (7.182), we see that the particle-field interaction not only gives rise to the amplitude variations of the diverging wave $(S(E) = 1$ without interaction), but also causes a time delay in the formation of the diverging part of the wave packet [7.4], given by

$$Q = -i\hbar \frac{d}{dE_0} \ln S(E_0) \quad . \tag{7.186}$$

The quantity Q describes the additional time (over the time of free flight) spent by the particle in the interaction region due to interaction. If $Q > 0$, the particle is retarded in the interaction region; if $Q < 0$, then the interaction "ejects" the particle.

Relation (7.186) may be rewritten in another form. Having calculated the derivative of the logarithm, we obtain

$$Q = -i\hbar S^{-1} \frac{d}{dE} S \quad . \tag{7.187}$$

With the scattering matrix expressed in terms of the phase shift, $S(E) = \exp[2i \, \delta(E)]$, (7.187) yields the Wigner formula for the time delay:

$$Q = 2\hbar \frac{d}{dE} \delta(E) \quad . \tag{7.188}$$

Without interaction, the time spent by the particle inside a sphere of radius R is $(2R + k^{-1})/v$, where v is the particle velocity (the term k^{-1}, equal to the wavelength, arises from the wave properties of the particle). In the presence of the field $V_r(r)$, the time spent by the particle at distances shorter than R is

$$\frac{(2R + k^{-1})}{v} + Q \geq 0 \quad . \tag{7.189}$$

This relation is in agreement with the inequality (4.113). As follows from (7.189), the inequality

$$Q \geq -\frac{2R + k^{-1}}{v} \tag{7.190}$$

is always satisfied.

7.5   Find the scattering matrix and consider quasistationary states for a square well potential surrounded by a square barrier,

$$V_r(r) = \begin{cases} 0 & , & r < r \quad , \\ V_{r0} & , & R < r < R' \quad , \\ 0 & , & R' < r \quad . \end{cases} \tag{7.191}$$

To be precise, we assume $k^2 < V_{r0}$. The Jost solution that satisfies the boundary condition (7.20) is given by

$$f(k, r) = \begin{cases} C_1\, e^{ikr} + C_2\, e^{-ikr} & , & r < R \quad , \\ C_1'\, e^{\kappa r} + C_2'\, e^{-\kappa r} & , & R < r < R' \quad , \\ e^{-ikr} & , & R' > r \quad , \end{cases} \tag{7.192}$$

where $\kappa^2 = V_{r0} - k^2$. The coefficients are determined by the joint conditions at the points $r = R$ and $r = R'$, so that

$$C_1 = \frac{i\kappa}{4k}\left(1 + \frac{k^2}{\kappa^2}\right)\left[1 - \exp\left(-2\kappa(R' - R)\right)\right]$$
$$\times \exp\left[-ik(R' + R) + \kappa(R' - R)\right] \quad ,$$

$$C_2 = -\frac{i\kappa}{4k}\left\{\left(1 - \frac{k^2}{\kappa^2}\right)\left[1 - \exp\left(-2\kappa(R' - R)\right)\right]\right.$$
$$\left. + 2i\frac{k}{\kappa}\left[1 + \exp\left(-2\kappa(R' - R)\right)\right]\right\}$$
$$\times \exp\left[-ik(R' - R) + \kappa(R' - R)\right] \quad ,$$

$$C_1' = \frac{1}{2}\left(1 - i\frac{k}{\kappa}\right)\exp\left(-ikR' - \kappa R'\right) \quad ,$$

$$C_2' = \frac{1}{2}\left(1 + i\frac{k}{\kappa}\right)\exp\left(-ikR' + \kappa R'\right) \quad .$$

The sum of the coefficients C_1 and C_2 determines the Jost function $f(k)$, and thus the scattering matrix in question is given by

$$S(k) = \exp\left(2ikR'\right)\frac{\kappa - ik}{\kappa + ik}\frac{\exp\left[-2\kappa(R' - R)\right] + \left((\kappa + ik)/(\kappa - ik)\right)\zeta(k)}{\exp\left[-2\kappa(R' - R)\right] + \left((\kappa - ik)/(\kappa + ik)\right)\zeta(k)} \quad , \tag{7.193}$$

where

$$\zeta(k) = \frac{k\cot kR + \kappa}{k\cot kR - \kappa} \quad . \tag{7.194}$$

Let us find the poles of the scattering matrix (7.193). If the barrier is sufficiently wide and the pole does not correspond to very small κ, then $\kappa(R'-R) \gg 1$ and the term $\exp\left[-2\kappa(R'-R)\right]$ is minor. However, we cannot disregard it in the numerator and denominator of (7.193) because otherwise (7.193) would have no poles. The small quantity $\exp\left[-\kappa(R'-R)\right]$ is important only in the vicinity of the zeros k_r of the function $\zeta(k)$. We expand $\zeta(k)$ about the zero in power series of $k - k_r$,

$$\zeta(k) \simeq (k - k_r) \left.\frac{d\zeta(k)}{dk}\right|_{k_r} \quad,$$

and require that the denominator of (7.193) must vanish. Then

$$\tilde{k}_r = k_r - \exp\left[-2\kappa_r(R'-R)\right] \frac{\left(\kappa_r + ik_r\right)^2}{\kappa_r^2 + k_r^2} \left[\left.\frac{d\zeta(k)}{dk}\right|_{k_r}\right]^{-1} \quad. \tag{7.195}$$

As could be anticipated, the pole of the scattering matrix (7.193) lies in the lower half-plane of complex k. The real and imaginary parts of $\tilde{k}_r$ determine the energy and the width of the quasistationary state, i.e.,

$$E_r \simeq \frac{\hbar^2 k_r^2}{2\mu} \quad,$$

$$\Gamma_r = 8E_r \frac{\kappa_r}{\kappa_r^2 + k_r^2} \left[\left.\frac{d\zeta(k)}{dk}\right|_{k_r}\right]^{-1} \exp\left[-2\kappa_r(R'-R)\right] \quad. \tag{7.196}$$

In the nearest vicinity of the pole, the scattering matrix is

$$S(k) \simeq e^{2i\varphi} \frac{k + \tilde{k}_r^*}{k - \tilde{k}_r} \quad, \tag{7.197}$$

or

$$S(E) \simeq e^{2i\varphi} \frac{E - E_r - (i/2)\,\Gamma_r}{E - E_r + (i/2)\,\Gamma_r} \quad, \tag{7.198}$$

where φ is a slowly varying function of k and E.

We use (7.187) to find the lifetime of the quasistationary state to be

$$Q = \frac{\Gamma_r^2}{\left(E - E_r\right)^2 + \frac{1}{4}\,\Gamma_r^2}\, Q_r \quad, \qquad Q_r = \frac{\hbar}{\Gamma_r} \quad. \tag{7.199}$$

It is clear that quasistationary states exist only if Q is much greater than $2R/v$, the time of particle free flight through the interaction region. To satisfy this requirement, the particle energy must be sufficiently close to the resonance energy,

$$|E - E_r| < \frac{1}{2}\,\Gamma_r \left(\frac{\hbar}{\Gamma_r} \frac{v}{2R}\right)^{1/2} \quad.$$

If the right edge of the potential barrier tends to infinity, then the pole $\tilde{k}_r$ tends to k_r (the imaginary part of $\tilde{k}_r$ vanishes as $R' \to \infty$), i.e., in the limiting case of $R' \to \infty$ the potential (7.191) reduces to the square well potential whose spectrum is discrete for $\kappa^2 < V_{r0}$. The stationary states of the square well are associated with the same points as the quasistationary states of the potential (7.191).

8. Dispersion Relations

8.1 Integral Representations of the Jost Functions

The Cauchy theorem enables one to obtain integral relations for the real and imaginary parts of any analytic function, which are referred to as dispersion relations.

Let us derive the dispersion relations for the Jost functions $f_l(k)$. We have already shown that $f_l(k)$ is an entire analytic function in the lower half-plane of the complex variable k. It is not difficult to show that within the context of (7.42) and boundary conditions (7.49) $f_l(k)$ tends to unity as $k \to \infty$ along any direction in the lower half-plane including the real axis. We apply the Cauchy theorem to the function $f_l(k) - 1$ to obtain

$$f_l(k) - 1 = \frac{1}{2\pi i} \int_C dk' \frac{f_l(k') - 1}{k' - k} \quad . \tag{8.1}$$

Integration in (8.1) must be carried out over any closed contour C which lies in the lower half-plane and encircles the point k. We take C to be the contour consisting of the real axis and a semicircle of infinite radius in the lower half-plane. As long as the integral over the semicircle vanishes, we have

$$f_l(k) = 1 - \frac{1}{2\pi i} \int_{-\infty}^{\infty} dk' \frac{f_l(k') - 1}{k' - k} \quad . \tag{8.2}$$

Formula (8.2) makes it possible to find $f_l(k)$ for any k from the lower half-plane by its given values on the real axis.

Suppose the point k moves to the real axis from below. Making use of the relation

$$\lim_{\mathrm{Im}\{k\} \to 0} \int_{-\infty}^{\infty} dk' \frac{f_l(k') - 1}{k' - k}$$

$$= \fint_{-\infty}^{\infty} dk' \frac{f_l(k') - 1}{k' - k} - i\pi \left\{ f_l(k) - 1 \right\} \quad , \tag{8.3}$$

where $\fint \ldots$ is the principle value of the integral, we find from (8.2) that

$$f_l(k) = 1 - \frac{1}{i\pi} \fint_{-\infty}^{\infty} dk' \frac{f_l(k') - 1}{k' - k} \quad , \qquad \mathrm{Im}\{k\} = 0 \quad . \tag{8.4}$$

Separating the real and imaginary parts in (8.4), we obtain the dispersion relations for the Jost function:

$$\text{Re}\left\{f_l(k)\right\} = 1 - \frac{1}{\pi} \fint_{-\infty}^{\infty} dk' \, \frac{\text{Im}\left\{f_l(k')\right\}}{k' - k} \quad , \tag{8.5}$$

$$\text{Im}\left\{f_l(k)\right\} = \frac{1}{\pi} \fint_{-\infty}^{\infty} dk' \, \frac{\text{Re}\left\{f_l(k')\right\} - 1}{k' - k} \quad . \tag{8.6}$$

Relation (8.5) allows one to find the real part of the Jost function by the given imaginary part; (8.6) determines the imaginary part provided the real part is known.

Since

$$\int_{-\infty}^{\infty} dk' \, \frac{\text{Im}\left\{f_l(k')\right\}}{k' - k + i0} = \fint_{-\infty}^{\infty} \frac{\text{Im}\left\{f_l(k')\right\}}{k' - k} - i\pi \, \text{Im}\left\{f_l(k)\right\} \quad ,$$

the dispersion relation (8.5) can be rewritten in a form more suitable for further calculations, i.e., as

$$f_l(k) = 1 - \frac{1}{\pi} \fint_{-\infty}^{\infty} dk' \, \frac{\text{Im}\left\{f_l(k')\right\}}{k' - k + i0} \quad . \tag{8.7}$$

This version of the dispersion relation enables one to regard it as an integral equation for the function $f_l(k)$. Indeed, because

$$f_l(k) = \left|f_l(k)\right| \exp\left[i\Delta_l(k)\right]$$

and

$$\begin{aligned}\text{Im}\left\{f_l(k)\right\} &= \left|f_l(k)\right| \sin \Delta_l(k) \\ &= f_l(k) \sin \Delta_l(k) \exp\left[-i\Delta_l(k)\right] \quad ,\end{aligned}$$

(8.7) reduces to the integral equation

$$f_l(k) = 1 - \frac{1}{\pi} \int_{-\infty}^{\infty} dk' \, \frac{e^{-i\Delta_l(k')} \sin \Delta_l(k')}{k' - k + i0} f_l(k') \quad , \tag{8.8}$$

the kernel of which is determined by the scattering phase shift $\Delta_l(k)$. However, the integral equation (8.8) has an unambiguous solution only if the dislocation of zeros of the function $f_l(k)$ in the lower half-plane, associated with bound states, is known.

We denote the zeros of $f_l(k)$ in the lower half-plane by $k_n = -i\kappa_n$ ($\kappa_n > 0$) and introduce a new function $\tilde{f}_l(k)$ defined by

$$\tilde{f}_l(k) = \left(\prod_{n=1}^{N} \frac{k - i\kappa_n}{k + i\kappa_n}\right) f_l(k) \quad . \tag{8.9}$$

The function $\tilde{f}_l(k)$ is analytic, has no zeros in the lower half-plane and tends to unity as $k \to \infty$. Therefore, $\ln \tilde{f}_l(k)$ satisfies a relation similar to (8.4), i.e.,

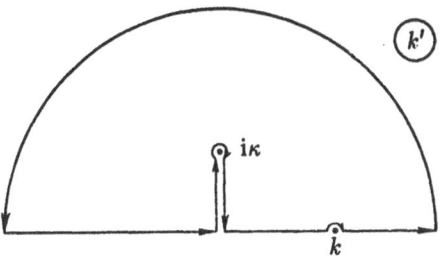

Fig. 8.1. Integration contour C in the complex k plane for the right-hand part of (8.11)

$$\ln \tilde{f}_l(k) = -\frac{1}{i\pi} \int_{-\infty}^{\infty} dk' \frac{\ln \tilde{f}_l(k')}{k' - k} \quad . \tag{8.10}$$

We separate the real part of (8.10) by using

$$\mathrm{Re}\left\{\ln \tilde{f}_l\right\} = \ln \left|\tilde{f}_l\right| = \ln \left|f_l\right| \quad ,$$

$$\mathrm{Im}\left\{\ln \tilde{f}_l\right\} = \Delta_l(k) - i \sum_{n=1}^{N} \left\{\ln\left(k - i\kappa_n\right) - \ln\left(k + i\kappa_n\right)\right\} \quad ,$$

then we have

$$\ln \left|f_l(k)\right| = -\frac{1}{\pi} \int_{-\infty}^{\infty} dk' \frac{\Delta_l(k')}{k' - k}$$

$$+ \frac{i}{\pi} \sum_{n} \int_{-\infty}^{\infty} dk' \frac{\ln\left(k' - i\kappa_n\right) - \ln\left(k' + i\kappa_n\right)}{k' - k} \quad . \tag{8.11}$$

The integral in the right-hand part of (8.11) can easily be calculated since

$$i \int_{C} dk' \frac{\ln\left(k' - i\kappa_n\right) - \ln\left(k' + i\kappa_n\right)}{k' - k} = 0$$

if integration is carried out over the contour C shown in Fig.8.1. The integral over the semicircle of infinite radius results in a zero contribution and the contribution of the integral over the small semicircle around the point k is pure imaginary. Thus the integral over the real axis reduces to the real part of the integral over the imaginary axis from zero to the branch point $i\kappa_n$ and vice versa, i.e.,

$$i \int_{-\infty}^{\infty} dk' \frac{\ln\left((k - i\kappa)/(k + i\kappa)\right)}{k' - k}$$

$$= \mathrm{Re}\left\{i \left(\int_{0}^{i\kappa} dk' \frac{\ln\left((k' - i\kappa)/(k' + i\kappa)\right) + 2\pi i}{k' - k}\right.\right.$$

$$\left.\left. + \int_{i\kappa}^{0} dk' \frac{\ln\left((k' - i\kappa)/(k' + i\kappa)\right)}{k' - k}\right)\right\}$$

$$= \pi \ln\left(1 + \frac{\kappa^2}{k^2}\right) \quad . \tag{8.12}$$

Substituting (8.12) into (8.11) and making use of the relation $\ln |f_l| = \ln f_l - i\Delta_l$, we find that

$$\ln f_l(k) = -\frac{1}{\pi} \int_{-\infty}^{\infty} dk' \frac{\Delta_l(k')}{k' - k + iO} + \sum_n \ln \left(1 + \frac{\kappa_n^2}{k^2} \right) \quad , \tag{8.13}$$

and therefore

$$f_l(k) = \prod_{n=1}^{N} \left(1 + \frac{\kappa_n^2}{k^2} \right) \exp \left[-\frac{1}{\pi} \int_{-\infty}^{\infty} dk' \frac{\Delta_l(k')}{k' - k + iO} \right] \quad . \tag{8.14}$$

Formula (8.14) gives the integral representation of the Jost function. We see that calculating the Jost function requires knowledge of both the energy dependence of the scattering phase shift and all the bound state energies. It is clear that the integral representation (8.14) may be regarded as a solution of the integral equation (8.8).

8.2 Levinson Theorem

Zero- and infinite-energy scattering phase shifts satisfy a relation which is a corollary of the scattering matrix analyticity [8.1]. This relation is usually referred to as the Levinson theorem; it describes the difference of zero- and infinite-energy scattering phase shifts in terms of the number of bound states. To derive this relation, we employ the analytic properties of the Jost function $f_l(k)$. According to (2.69), the logarithmic derivative of the scattering matrix is a function of the scattering phase shift derivative, i.e.,

$$\frac{S_l'(k)}{S_l(k)} = 2i \, \delta_l'(k) \quad . \tag{8.15}$$

Let us consider the integral

$$I \equiv \int_{-\infty}^{\infty} dk \, \frac{S_l'(k)}{S_l(k)} \quad . \tag{8.16}$$

Substituting (8.15) into (8.16) and bearing in mind that the phase shift $\delta_l(k)$ is an odd function of k, we obtain

$$I = 4i \int_{0}^{\infty} dk \, \delta_l'(k) = 4i \left\{ \delta_l(\infty) - \delta_l(0) \right\} \quad . \tag{8.17}$$

Here the phase shifts are regarded as continuous functions of k and are not referred to any finite interval.

On the other hand, the scattering matrix $S_l(k)$ may be expressed in terms of the Jost functions $f_l(k)$ and $f_l(-k)$. Then the integral (8.16) reduces to

$$I = \int_{-\infty}^{\infty} dk \frac{d}{dk} \left\{ \ln f_l(k) - \ln f_l(-k) \right\} = 2 \int_{-\infty}^{\infty} dk \frac{d}{dk} \ln f_l(k) \quad . \qquad (8.18)$$

The function $\ln f_l(k)$ vanishes as $k \to \infty$ along any direction in the lower half-plane, hence in order to calculate the integral in the right-hand part of (8.18) we close the contour from below. Since $f_l(k)$ is regular in the lower half-plane, only simple isolated zeros at the points $k_n = -i\,\kappa_n$ contribute to the integral. Therefore,

$$I = -4\pi i\, N_l \quad , \qquad (8.19)$$

where N_l is the number of bound states with the angular momentum quantum number l associated with the zeros of the function $f_l(k)$.

We compare (8.17) to (8.19) and thus obtain the Levinson theorem:

$$\delta_l(0) - \delta_l(\infty) = \pi\, N_l \quad . \qquad (8.20)$$

If $f_l(k) = 0$ for $k = 0$ and $l = 0$, then formula (8.20) is modified so that

$$\delta_l(0) - \delta_l(\infty) = \pi \left(N_0 + \frac{1}{2} \right) \quad , \qquad f_0(0) = 0 \quad . \qquad (8.21)$$

If the potential range is finite, then $\delta_l(\infty) = 0$, and the zero-energy scattering phase shift is determined by the number of bound states.

By virtue of (8.21), for $l = 0$ and $f_0(0) = 0$ we have

$$S_0(0) = \exp\left[2i\, \delta_0(0) \right] = -1 \quad . \qquad (8.22)$$

For all other cases, $S_l(0) = 1$. Therefore, the partial scattering amplitude

$$f_l(k) = \frac{1}{2ik} \left\{ S_l(k) - 1 \right\} \qquad (8.23)$$

is finite at $k = 0$ except for the case when $l = 0$, $f_0(0) = 0$. This case is referred to as the zero-energy resonance.

8.3 Complex Energy Shell

It is sometimes convenient to treat the scattering matrix as a function of energy E rather than of the wave vector k. The scattering matrix $S_l(E)$, defined for real energies, can be analytically continued to complex E. In order to find the correspondence between $S_l(E)$ and $S_l(k)$, we must take into account that by virtue of the relation $E = \hbar^2 k^2 / 2\mu$, the passage from complex k to complex energies E maps the k plane onto the two-sheeted energy shell. The upper half-plane of k becomes the entire plane, and the lower half-plane of k develops into the second sheet of the Riemann E surface; the cut must be made along the positive part of the real E axis. The upper sheet of the Riemann surface is referred to as the physical sheet $\mathrm{Im}\left\{ E^{1/2} \right\} > 0$. The bound states are associated with the poles of the scattering matrix $S_l(E)$ on the left real semi-axis of the physical sheet. The scattering matrix poles on the nonphysical sheet $\text{Im}\left\{E^{1/2}\right\} < 0$ correspond to the quasistationary states.

To define the scattering matrix $S_l(E)$ in the complex E plane, we introduce the function $D_l(E)$ which for real positive E coincides with the Jost function $f_l(-k)$, where $k = \left(2\mu E/\hbar^2\right)^{1/2}$. In order to continue $D_l(E)$ analytically to the complex E plane, we employ the dispersion relation (8.7) for the function $f_l(-k)$,

$$f_l(-k) = 1 + \frac{1}{\pi} \int_{-\infty}^{\infty} dk' \frac{\text{Im}\left[f_l(-k)\right]}{k' - k - iO} \ . \tag{8.24}$$

We multiply both the numerator and the denominator of the integrand in (8.24) by $(k' + k + iO)$. Then, since $\text{Im}\left\{f_l(-k)\right\}$ is an odd function of k, we obtain

$$f_l(-k) = 1 + \frac{1}{\pi} \int_0^{\infty} dE' \frac{\text{Im}\left\{f_l(-k)\right\}}{E' - E - iO} \ , $$

or

$$D_l(E) = 1 + \frac{1}{\pi} \int_0^{\infty} dE' \frac{\text{Im}\left\{D_l(E')\right\}}{E' - E - iO} \ . \tag{8.25}$$

Relation (8.25) determines $D_l(E)$ for the whole complex energy plane with a cut along the real axis from zero to infinity. To obtain the physical value of $D_l(E)$, we approach the real axis from above, i.e.,

$$D_l(E + iO) = f_l(-k) \qquad (k \text{ is real}) \ . \tag{8.26}$$

The phase $D_l(E)$ on the real axis is by definition equal to $-\Delta_l(E)$. When approaching the real axis from below (on the lower edge of the cut), we have

$$D_l(E - iO) = f_l(k) \ . \tag{8.27}$$

Note that $D_l(E - iO) = D_l^*(E + iO)$, i.e., $D_l(E)$ takes complex conjugate values on the upper and lower edges of the cut.

On the physical sheet, $D_l(E)$ is given by the relation

$$D_l^1(E) = 1 + \frac{1}{\pi} \int_0^{\infty} dE' \frac{\text{Im}\left\{D_l(E')\right\}}{E' - E + iO} \ . \tag{8.28}$$

Equations (8.25, 28) determine the single-valued function $D_l(E)$ on the two-sheeted surface. The scattering matrix $S_l(E)$ for physical (real) energies E is given by

$$S_l(E) = (-1)^l \frac{D_l(E - iO)}{D_l(E + iO)} \ . \tag{8.29}$$

If energy is complex, the scattering matrix is described by the relation

$$S_l(E) = (-1)^l \, \frac{D_l^1(E)}{D_l(E)} \quad , \tag{8.30}$$

i.e., it is determined by the ratio of nonphysical to physical values of the function $D_l(E)$. The partial scattering amplitude is expressed in terms of the scattering matrix by means of the relation

$$f_l(E) = \frac{\hbar}{2\mathrm{i}\,(2\mu E)^{1/2}} \left\{ S_l(E) - 1 \right\} \quad . \tag{8.31}$$

Regarding the partial scattering amplitude as a function of complex energy on the whole physical sheet, we find the discrete energy levels to be its simple poles.

8.4 Analyticity of the Scattering Matrix and the Causality Principle

We have already seen that the scattering matrix is analytic for arbitrary potentials in both complex k and E planes. Analyticity of the scattering matrix is a corollary of the causality principle which infers that the cause must preceed the effect. We shall show by a simple example that the physical requirement of a causal relationship between the events indeed leads to the analyticity property of the scattering matrix.

Let us consider scattering by a finite-range potential. For given energy E, the wave function off the interaction region is

$$\psi_E(r,t) = \frac{1}{r} \left\{ e^{-ikr} - S(E)\, e^{ikr} \right\} e^{(-i/\hbar)Et} \quad , \tag{8.32}$$

$$E = \frac{\hbar^2 k^2}{2\mu} \quad , \qquad r > R \quad .$$

The first term describes the wave incident on the scattering center, the second one is associated with the scattered wave. (For the sake of simplicity we consider only S-scattering.) We consider a wave packet localized in time and space,

$$\psi(r,t) = \int_0^\infty dE' \, a(E') \, \psi_{E'}(r,t) \quad , \tag{8.33}$$

with the shape determined by the amplitude $a(E')$. It is clear that

$$\psi(r,t) = \Phi^{(-)}(r,t) - \Phi^{(+)}(r,t) \quad , \tag{8.34}$$

where $\Phi^{(-)}(r,t)$ is the incident wave packet,

$$\Phi^{(-)}(r,t) = \int_0^\infty dE' \, a(E') \frac{1}{r} \exp\left[-ik'r - \frac{i}{\hbar} E't \right] \quad ,$$

and $\Phi^{(+)}(r,t)$ is the packet of diverging waves,

$$\Phi^{(+)}(r,t) = \int_0^\infty dE' \, a(E') \, S(E') \frac{1}{r} \exp\left[ik'r - \frac{i}{\hbar} E't\right] \quad .$$

The diverging wave amplitude is completely determined by the amplitude of the incident wave, therefore, the relation between $\Phi^{(+)}(r,t)$ and $\Phi^{(-)}(r,t)$ must be linear. The most general form of linear dependence between these functions for all previous times is given by the integral relation

$$\Phi^{(+)}(r,t) = \int_0^\infty d\tau \, F(\tau) \, \Phi^{(-)}(r, t - \tau) \quad , \tag{8.35}$$

where $F(\tau)$ is a finite function of time determined by the properties of the system. According to the causality principle, integration in (8.35) extends only over times preceeding t.

We multiply (8.35) by $\exp[(i/\hbar)Et]$ and carry out integration over t from $-\infty$ to ∞. Thus we obtain

$$S(E) \, e^{2ikr} = \int_0^\infty d\tau \, F(\tau) \, e^{(i/\hbar)E\tau} \tag{8.36}$$

This relation determines the scattering matrix as a function of real energy, but allows extensions to complex E.

Regarding E as a complex variable, we can consider the properties of the function $S(E)$ in the upper half-plane of E. According to (8.36), the scattering matrix $S(E)$ in the upper half-plane is a single-valued function and does not tend to infinity, i.e., it has no singular points. Indeed, for Im $\{E\} > 0$, the integrand of (8.36) contains a decreasing exponential factor $\exp[-\text{Im}\{E\tau\}/\hbar]$ $(\tau > 0)$, and since the function $F(\tau)$ is finite in the whole integration region, the integral is convergent. In the lower half-plane, the integral is divergent and hence the function $S(E)$ can be defined only as the analytic continuation of (8.36).

Thus, we have shown that analyticity of the scattering matrix in the upper half-plane of E directly follows from the causality principle. This conclusion is drawn irrespective of potential properties in the interaction region. The analyticity region obtained corresponds to the first quadrant of the complex k plane. By virtue of the symmetry properties of the scattering matrix (7.80, 82), the analyticity region can be extended to the whole k plane with the exception of isolated poles.

The factor $\exp(2ikr)$ in (8.36) describes the phase difference of the wave travelling through the scattering field center and the wave reflected from the surface of the sphere of radius r (we consider the case of S-scattering). In the general case, when a plane wave is scattered at an angle ϑ, the path difference for the rays passing through the center and reflected from the surface of radius r is equal to $2r \sin(\vartheta/2)$. Hence, it is not difficult to show that the quantity $f(k,\vartheta) \exp[2ikr \sin(\vartheta/2)]$, rather than the scattering amplitude, is analytic in the upper half-plane of E. Therefore, the amplitude of forward direction scattering $f(k,0)$ is analytic in the upper half-plane of E and possesses the simplest analytic properties.

8.5 Dispersion Relations
for the Forward Direction Scattering Amplitude

Up to now, we have considered the analytic properties of the scattering matrix S_l for states with angular momentum quantum number l. According to (3.77), the partial scattering amplitude can be expressed in terms of S_l. Therefore, singularities of the scattering matrix at the same time are singularities of the partial amplitude. Obviously, the total amplitude $f(k, k')$, regarded as a function of k or E for given scattering angles, also possesses the same singularities. The above study of analytic properties of the scattering matrix involved some difficulties because of extra poles and singularities which could occur at infinity. We shall show in this section that analytic properties of the forward direction scattering amplitude are simpler.

According to (3.15), the amplitude of zero-angle scattering is given by the expression

$$f(k, 0) \equiv f(k, k') = -\frac{\mu}{2\pi\hbar^2} \int dr \, e^{-ikr} \, V(r) \, \psi_k^{(+)}(r) \quad , \tag{8.37}$$

where $\psi_k^{(+)}(r)$ is the exact wave function of the scattering problem. At large distances from the scatterer, the wave function consists of the two parts — the incident and diverging waves. The diverging wave is proportional to $\exp(ikr)$, therefore the integrand contains the factor $\exp[ikr(1 - \cos\theta)]$, where θ is the angle between the vectors k and r. Since $\cos\theta \leq 1$, this integral is convergent if k lies in the upper half-plane. The incident wave is proportional to $\exp(ikr)$, hence the exponential factors in the integrand cancel out and this part is also convergent.

Since the function $\psi_k^{(+)}(r)$ for complex k (in the upper half-plane) is unambiguously determined by the solution of the Schrödinger equation which contains the plane wave and the part that exponentially decreases as $r \to \infty$, the amplitude (8.37) is also determined unambiguously.

The singularities of the amplitude (8.37) can arise only if the function $\psi_k^{(+)}(r)$ itself becomes infinite. This occurs for k such that $f_l(-k) = 0$. Indeed, as long as $\psi_k^{(+)}(r)$ is normalized to the unit amplitude of the plane wave, the coefficients of its expansion in terms of the function $\Phi_l(k, r)$ are inversely proportional to $f_l(-k)$, i.e.,

$$\psi_k^{(+)}(r) = \sum_l \frac{(-1)^l (2l + 1)}{f_l(-k)} \frac{1}{r} \Phi_l(k, r) \, P_l(\cos\vartheta) \quad , \tag{8.38}$$

and therefore $\psi_k^{(+)}(r)$ tends to infinity at the zeros of $f_l(-k)$. We have already shown that these zeros are associated with the bound states of the system.

If the potential range is finite, then the amplitude (8.37) remains finite as $k \to \infty$. Indeed, for $k \to \infty$ the potential in the Schrödinger equation can be disregarded. Then $\psi_k^{(+)}(r)$ reduces to the plane wave. As a result, we find that the amplitude can be described by

$$f(\infty, 0) = -\frac{\mu}{2\pi\hbar^2} \int dr\, V(r) \quad , \tag{8.39}$$

which coincides with the Born forward direction scattering amplitude. Note that the Born amplitude is real.

Thus, we have shown that the forward direction scattering amplitude (8.37) is analytic in the upper half-plane of k and has simple poles on the imaginary axis associated with the bound states of the system.

Let us consider the integral

$$\frac{1}{2\pi i} \int_C dk' \frac{f(k', 0) - f(\infty, 0)}{k' - k} \quad .$$

Suppose the point k lies in the upper half-plane. We carry out integration over the closed contour C consisting of the real axis and a semicircle of infinitely large radius in the upper half-plane. The integral over the semicircle vanishes and by virtue of the Cauchy theorem, we have

$$\frac{1}{2\pi i} \int dk' \frac{f(k', 0) - f(\infty, 0)}{k' - k} = f(k, 0) - f(\infty, 0) + \sum \quad , \tag{8.40}$$

where $\sum$ is the sum of residues of the integrand at the poles of the amplitude $f(k', 0)$.

The amplitude $f(k, 0)$ and the scattering matrix $S_l(k)$ satisfy the relation

$$f(k, 0) = \frac{1}{2k} \sum_l (2l + 1) \left\{ 1 - S_l(k) \right\} \quad , \tag{8.41}$$

therefore, within the context of (7.76), the sum of residues entering (8.40) is given by

$$\sum = \sum_{l,n} (-1)^l (2l + 1) \frac{1}{2i\,\kappa_{ln}} \frac{\beta_{ln}^2}{k - i\,\kappa_{ln}} \quad . \tag{8.42}$$

Here β_{ln} is the coefficient before $\exp(-\kappa_{ln} r)$ in the asymptotic of the normalized bound state wave function; the summation extends over all bound states of the system (all angular momentum quantum numbers l associated with bound states, and all states n for fixed l).

If the imaginary part of k in (8.40) tends to zero, then in order to calculate the integral over the real axis, the pole $k' = k$ must be encircled from below. Since

$$\int_{-\infty}^{\infty} dk' \frac{F(k')}{k' - k - i0} = \int_{-\infty}^{\infty} dk' \frac{F(k')}{k' - k} + i\pi F(k) \quad , \tag{8.43}$$

the relation (8.40) with $\mathrm{Im}\,\{k\} = 0$ may be rewritten as

$$f(k,0) = f(\infty, 0) + \frac{1}{i\pi} \int_{-\infty}^{\infty} dk' \frac{f(k',0) - f(\infty,0)}{k' - k}$$

$$- \sum_{l,n} (-1)^l (2l+1) \frac{1}{i\kappa_{ln}} \frac{\beta_{ln}^2}{k - i\kappa_{ln}} \quad . \tag{8.44}$$

Separating the real and imaginary parts of this relation, we obtain the dispersion relations for the forward direction scattering amplitude:

$$\text{Re}\,\{f(k,0)\} = f(\infty,0) + \frac{1}{\pi} \int_{-\infty}^{\infty} dk' \frac{\text{Im}\{f(k',0)\}}{k' - k}$$

$$+ \sum_{l,n} (-1)^{l+1} (2l+1) \frac{\beta_{ln}^2}{k^2 + \kappa_{ln}^2} \quad , \tag{8.45}$$

$$\text{Im}\,\{f(k,0)\} = -\frac{1}{\pi} \int_{-\infty}^{\infty} dk' \frac{\text{Re}\{f(k',0)\} - f(\infty,0)}{k' - k}$$

$$+ \sum_{l,n} (-1)^l (2l+1) \frac{k}{\kappa_{ln}} \frac{\beta_{ln}^2}{k^2 + \kappa_{ln}^2} \quad . \tag{8.46}$$

The dispersion relation (8.45) gives the real part of the forward direction scattering amplitude for Im $\{k\} = 0$ as a function of the imaginary part of the scattering amplitude, bound state energies, and normalization coefficients of all the bound state wave functions. As follows from the optical theorem (5.8), the imaginary part of the forward direction scattering amplitude is determined by the total scattering cross section, i.e.,

$$\text{Im}\,\{f(k,0)\} = \frac{k}{4\pi} \sigma(k) \quad . \tag{8.47}$$

Hence, relation (8.45) enables us to express the scattering amplitude $f(k,0)$ only in terms of observable quantities, so that

$$f(k,0) = \text{Re}\,\{f(k,0)\} + i \frac{k}{4\pi} \sigma(k) \quad , \tag{8.48}$$

$$\text{Re}\,\{f(k,0)\} = f(\infty,0) + \frac{1}{4\pi^2} \int_{-\infty}^{\infty} dk' \frac{k' \sigma(k')}{k' - k}$$

$$+ \sum_{l,n} (-1)^{l+1} (2l+1) \frac{\beta_{ln}^2}{k^2 + \kappa_{ln}^2} \quad . \tag{8.49}$$

The dispersion relation (8.46) describes the imaginary part of the forward direction scattering amplitude for Im $\{k\} = 0$ in terms of its real part.

Making use of (8.43), we rewrite the dispersion relation (8.45) as

$$f(k,0) = f(\infty,0) + \frac{1}{\pi} \int_{-\infty}^{\infty} dk' \frac{\text{Im}\{f(k',0)\}}{k' - k - iO}$$

$$+ \sum_{l,n} (-1)^{l+1} (2l+1) \frac{\beta_{ln}^2}{k^2 + \kappa_{ln}^2} \quad . \tag{8.50}$$

The advantage of the dispersion relation written in this form is that it determines the amplitude $f(k, 0)$ for any complex k from the upper half-plane.

If the scattering amplitude is treated as a function of energy, then the dispersion relation (8.50) takes the form

$$
f(E, 0) = f(\infty, 0) + \frac{1}{\pi} \int_0^\infty dE' \, \frac{\mathrm{Im}\,\{f(E', 0)\}}{E' - E - iO}
$$
$$
+ \sum_{l,n} (-1)^{l+1} (2l+1) \frac{\hbar^2}{2\mu} \frac{\beta_{ln}^2}{E - E_{ln}} \quad . \tag{8.51}
$$

Formula (8.51) expresses the amplitude $f(E, 0)$ at any point of the physical sheet (for positve energies) in terms of its imaginary part. We see that the dispersion relations impose stringent restrictions on the energy dependence of the scattering amplitude.

Note that generally speaking, the dispersion relations are corollaries of the causality principle. Indeed, the derivation only involved the analytic properties of the amplitude. Since analyticity of the scattering matrix, and hence the scattering amplitude, is implied by the causality principle, the latter is also responsible for the occurence of the dispersion relations.

8.6 Dispersion Relations
for the Arbitrary Direction Scattering Amplitude

If the scattering angle is not equal to zero, then the analytic properties of the scattering amplitude in the complex k plane essentially depend on how the potential decreases towards large distances. Dispersion relations of the type (8.50) or (8.51) hold only if the momentum transfer and the interaction range satisfy certain relations.

In order to derive dispersion relations for arbitrary scattering angles and the relevant applicability conditions, we first study the analytic properties of the scattering amplitude in the complex k plane for $k' \neq k$. We start from the general definition of the scattering amplitude,

$$
f(k, k') = -\frac{\mu}{2\pi\hbar^2} \int dr \, e^{-ik'r} V(r) \psi_k^{(+)}(r) \quad . \tag{8.52}
$$

If the potential $V(r)$ is centrally symmetric, then the scattering amplitude depends only on k and the scattering angle ϑ (the angle between the vectors k and k'). However, the angle variable is inappropriate for analytic continuation of the scattering amplitude into the region of complex k. A more suitable independent variable is the momentum transfer

$$
q = k - k' \quad , \tag{8.53}
$$

the absolute value of which is related to the scattering angle by

$$q = 2k \sin \frac{\vartheta}{2} = \sqrt{2k^2(1 - \cos \vartheta)} \quad . \tag{8.54}$$

In what follows, we shall regard the scattering amplitude (8.52) as a function of independent variables k and q, $f(k, k') \equiv f(k, q)$; we assume that k is complex and lies in the upper half-plane, q is real and fixed. We separate out in the amplitude (8.52) the part corresponding to the first Born approximation,

$$f_B(q) = -\frac{\mu}{2\pi\hbar^2} \int d\mathbf{r} \, e^{i\mathbf{q}\mathbf{r}} \, V(r) \quad . \tag{8.55}$$

Due to the spherical symmetry of the potential, the Born scattering amplitude is real and depends only on the magnitude of the momentum transfer q. The rest of the amplitude may be written as

$$f(k, q) - f_B(q) = -\frac{\mu}{2\pi\hbar^2} \int d\mathbf{r} \, e^{i\mathbf{q}\mathbf{r} - i\mathbf{k}\mathbf{r}} \, V(r) \left(\psi_k^{(+)}(\mathbf{r}) - e^{i\mathbf{k}\mathbf{r}} \right) . \tag{8.56}$$

[The vector k' in the right-hand part of (8.56) is expressed in terms of k and q.] Let us investigate the convergence of the integral in the right-hand part of (8.56). As long as q is fixed, we have to express the angle between the vectors k and r in the scalar product $k\mathbf{r}$ entering the exponent, in terms of the angles between each of these vectors and q. Then we have

$$k\mathbf{r} = kr \left[\cos(k, q) \cos(r, q) + \sin(k, q) \sin(r, q) \cos \varphi \right] \quad , \tag{8.57}$$

where φ is the angle between the planes containing the vectors k, q and r, q. Since

$$\cos(k, q) = \frac{1}{2} \frac{q}{k} \quad \text{and} \quad \sin(k, q) = \sqrt{1 - \frac{q^2}{4k^2}} \quad ,$$

the scalar product (8.57) may be rewritten as

$$k\mathbf{r} = \frac{1}{2} qr + r \sqrt{k^2 - \frac{q^2}{4}} \sin(r, q) \cos \varphi \quad . \tag{8.58}$$

The difference $[\psi_k^{(+)}(\mathbf{r}) - \exp(i\mathbf{k}\mathbf{r})]$ at large distances corresponds to the diverging wave. Therefore, the integrand in (8.56) for large r contains the exponential factor

$$\exp \left[\frac{i}{2} qr + i \left(k - \sqrt{k^2 - \frac{q^2}{4}} \sin(r, q) \cos \varphi \right) r \right] \quad . \tag{8.59}$$

It is not difficult to show that the real part of the exponent is not negative definite in the upper half-plane of complex k. Therefore, the integral in the right-hand part of (8.56) in general can be divergent. The requirement that the integral must converge imposes the following condition on the potential: as r grows, $V(r)$ must not decrease more slowly than the exponential. Indeed, assuming that the potential satisfies the condition

$$\int_0^\infty dr\, r^2 e^{mr}\, |V(r)| < \infty \quad , \tag{8.60}$$

where m is some positive quantity, we find that the inequality

$$\text{Im}\,\{k\} + m \geq \text{Im}\,\left\{\sqrt{k^2 - \frac{q^2}{4}}\right\}$$

occurs if

$$q \leq 2m \quad . \tag{8.61}$$

Therefore, the integral under consideration is convergent for all k except at the points for which the function $\psi_k^{(+)}(r)$ tends to infinity.

Thus, if the potential satisfies the condition (8.60), then for $q \leq 2m$ the scattering amplitude $f(k, q)$ is analytic in the upper half-plane of complex k. It possesses singularities – simple poles – located on the imaginary axis and associated with the bound states of the system. On the real axis, the scattering amplitude is continuous except for the branch point $k = \pm q/2$. As $|k| \to \infty$ along any direction, the amplitude (8.56) tends to zero.

In view of the above properties of the scattering amplitude, we choose the integration contour in the complex plane in a manner similar to the case of zero-angle scattering. Thus, we obtain the dispersion relation for the amplitude of scattering at arbitrary angles, i.e.,

$$f(k, q) = f_B(q) + \frac{1}{\pi}\int_{-\infty}^\infty dk'\, \frac{\text{Im}\,\{f(k', q)\}}{k' - k - iO}$$
$$+ \sum_{l,n}(-1)^{l+1}\,(2l+1)\,\frac{\beta_{ln}^2}{k^2 + \kappa_{ln}^2}\,P_l\left(1 + \frac{q^2}{2\kappa_{ln}^2}\right) \quad . \tag{8.62}$$

If the scattering amplitude is regarded as a function of energy and momentum transfer, the dispersion relation takes the form[1]

$$f(E, q) = f_B(q) + \frac{1}{\pi}\int_0^\infty dE'\, \frac{\text{Im}\,\{f(E', q)\}}{E' - E - iO}$$
$$+ \sum_{l,n}(-1)^{l+1}\,(2l+1)\,\frac{\hbar^2}{2\mu}\,\frac{\beta_{ln}^2}{E - E_{ln}}\,P_l\left(1 - \frac{\hbar^2 q^2}{4\mu\,E_{ln}}\right) \quad . \tag{8.63}$$

Both dispersion relations (8.62, 63) are valid under the condition $q \leq 2m$. It is clear that if the potential decreases with growing distance faster than the exponential function (say, by the Gauss law or faster), then the dispersion relations (8.62, 63) hold for an arbitrary momentum transfer q.

Note that the integrals in the right-hand parts of (8.62, 63) partially extend over the nonphysical region. Indeed, elastic scattering cannot produce particle

---

[1] The dispersion relations for the arbitrary direction scattering amplitude were derived by *Khuri* [8.2]. A simple derivation of the dispersion relations is given in [8.3].

momentum change greater than twice the initial value. At the same time, for the momentum tranfer q fixed, integration in (8.62, 63) extends over all k' and E'. The integration ranges $-q/2$ to $q/2$ in (8.62) or 0 to $\hbar^2 q^2/4\mu$ in (8.63) are obviously nonphysical. Therefore, in order for the dispersion relations (8.62, 63) to be employed in practice, we must propose the rule for the analytic continuation of the imaginary part of the scattering amplitude to the nonphysical region.

If the potential $V(r)$ satisfies the condition (8.60), then it is not difficult to show (problem 1) that the imaginary part of the scattering amplitude is an analytic function of the variable $z = \cos \vartheta$ within an ellipse in the complex z plane with focal points ± 1 and semimajor axis $(1 + 2m^2/k^2)$. A function that is analytic within an ellipse in the complex z plane can be expanded in terms of the Legendre polynomials; the series converges in the entire region of analyticity. Therefore, substituting the partial wave expansion for the scattering amplitude, we can perform analytic continuation of its imaginary part to the nonphysical region contained in the integrals (8.62, 63). For real z, the analyticity region boundary for the imaginary part of the scattering amplitude is determined by the inequality

$$q^2 < 4(m^2 + k^2) \quad . \tag{8.64}$$

Since the dispersion relations hold only for $q \leq 2m$, one can analytically continue the imaginary part of the scattering amplitude and thus define the integrands for the entire integration range.

In the case of forward direction scattering, the dispersion relations (8.62, 63) reduce to (8.50, 51) and the nonphysical regions disappear. Note also that the dispersion relations for the forward direction scattering are valid if the potential satisfies the condition (7.3). This is appreciably less stringent than (8.60) which ensures validity of (8.62, 63).

Problems

8.1 Prove that if the potential satisfies the condition (8.60), then the imaginary part of the scattering amplitude (regarded as a function of independent variables k and $z = \cos \vartheta$) is analytic in the complex z plane within the ellipse with focal points ± 1 and semimajor axis $(1 + 2m^2/k^2)$ [8.4].

Making use of (8.56) and (3.61), we can express the difference between the total and Born scattering amplitudes in terms of the exact Green function (3.66). We obtain

$$f(k, z) - f_B(k, z) = -\frac{\mu}{2\pi\hbar^2} \int d\mathbf{r} \int d\mathbf{r}' \, e^{-\mathbf{k}'\mathbf{r}} V(r)$$
$$\times G^{(+)}(E; \mathbf{r}, \mathbf{r}') V(r') e^{i\mathbf{k}\mathbf{r}'} \quad . \tag{8.65}$$

Since the Born amplitude $f_B(k, z)$ is real, the imaginary part of the scattering amplitude $f(k, z)$ is determined by the imaginary part of the expression in the right-hand part of (8.65). According to the definition (4.79), the Green function $G(E; r, r')$ depends on the absolute values and relative orientation of the vectors r and r' and hence integration over one of the angles can be carried out explicitly. We introduce the coordinate system

$$k = k(1, 0, 0) \quad ,$$

$$k' = k(\cos \vartheta, \sin \vartheta, 0) \quad ,$$

$$r = r(\cos \alpha \sin \beta, \sin \alpha \sin \beta, \cos \beta) \quad , \tag{8.66}$$

$$r' = r'(\cos \alpha' \sin \beta', \sin \alpha' \sin \beta', \cos \beta') \quad ,$$

where ϑ is the scattering angle; α and β (α' and β') are the usual spherical angles, the polar axis being perpendicular to the plane of the vectors k and k'. The Green function $G(E; r, r')$ depends on r, r', and the cosine of the angle θ between the vectors r and r' which is given by

$$\cos \theta = \cos \beta \cos \beta' + \sin \beta \sin \beta' \cos(\alpha - \alpha') \quad .$$

Hence, if we introduce new variables $\chi = \alpha - \alpha'$ and α instead of α and α', then the α-dependence of the integral I that enters (8.65), as well as the dependence on the angle ϑ, is contained in the exponent so that

$$I = \int_0^\infty dr\, r^2 \int_0^\infty dr'\, r'^2 \int_0^\pi d\beta \sin \beta \int_0^\pi d\beta' \sin \beta' \int_0^{2\pi} d\chi \int_0^{2\pi} d\alpha$$

$$\times \exp \left[-ikr \sin \beta \cos(\vartheta - \alpha) \right] V(r)\, G^{(+)}(E; r, r', \theta)\, V(r')$$

$$\times \exp \left[ikr' \sin \beta' \cos(\chi - \alpha) \right] \quad . \tag{8.67}$$

After integration over α is carried out, we have

$$I = 2\pi \int_0^\infty dr\, r^2 \int_0^\infty dr'\, r'^2 \int_0^\pi d\beta \sin \beta \int_0^\pi d\beta' \sin \beta' \int_0^{2\pi} d\chi$$

$$\times J_0 \left(k \sqrt{2rr' \sin \beta \sin \beta' (\zeta - \cos(\vartheta - \chi))} \right) \tag{8.68}$$

$$\times V(r)\, G^{(+)}(E; r, r', \theta)\, V(r') \quad ,$$

where

$$\zeta = \frac{r^2 \sin^2 \beta + r'^2 \sin^2 \beta'}{2rr' \sin \beta \sin \beta'} \quad .$$

Let us investigate the convergence of the integral (8.68). The quantity ζ takes its minimum value $\zeta = 1$ for $r = r'$ and $\beta = \beta' = \pi/2$. In the region of complex $\cos \vartheta$, the quantity $\cos(\vartheta - \chi)$ can be greater than unity. Then for large r the radical $\sqrt{1 - \cos(\vartheta - \chi)}$ becomes imaginary and the Bessel function J_0 grows exponentially, i.e.,

$$J_0 \left(kr \sqrt{2(1 - \cos(\vartheta - \chi))} \right) \underset{r \to \infty}{\sim} \exp \left[\sqrt{2}\, kr \left| \mathrm{Re}\, \{\cos(\vartheta - \chi) - 1\} \right| \right]$$

Evidently, the integral is convergent if this growth is compensated by the decrease of the integrand due to the radial dependence of the potential $V(r)$. If the potential satisfies the condition (8.60), then the requirement that the integral must be convergent imposes the condition

$$\sqrt{2}\, k \left| \mathrm{Re} \left\{ \sqrt{\cos(\vartheta - \chi) - 1} \right\} \right| < 2m \quad . \tag{8.69}$$

It is convenient to rewrite this inequality in another form. Introducing the notation $\cos \vartheta = \cosh(\xi + i\eta)$, we have

$$\cos(\vartheta - \chi) = \cosh(\xi + i\psi) \quad , \qquad \psi = \eta + \chi \quad . \tag{8.70}$$

Substituting (8.70) into (8.69), after some algebra we obtain

$$\sinh^2 \frac{\xi}{2} \cos^2 \frac{\psi}{2} < \frac{m^2}{k^2} \quad .$$

The most stringent restriction for ξ is imposed for $\psi = 0$, thus we have

$$\cosh \xi < 1 + \frac{2m^2}{k^2} \quad . \tag{8.71}$$

The condition (8.71) determines the analyticity region for the imaginary part of the scattering amplitude $\mathrm{Im}\{f(k, z)\}$ with respect to the variable $z \equiv \cos \vartheta \equiv \cosh(\xi + i\eta)$. In the complex z plane, this region is an ellipse with focal points ± 1, semimajor axis $(1 + 2m^2/k^2)$, and semiminor axis $(2m/k)(1 + m^2/k^2)^{1/2}$.

8.2 Find the high energy elastic scattering amplitude making use of the analyticity and unitarity conditions [8.5].

Suppose the energy dispersion relation (8.63) for the scattering amplitude is valid for all values of the momentum transfer. For the sake of simplicity, we also assume that the bound states do not occur. The Fourier–Bessel transformation of the amplitude $f(k, q)$ yields

$$f(k, q) = \int_0^\infty db \, bh(k, b) \, j_0(qb) \quad . \tag{8.72}$$

Multiplying this relation by $q J_0(qb)$ and integrating over q from 0 to ∞, we obtain

$$h(k, b) = h_B(b) + \frac{1}{\pi} \int_{-\infty}^{\infty} dk' \, \frac{\mathrm{Im}\{h(k', b)\}}{k' - k - i0} \quad ,$$

$$h_B(b) = -\int_0^\infty dz \, v\left(\sqrt{b^2 + z^2}\right) \quad . \tag{8.73}$$

For b fixed, the function $h(k, b)$ is analytic in the upper half-plane of k. To transform the dispersion relation (8.73) into an equation, we employ the unitarity condition (5.9) and express Im $\{h(k, b)\}$ in terms of $h(k, b)$. Then after some algebra we have

$$
\text{Im} \{h(k, b)\} = \int_0^\infty db' \, b'
$$

$$
\times \int_0^\infty db'' \, b'' \, h^*(k, b') \, h(k, b'') \, \mathcal{Y}(k, b; b', b'') \quad , \tag{8.74}
$$

where

$$
\mathcal{Y}(k, b; b', b'') = \frac{4k^2}{\pi} \int_0^\pi d\varphi \int_0^\pi d\vartheta \, \sin \vartheta \, J_0 \left(2kb \, \sin \frac{\vartheta}{2} \right)
$$

$$
\times \frac{J_1 \left[2k \, \beta(\vartheta, \varphi) \right]}{\beta(\vartheta, \varphi)} \quad , \tag{8.75}
$$

$$
\beta^2(\vartheta, \varphi) = b'^2 + b''^2 - 2b'b'' \cos \frac{\vartheta}{2} \cos \varphi \quad .
$$

For sufficiently high energies, the expression for $\mathcal{Y}(k, b; b', b'')$ is simplified to

$$
\mathcal{Y}(k, b; b', b'') \simeq \frac{1}{2kb^2} \, \delta(b - b') \, \delta(b - b'') \quad , \tag{8.76}
$$

and the unitarity condition (8.74) takes the form

$$
\text{Im} \{h(k, b)\} = \frac{1}{2k} \, |h(k, b)|^2 \quad . \tag{8.77}
$$

Substituting (8.75) into (8.73) yields an equation for the amplitude $h(k, b)$. If we employ, for all energies, the unitarity condition (8.77) which is valid only for high energies, then (8.73) yields a simple equation for $h(k, b)$. However, instead of solving this equation, we carry out the following procedure. We introduce the function

$$
L(k, b) \equiv \frac{h_B(b)}{h(k, b)} \tag{8.78}
$$

which is obviously analytic in the upper half-plane of k and tends to unity as $|k| \to \infty$ [$h(k, b)$ is assumed to have no zeros]. The explicit analyticity condition for this function is

$$
L(k, b) = 1 + \frac{1}{\pi} \int_{-\infty}^\infty dk' \, \frac{\text{Im} \{L(k', b)\}}{k' - k - iO} \quad , \tag{8.79}
$$

where

$$
\text{Im} \{L(k, b)\} = -\frac{h_B(b)}{|h(k, b)|^2} \, \text{Im} \{h(k, b)\} \quad . \tag{8.80}
$$

Making use of the unitarity condition (8.77), we obtain

$$
\text{Im} \{L(k, b)\} = -\frac{1}{2k} \, h_B(b) \quad ; \tag{8.81}
$$

substituting this relation into (8.79) we find that

$$L(k, b) = 1 + \frac{1}{2ik} h_B(b) \quad , \tag{8.82}$$

and then

$$h(k, b) = \frac{h_B(b)}{1 + (1/2ik) h_B(b)} \quad . \tag{8.83}$$

The function $h(k, b)$ thus obtained, possesses the required analytic properties and satisfies the high energy unitarity condition (8.77).

The scattering amplitude in the approximation under consideration is given by

$$f(k, q) = \int_0^\infty db\, b J_0(qb) \frac{h_B(b)}{1 + (1/2ik) h_B(b)} \quad ,$$

or

$$f(k, q) = ik \int_0^\infty db\, b J_0(qb) \left\{ 1 - \exp\left[2i\, \delta(k, b)\right] \right\} \quad , \tag{8.84}$$

where

$$\delta(k, b) = \arctan\left(\frac{1}{2k} h_B(b)\right)$$

$$\equiv -\arctan\left(\frac{1}{2k} \int_0^\infty dz\, v\left(\sqrt{b + z^2}\right)\right) \quad . \tag{8.85}$$

To pass to the high energy approximation considered in Sect. 3.8, we have to substitute the first term of the arctangent expansion for the function $\delta(k, b)$.

It is of interest to note that the amplitude $h(k, b)$ and the partial amplitude $f_l(k)$ are related. Since at high energies small scattering angles and large l are the most important, it is not difficult to derive the approximate relation

$$f_l(k) \simeq \frac{1}{2k^2} h\left(k, \frac{l}{k}\right) \quad . \tag{8.86}$$

It is clear that in the limiting case of large l and high energies, the quantity b becomes the classical impact parameter.

9. Complex Angular Momenta

9.1 Analytic Properties of the Scattering Matrix in the Complex Angular Momentum Plane

In the previous chapters we studied the important properties of the scattering matrix associated with its singularities in the plane of complex k. Since the scattering matrix $S_l(k)$ depends, along with the wave vector k, on the angular momentum quantum number l, the singularities of the scattering matrix can occur not only for complex k and integer physical values of l, but also for physical values of k and complex l [9.1].

To consider the analytic properties of the scattering matrix in the complex moment plane, we introduce instead of l a quantity λ, defined by

$$\lambda = l + \tfrac{1}{2} \quad . \tag{9.1}$$

Then equation (7.42) for the radial function may be rewritten as

$$u''(\lambda, k, r) + \left\{ k^2 - \frac{\lambda^2 - 1/4}{r^2} - V_r(r) \right\} u(\lambda, k, r) = 0 \quad . \tag{9.2}$$

Let $\Phi(\lambda, k, r)$ be the solution of (9.2) regular at the point $r = 0$. It satisfies the boundary condition

$$\lim_{r \to 0} r^{-[\lambda + (1/2)]} \Phi(\lambda, k, r) = 1 \quad . \tag{9.3}$$

For the second independent solution of (9.2), we take $\Phi(-\lambda, k, r)$; it satisfies the condition

$$\lim_{r \to 0} r^{[\lambda - (1/2)]} \Phi(-\lambda, k, r) = 1 \quad . \tag{9.4}$$

It is not difficult to verify that

$$W\left\{ \Phi(\lambda, k, r), \Phi(-\lambda, k, r) \right\} = -2\lambda \quad . \tag{9.5}$$

We denote the Jost solutions of (9.2) by $f(\lambda, k, r)$ and $f(\lambda, -k, r)$; the boundary conditions for these solutions are

$$\lim_{r \to \infty} e^{ikr} f(\lambda, k, r) = 1 \quad . \tag{9.6}$$

In what follows we assume that the quantity λ takes arbitrary complex values and consider the behavior of various solutions of (9.2) in the complex λ plane. Since the coefficient in (9.2) is an entire function of λ, and the boundary condition (9.6) does not depend on λ, $f(\lambda, k, r)$ must be an entire function of λ. Note that for real λ, $f(\lambda, k, r)$ is an even function of λ. If the potential is real, then (9.2, 6) yield

$$f(\lambda, k, r) = f^*(\lambda^*, -k^*, r) \quad . \tag{9.7}$$

It is not difficult to show that the regular solution $\Phi(\lambda, k, r)$ is analytic in the right half-plane Re $\{\lambda\} > 0$. To do this, we introduce, instead of r and u, the new variables ϱ and w:

$$r = r_0 e^{-\varrho} \quad , \qquad u = w e^{-\varrho/2} \tag{9.8}$$

(r_0 is a constant of length dimensionality). Then (9.2, 3) reduce to

$$\frac{d^2 w}{d\varrho^2} - \left\{ \lambda^2 + \left[V_r(e^{-\varrho}) - k^2 \right] r_0^2 e^{-2\varrho} \right\} w = 0 \quad , \tag{9.9}$$

$$\lim_{\varrho \to \infty} e^{\lambda \varrho} w(\varrho) = 1 \quad . \tag{9.10}$$

Comparing (9.9) to (9.2) and (9.10) to (9.6), we find that

$$w(\varrho) = f(\tfrac{1}{2}, -i\lambda, \varrho) \tag{9.11}$$

if the quantity $\left[V_r(e^{-\varrho}) - k^2 \right] r_0^2 e^{-2\varrho}$ is taken as the potential. From this analysis follows that the solution (9.11) is an analytic (regular) function for Im $\{-i\lambda\} < 0$ (i.e., for Re $\{\lambda\} > 0$) and therefore, the solution $\Phi(\lambda, k, r)$ is an analytic (regular) function of λ for Re $\{\lambda\} > 0$.

Thus we have shown that the domains of existence of the solutions $\Phi(\lambda, k, r)$ and $f(\lambda, k, r)$ are determined by the conditions

$$\pm\text{Re}\,\{\lambda\} > 0 \quad , \quad k \text{ is finite} \quad \text{for} \quad \Phi(\pm\lambda, k, r) \quad ,$$

$$\lambda \text{ is finite} \quad , \quad \pm\text{Im}\,\{k\} < 0 \quad \text{for} \quad f(\lambda, \pm k, r) \quad .$$

To extend these domains, we must impose some restrictions on the potential $V_r(r)$.

We introduce the Jost function

$$f(\lambda, k) \equiv W\{f(\lambda, k, r), \, \Phi(\lambda, k, r)\} \quad , \tag{9.12}$$

then the regular solution of (9.2) can be expressed in terms of the Jost solutions, so that

$$\Phi(\lambda, k, r) = -\frac{1}{2ik} \left[f(\lambda, -k) f(\lambda, k, r) - f(\lambda, k) f(\lambda, -k, r) \right] \quad . \tag{9.13}$$

Note that the functions $f(\pm\lambda, \pm k)$ completely determine the relationship between the solutions associated with the boundary conditions at $r = 0$ and for $r \to \infty$.

Since $f(\lambda, k, r)$ and $f'(\lambda, k, r)$ are entire functions of λ and $\Phi(\lambda, k, r)$ and $\Phi'(\lambda, k, r)$ are analytic (regular) functions of λ for $\operatorname{Re}\{\lambda\} > 0$, the function $f(\lambda, k)$ is analytic (regular) for $\operatorname{Re}\{\lambda\} > 0$ and fixed k. Earlier we have shown that $f(\lambda, k)$ is an analytic (regular) function of k in the half-plane $\operatorname{Im}\{k\} < 0$ for fixed λ. Therefore, being treated as a function of two variables λ and k, the function $f(\lambda, k)$ is analytic and has no singularities in the direct product of the domains $\operatorname{Im}\{k\} < 0$ and $\operatorname{Re}\{\lambda\} > 0$.

Comparing the asymptotics of (9.13) to (4.24), we obtain

$$S(\lambda, k) = \exp\left[i\pi\left(\lambda - \frac{1}{2}\right)\right] \frac{f(\lambda, k)}{f(\lambda, -k)} \quad . \tag{9.14}$$

The scattering matrix $S(\lambda, k)$ is a meromorphic function of λ and k in the direct product of the domains $\operatorname{Re}\{\lambda\} > 0$ and $\operatorname{Im}\{k\} > 0$. The relation (9.7) yields the complex unitarity condition for the scattering matrix,

$$S^*(\lambda^*, k^*) = S^{-1}(\lambda, k) \quad . \tag{9.15}$$

This condition is satisfied if the potential $V_r(r)$ is real for real positive values of r.

The complex unitarity condition (9.15) imposes certain restrictions on the dislocation of the scattering matrix poles in the λ plane. If k is real, then (9.15) takes the form

$$S^*(\lambda^*, k) = S^{-1}(\lambda, k) \tag{9.16}$$

and we immediately find that if the function $S(\lambda, k)$ has a pole at the point $\lambda = \lambda_0$ for real k, then it also has a zero at the point $\lambda = \lambda_0^*$. If λ is real, too, then

$$S(\lambda, k) = \exp\left[2i\,\delta(\lambda, k)\right] \quad , \tag{9.17}$$

where $\delta(\lambda, k)$ is the real scattering phase shift. Equation (9.17) also holds for complex λ, however, the phase shift $\delta(\lambda, k)$ is then complex.

Let us express the Jost functions $f(\lambda, k)$ and $f(\lambda, -k)$ in terms of the phase shift $\delta(\lambda, k)$. To do this, we note that $f(\lambda, k)$ can always be written as

$$f(\lambda, k) = \exp\left[i\theta(\lambda, k)\right] \quad ,$$

where $\theta(\lambda, k)$ is some phase shift which is in general complex for complex λ. Obviously, the function $f(\lambda, -k)$ may be written as

$$f(\lambda, -k) = \exp\left[i\theta(\lambda, -k)\right] \quad .$$

The relation between $\theta(\lambda, k)$ and $\delta(\lambda, k)$ follows from (9.14), we have

$$\theta(\lambda, k) - \theta(\lambda, -k) + \pi\left(\lambda - \tfrac{1}{2}\right) = 2\delta(\lambda, k) \quad . \tag{9.18}$$

For k real, it is convenient to divide the complex phase shift $\theta(\lambda, k)$ into two parts $\tilde{\Delta}(\lambda, k)$ and $\Delta(\lambda, k)$, so that

$$\theta(\lambda, k) = \tilde{A}(\lambda, k) + \Delta(\lambda, k) \quad ,$$

where $\tilde{A}(\lambda, k)$ is an even function of k, $\Delta(\lambda, k)$ is an odd function of k. Relation (9.18) immediately determines the odd part of the phase shift $\theta(\lambda, k)$:

$$\Delta(\lambda, k) = \delta(\lambda, k) - \frac{\pi}{2}\left(\lambda - \frac{1}{2}\right) \quad .$$

Thus we find that the relations between the Jost functions $f(\lambda, k)$ and $f(\lambda, -k)$, and the phase shift $\delta(\lambda, k)$ are given by

$$f(\lambda, k) = \tau(\lambda, k) \, \exp\left[i\delta(\lambda, k) - i\frac{\pi}{2}\left(\lambda - \frac{1}{2}\right)\right] \quad ,$$

(9.19)

$$f(\lambda, -k) = \tau(\lambda, k) \, \exp\left[-i\delta(\lambda, k) + i\frac{\pi}{2}\left(\lambda - \frac{1}{2}\right)\right] \quad ,$$

where $\tau(\lambda, k) = \exp\left[i\tilde{A}(\lambda, k)\right]$ for complex λ is complex. Note that for complex λ the complex scattering phase shift $\delta(\lambda, k)$ contains both even and odd parts with respect to k. By using (9.18), the odd part of the scattering phase shift $\delta(\lambda, k)$ can be shown to coincide with $\Delta(\lambda, k)$, and the even part of $\delta(\lambda, k)$ to be equal to $\frac{\pi}{2}\left(\lambda - \frac{1}{2}\right)$. If λ is real, then the phase shift $\delta(\lambda, k)$ is real and $\tau(\lambda, k) = |f(\lambda, k)|$.

Let us show that the scattering matrix has no poles if $\text{Im}\{\lambda\} < 0$ and k is real positive. We multiply (9.2) by u^* and subtract the complex conjugate from the equation obtained. The result is

$$\frac{d}{dr}\left(u^*u' - uu'^*\right) = 2i \, \text{Im}\{\lambda^2\} \frac{|u|^2}{r^2} \quad . \tag{9.20}$$

Taking u to be the regular solution Φ and integrating (9.20) over r from 0 to ∞ accounting for (9.13, 19), we find that

$$2 \, \text{Im}\{\lambda^2\} \int_0^\infty dr \, \frac{|\Phi|^2}{r^2} = \frac{1}{2k}\left[|f(\lambda, k)|^2 - |f(\lambda, -k)|^2\right]$$

$$= \frac{1}{2k} |\tau(\lambda, k)|^2 \left\{\exp\left[-2\left(\text{Im}\{\delta\} - \frac{\pi}{2}\text{Im}\{\lambda\}\right)\right]\right.$$

$$\left. - \exp\left[2\left(\text{Im}\{\delta\} - \frac{\pi}{2}\text{Im}\{\lambda\}\right)\right]\right\} \quad . \tag{9.21}$$

As follows from (9.21), $\text{Im}\{\delta\} > (\pi/2)\text{Im}\{\lambda\}$ if $\text{Im}\{\lambda\} < 0$. Therefore, the imaginary part of δ is limited from below for $\text{Im}\{\lambda\} < 0$ and hence the quantity $S(\lambda, k) = \exp\left[2i\,\delta(\lambda, k)\right]$ is limited from above. Thus, the scattering matrix has no poles in the domain $\text{Im}\{\lambda\} < 0$. In the same manner, $S(\lambda, k)$ can be shown to have no zeros in the domain $\text{Im}\{\lambda\} > 0$ by virtue of the relation (9.16).

Let us clarify the physical meaning of the fact that $S(\lambda, k)$ has no poles for $\text{Im}\{\lambda\} < 0$. First of all we note that for complex λ the centrifugal potential $\left(\lambda^2 - \frac{1}{4}\right)r^{-2}$ is also complex. The sign of the imaginary part of λ^2 determines absorptive $(\text{Im}\{\lambda\} < 0)$ or emissive $(\text{Im}\{\lambda\} > 0)$ nature of the potential. Suppose Φ is a regular solution of (9.2) associated with a pole. Then $f(\lambda, -k) = 0$ and the asymptotic is given by

$$\Phi \rightarrow \frac{f(\lambda, k)}{2ik} e^{ikr} \quad , \qquad r \rightarrow \infty \quad .$$

Making use of this expression we can verify that for $k > 0$ the probability flux density is positive at large distances, i.e., the probability flux is directed outwards. Since the wave function Φ is regular and describes a stationary state (E is real), the probability losses can be compensated only at the expense of some source, say, a centrifugal potential corresponding to complex angular momentum with $\mathrm{Im}\{\lambda\} > 0$. For $\mathrm{Im}\{\lambda\} < 0$, compensation is impossible, hence no poles occur in the domain under consideration.

The above mentioned restrictions imposed on the dislocation of the scattering matrix poles in the complex λ plane follow only from complex unitarity of the scattering matrix. If we consider the law of potential decrease at large distances, further restrictions for the pole positions can be obtained. For example, for the Yukawa potential which satisfies the condition $|V_r(iy)| < P/y$, the poles can lie only in the region $\mathrm{Re}\{\lambda\} < P/k$, i.e., a line exists in the λ plane, to the right of which no poles occur.

9.2 Poles of the Scattering Matrix
in the Complex Angular Momentum Plane

Let us reveal the physical meaning of the scattering matrix poles in the plane of complex angular momentum λ for physical, i.e., real energies E, which usually are referred to as the Regge poles. According to (9.14), the Regge pole is determined by the condition

$$f(\lambda, -k) = 0 \quad . \tag{9.22}$$

Suppose $\lambda(k)$ is a solution of (9.22).[1] It is clear that (9.22) is satisfied for

$$\lambda = \lambda(k) \quad .$$

The quantity $\lambda(k)$ determines the position of the Regge pole and is an analytic function of k. If the energy E is fixed, a number of poles $\lambda_1(k)$, $\lambda_2(k)$, ... can occur. If energy varies continuously, then each point $\lambda_i(k)$ moves along some curve in the complex λ plane. These curves are referred to as the Regge trajectories. We introduce the notation

$$\lambda(k) = \alpha(k) + \tfrac{1}{2} \quad . \tag{9.23}$$

---

[1] It is not difficult to show that if $f(\lambda, -k)$ is analytic with respect to both arguments in a limited domain and $\partial f / \partial \lambda \neq 0$ in the general case, then the equation $f(\lambda, -k) = 0$ has a unique solution $\lambda = \lambda(k)$ which is an analytic function of k and takes definite values $\bar{\lambda}$ at given points $\bar{k}$, so that $f(\bar{\lambda}, -\bar{k}) = 0$.

Obviously, the physical values of the quantity $\alpha(k)$ are positive integers associated with the values of the angular momentum quantum number l of the system.

Note that the difference between the Regge poles and the ones considered in Chap. 7 is purely formal. There we considered the poles with respect to the variable k (or E) for fixed quantum number l. Now we fix the variable E and study the poles with respect to λ (or l). In both cases, the pole is determined by λ and k, for which $f(\lambda, -k) = 0$.

The poles of the scattering matrix $S(\lambda, k)$ in the plane of complex $\lambda(k) = \alpha(k) + \frac{1}{2}$ describe either bound states or resonances in the system. If the function $\alpha(k)$ is real and equal to the positive integer l, then a bound state occurs. If $\alpha(k)$ is complex, the real part being close to a positive integer l and the imaginary part being small, $\text{Im}\,\{\alpha(k)\} \ll 1$, then the Regge pole is associated with a resonance.

Let us consider the properties of the Regge trajectory which occur as E varies from $-\infty$ to ∞. We assume that the trajectory under consideration does not intersect other trajectories.

If energy is negative, $E > 0$, then the wave vector is imaginary, $k = i\kappa$. We coonsider the values of E on the physical sheet, in this case $\kappa > 0$. As long as the regular solution (9.13) is real, the function $f(\lambda, -k)$ is real, too. Therefore, the equation $f(\lambda, -k) = 0$ has either real or complex conjugate solutions.

Let us show that for $E < 0$ solutions of (9.22) on the physical sheet are real. Within the context of (9.22), the asymptotic of the regular solution (9.13) is given by

$$\Phi_{\alpha(k)}(k, r) \xrightarrow[r \to \infty]{} -\frac{1}{2\kappa}\, f_{\alpha(k)}(i\kappa)\, e^{-\kappa r} \quad . \tag{9.24}$$

Since Φ vanishes at $r = 0$ and for $r \to \infty$, we obtain from (9.20)

$$\text{Im}\,\{\alpha(k)\,(\alpha(k) + 1)\} \int_0^\infty dr\, \frac{|\Phi|^2}{r^2} = 0 \quad , \tag{9.25}$$

and therefore, $\alpha(k)$ is real. The convergence of the integral is ensured because $\alpha(k) > -\frac{1}{2}$. The quantity $\alpha(k)$ can take physical values, i.e., be equal to the integer l,

$$\alpha(k) = l \quad . \tag{9.26}$$

This condition yields $k_l = i\,\kappa_l$ and hence

$$E_l = -\frac{\hbar^2 \kappa_l^2}{2\mu} \quad . \tag{9.27}$$

Since the wave function Φ is square integrable, the energy E for which the Regge trajectory goes through a physical point l is associated with a bound state.

It is not difficult to show that $\alpha(k)$ is an ascending function of k^2. To do this, we differentiate (9.2) with respect to k^2 and multiply by Φ, then subtract the result obtained from (9.2) multiplied by $\partial \Phi/\partial k^2$, to obtain

$$\frac{\partial}{\partial r}\left(\Phi'\frac{\partial\Phi}{\partial k^2}-\Phi\frac{\partial\Phi'}{\partial k^2}\right)=\left\{1-\frac{1}{r^2}\frac{\partial}{\partial k^2}\,\alpha(k)\,[\alpha(k)+1]\right\}\Phi^2\quad. \tag{9.28}$$

Integrating this equality over r from 0 to ∞ and bearing in mind that the function Φ vanishes at $r=0$ and for $r\to\infty$, we find that

$$\frac{\partial}{\partial k^2}\,\alpha(k)\,[\alpha(k)+1]=\frac{\int_0^\infty dr\,\Phi^2}{\int_0^\infty dr\,(\Phi^2/r^2)}\quad. \tag{9.29}$$

Since Φ is real, this result implies that the derivative $\partial\alpha(k)/\partial k^2$ is positive, i.e., $\alpha(k)$ is an ascending function. As energy E increases, the Regge pole $\alpha(k)$ moves to the right along the real axis in the complex l plane. It is clear that several bound states with discrete energies $E_l < 0$, up to the threshold value $E = 0$, can lie on the same Regge trajectory. For energies above the threshold (i.e. positive), the Regge trajectory abandons the real axis and no new bound states appear because it does not go through the points associated with integers. Note that since $\alpha(k)$ is an ascending function, the quantity $\alpha(0)$ determines the maximum angular momentum for the bound states associated with a given Regge trajectory.

Let us consider the Regge poles in the range of positive energies $E > 0$. In this case the function $f(\lambda, -k)$ possesses an imaginary part, hence $\alpha(k)$ acquires an imaginary part, too, so that

$$\alpha(k)=\mathrm{Re}\left\{\alpha(k)\right\}+i\,\mathrm{Im}\left\{\alpha(k)\right\}\quad. \tag{9.30}$$

Let us show that for $E > 0$ the Regge pole $\alpha(k)$ with the real part close to a positive integer l and sufficiently small imaginary part describes a resonance. Suppose

$$\mathrm{Re}\left\{\alpha(k)\right\}\simeq l\quad,\qquad\mathrm{Im}\left\{\alpha(k)\right\}\ll 1\quad, \tag{9.31}$$

and, more precisely, $k > 0$. We emphasize that the Regge pole (9.31) corresponds to a physical value of energy and a complex value of the angular momentum.

We pass to the complex k plane and choose some complex k_0 for which

$$\mathrm{Re}\left\{\alpha(k_0)\right\}=l\quad,\qquad\mathrm{Im}\left\{\alpha(k_0)\right\}=0\quad. \tag{9.32}$$

Since $\alpha(k)$ is an analytic function, we can expand it in a power series of the difference $(k-k_0)$ and retain the linear term. Then we have

$$\mathrm{Re}\left\{\alpha(k)\right\}+i\,\mathrm{Im}\left\{\alpha(k)\right\}=l+(k-k_0)\left.\frac{d\,\alpha(k)}{dk}\right|_{k_0}+\dots$$

and thus

$$k_0=k_0'+ik_0''\quad,$$

$$k_0'=k+\frac{l-\mathrm{Re}\left\{\alpha(k)\right\}}{(d\,\alpha(k)/dk)\big|_{k_0}}\quad,\qquad k_0''=-\frac{\mathrm{Im}\left\{\alpha(k)\right\}}{(d\,\alpha(k)/dk)\big|_{k_0}}\quad. \tag{9.33}$$

Since $\text{Im}\{\alpha(k)\} > 0$ and $d\,\alpha(k)/dk > 0$, we have $k_0'' < 0$. The complex value of k_0 corresponds to the complex energy

$$W = E_0 - \frac{i}{2}\,\Gamma_0 \ ,$$

$$E_0 = \frac{\hbar^2}{2\mu}\left(k_0'^2 - k_0''^2\right) \ , \qquad \Gamma_0 = \frac{2k_0'}{\mu}\,\frac{\text{Im}\{\alpha(k)\}}{(d\,\alpha(k)/dk)\big|_{k_0}} \ . \qquad (9.34)$$

Thus, we have found the scattering matrix pole corresponding to the physical value of the angular momentum (9.32), but the energy (9.34) is complex. The wave function at the point k_0 in the complex k plane is given by

$$\Phi \to \frac{f_l(k_0)}{2ik_0}\,\exp\left[ik_0'r - k_0''r\right] \qquad (r \to \infty) \qquad (9.35)$$

and describes an unbound state.

So, we have shown that the scattering matrix pole (9.31) in the complex l plane, associated with physical energy, can correspond to the pole (9.33) in the complex k plane, associated with physical angular momentum. Therefore, if the Regge trajectory approaches positive integer l, then a resonance appears in the system. If the function $\alpha(k)$ comes close to several integers l, the same Regge trajectory can go through several resonances.

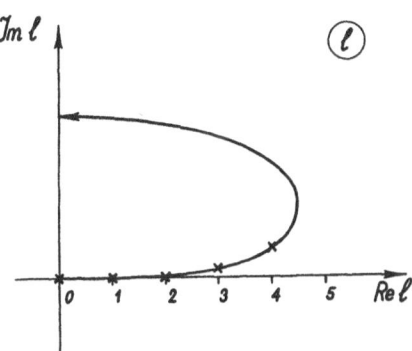

Fig. 9.1. Regge trajectory for a potential exponentially decreasing towards infinity

In the general case, both bound states and resonances can lie on the same Regge trajectory. Together they form a family of states with equal intrinsic quantum numbers and may be classified according to their Regge trajectory.

Figure 9.1 shows a typical Regge trajectory for a potential that decreases with growing distance by the exponential law. This trajectory determines three bound states with $l = 0, 1$, and 2, abandons the real axis in the interval between the points $l = 2$ and $l = 3$, describes two resonances with $l = 3$ and $l = 4$, and then ascends and turns to the left. If E is large, the behavior of the trajectory depends on the form of the potential.

To illustrate the relationship between the scattering matrix poles in the complex l plane and the bound states, we consider the example of a Coulomb field. According to (4.57), the scattering matrix for the Coulomb potential is given by

$$S_l(k) = \frac{\Gamma(l+1+i\xi)}{\Gamma(l+1-i\xi)} \quad , \qquad \xi = \frac{Ze^2\mu}{\hbar^2 k} \quad . \tag{9.36}$$

The poles of the function (9.36) in the complex l plane correspond to the values of $\alpha(k)$ for which the argument of the gamma function $\Gamma\left[\alpha(k)+1+i\xi\right]$ is equal to a negative integer or zero, i.e., when

$$\alpha(k) + 1 + i\xi = -n_r \quad , \tag{9.37}$$

where $n_r = 0, 1, 2, \ldots$. The number n_r determines the pole of the scattering matrix in the complex l plane and the corresponding Regge trajectory.

If $a(k)$ takes physical values $0, 1, 2, \ldots$, then the Regge poles correspond to bound states. Their energies lie on the Regge trajectory (9.37) and are given by

$$E_{n_r l} = -\frac{Ze^4\mu}{2\hbar^2} \frac{1}{\left(n_r + l + 1\right)^2} \quad . \tag{9.38}$$

Thus each Regge trajectory determined by the number n_r is associated with an inifinite number of bound states with different angular momentum quantum numbers l. Figure 9.2 shows the Regge trajectories $n_r = 0, 1$, and 2 for the Coulomb field.

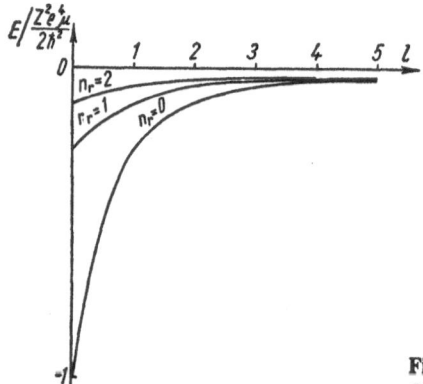

Fig. 9.2. Regge trajectories $n_r = 0, 1$, and 2 for a Coulomb field

Equation (9.38) describes the energy levels of the hydrogen-like atom. The number n_r coincides with the number of nodes of the radial wave function. We see that the two classifications of the hydrogen-like atom bound states — by the number of nodes of the radial wave function and by the Regge trajectories — are equivalent. Each trajectory contains an infinite number of bound states with the same radial quantum number n_r (the number of nodes of the radial wave function) and different angular momentum quantum numbers l.

9.3 Analytic Properties of the Scattering Amplitude in the Complex z Plane

The complex angular momentum approach enables one to consider the general analytic properties of the scattering amplitude in the plane of complex cosine of the scattering angle $z = \cos \vartheta$. Expanding the scattering amplitude in terms of the Legendre polynomials,

$$f(k, z) = \frac{1}{2ik} \sum_l (2l+1) \left\{ S_l(k) - 1 \right\} P_l(z) \quad , \tag{9.39}$$

is convenient in the range of physical scattering angles $-1 \leq z \leq 1$. However, this expansion is not valid for nonphysical z since the series (9.39) is convergent only within a limited region of the complex z plane. To ensure an adequate analytic continuation of the scattering amplitude to the entire complex z plane, we introduce the Watson-Sommerfeld transformation.

Let us consider the function $(\sin l\pi)^{-1}$ which has an infinite number of poles corresponding to each integer l; the corresponding residues are $(-1)^l/\pi$. Having encircled a fixed point l_0 in the complex l plane by a closed contour C_0, we find that according to the Cauchy formula,

$$\frac{1}{2i} \int_{C_0} dl \frac{(-1)^l}{\sin l\pi} = 1 \quad . \tag{9.40}$$

Hence, the sum (9.39) can be reduced to the integral

$$f(k, z) = \frac{1}{4k} \int_C dl(2l+1) \left\{ S_l(k) - 1 \right\} P_l(-z) \frac{1}{\sin l\pi} \tag{9.41}$$

over the contour C that encircles the positive real semi-axis. All physical values $l = 0, 1, 2, \ldots$ lie within the contour (Fig. 9.3).

The integral representation (9.41) of the scattering amplitude is suitable for the study of analytic properties of the scattering amplitude regarded as a function of z.

Let us consider a potential $V(r)$ which satisfies the condition (8.60) and can be written as a superposition of the Yukawa potentials, i.e.,

$$V(r) = \int_m^\infty d\mu\, \sigma(\mu) \frac{e^{-\mu r}}{r} \quad ; \tag{9.42}$$

the function $\sigma(\mu)$ is restricted with the exception of a finite number of singularities. Substituting $\sigma(\mu) = V_0\, \sigma(\mu - \lambda)$ into (9.42), we obtain the usual Yukawa potential

$$V(r) = V_0 \frac{e^{-\lambda r}}{r} \quad . \tag{9.43}$$

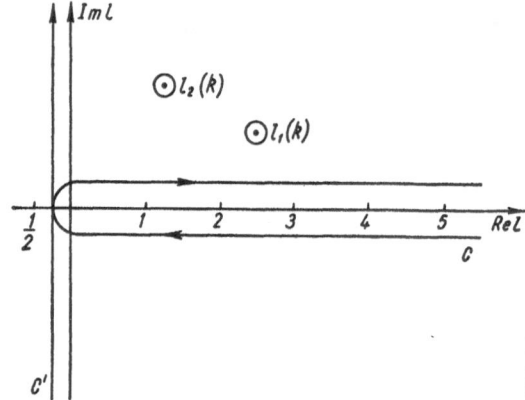

Fig. 9.3. Integration contour in the complex angular momentum plane for the Watson–Sommerfeld transformation

First of all we analyze the convergence of the integral (9.41) for large l. The scattering matrix $S_l(k)$ for $|l| \gg 1$ is majorized by the Born approximation, the difference for $|l| \to \infty$ can be as small as is required irrespective of k, i.e.,

$$S_l(k) - 1 \simeq -\frac{ik}{2\pi} \int_0^\infty dr\, r^2\, V_r(r)\, j_l^2(kr) \quad . \tag{9.44}$$

Substituting (9.42) for the potential $V_r(r)$ into (9.44) and bearing in mind that

$$\int_0^\infty dr\, r\, e^{-\mu r}\, j_l^2(kr) = \frac{1}{2k^2}\, Q_l\!\left(1 + \frac{\mu^2}{2k^2}\right) \quad ,$$

where $Q_l(z)$ is a Legendre function of the second-kind, we find that

$$S_l(k) - 1 \simeq -\frac{i}{4\pi k} \int_m^\infty d\mu\, \sigma(\mu)\, Q_l\!\left(1 + \frac{\mu^2}{2k^2}\right) \quad . \tag{9.45}$$

In order to derive the asymptotics for large z of the Legendre functions of the first- and second-kind $P_l(z)$ and $Q_l(z)$, we employ the following representations:

$$P_l(z) = \frac{1}{\sqrt{2\pi}} \frac{\Gamma(l+1)}{\Gamma(l+(3/2))} \left(z^2 - 1\right)^{-1/4}$$
$$\times \left\{ \left(z + \sqrt{z^2-1}\right)^{l+1/2} F\!\left(\frac{1}{2}, \frac{1}{2}, l+\frac{3}{2}; \frac{z+\sqrt{z^2-1}}{2\sqrt{z^2-1}}\right) \right.$$
$$\left. + i\left(z - \sqrt{z^2-1}\right)^{l+1/2} F\!\left(\frac{1}{2}, \frac{1}{2}, l+\frac{3}{2}; -\frac{z-\sqrt{z^2-1}}{2\sqrt{z^2-1}}\right) \right\} \quad , \tag{9.46}$$

$$Q_l(z) = \sqrt{\frac{\pi}{2}} \frac{\Gamma(l+1)}{\Gamma(l+(3/2))} \left(z^2 - 1\right)^{-1/4} \left(z - \sqrt{z^2-1}\right)^{l+(1/2)}$$
$$\times F\!\left(\frac{1}{2}, \frac{1}{2}, l+\frac{3}{2}; -\frac{z-\sqrt{z^2-1}}{2\sqrt{z^2-1}}\right) \quad , \tag{9.47}$$

where $F(a, b, c; x)$ is the hypergeometric function. We note that $F(a, b, c; x) \to 1$ as $\mathrm{Re}\,\{c\} \to \infty$, and $\lim_{|l| \to \infty} [\Gamma(l + 1)/\Gamma(l + \tfrac{3}{2})] \to l^{-1/2}$. Then for fixed $z = \cos\vartheta$ and $|l| \to \infty$ $(\mathrm{Re}\,\{l\} \geq 0)$ (9.46, 47) reduce to the asymptotic formulas

$$P_l(z) \to \frac{1}{\sqrt{2\pi l}} \frac{1}{\sqrt{\sin\vartheta}} \left\{ \exp\left[i\left(l + \frac{1}{2}\right)\vartheta\right] + i \exp\left[-i\left(l + \frac{1}{2}\right)\vartheta\right] \right\}$$
$$\times \exp\left(-i\frac{\pi}{4}\right) \quad , \tag{9.48}$$

$$Q_l(z) \to \sqrt{\frac{\pi}{2l}} \frac{1}{\sqrt{\sin\vartheta}} \exp\left[-i\left(l + \frac{1}{2}\right)\vartheta - i\frac{\pi}{4}\right] \quad . \tag{9.49}$$

Having substituted the asymptotics (9.48, 49) into (9.41, 45) one can easily verify that the integrand in (9.41) is smaller than

$$N\,|l|^{1/2} \exp\left[-|\mathrm{Re}\,\{\vartheta\}|\,\mathrm{Im}\,\{l\}| + (|\mathrm{Im}\,\{\vartheta\}| - \alpha)\,\mathrm{Re}\,\{l\}\right] \quad , \tag{9.50}$$

where N is a certain l-independent quantity; α is defined by

$$\cosh\alpha = 1 + \frac{m^2}{2k^2} \quad . \tag{9.51}$$

If the contour C in (9.41) is sufficiently close to the real axis, then we can put $\mathrm{Im}\,\{l\} = 0$ in (9.50), and the criterion of absolute convergence for the integral in (9.41) is $|\mathrm{Im}\,\{\vartheta\}| < \alpha$. The convergence domain in the complex ϑ plane is bounded by the straight lines

$$\mathrm{Im}\,\{\vartheta\} = \pm\alpha \quad . \tag{9.52}$$

It is not difficult to verify that the convergence domain boundary for the integral (9.41) in the z plane is an ellipse with the focal points ±1. Indeed, separating the imaginary and real parts of z and ϑ,

$$z = x + iy = \cos\left(\vartheta_1 + i\vartheta_2\right) \quad , \qquad 0 \leq \vartheta_1 \leq \pi \quad ,$$

we have

$$x = \cos\vartheta_1 \cosh\vartheta_2 \quad , \qquad y = -\sin\vartheta_1 \sinh\vartheta_2 \quad .$$

The boundary condition (9.52) corresponds to the ellipse equation in the z plane, i.e.,

$$\frac{x^2}{(\cosh\alpha)^2} + \frac{y^2}{(\sinh\alpha)^2} = 1 \quad . \tag{9.53}$$

Therefore, the boundary of the domain of absolute convergence in the complex z plane for the integral (9.41) is an ellipse with the focal points ±1. It is usually referred to as the Lehmann ellipse (Fig. 9.4). For real k, i.e., positive energies, the semimajor and semiminor axes of the ellipse are

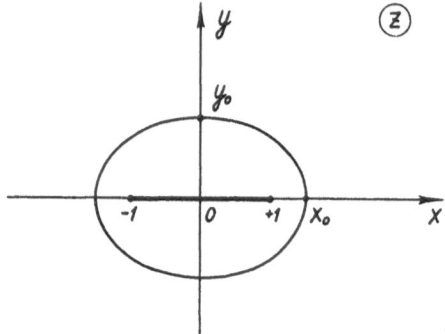

Fig. 9.4. The Lehmann ellipse in the complex z plane that determines the absolute convergence domain for the Watson–Sommerfeld integral

$$x_0 = 1 + \frac{m^2}{2k^2} \quad , \qquad y_0 = \frac{m}{k}\sqrt{1 + \frac{m^2}{4k^2}} \quad . \tag{9.54}$$

Thus we have shown that the integral in (9.41), and hence the partial wave expansion (9.39) of the scattering amplitude, absolutely converges within the Lehmann ellipse. Therefore the scattering amplitude $f(k, z)$ is analytic within the Lehmann ellipse in the complex z plane. Assuming z to lie within the Lehmann ellipse, we modify the integration contour C in (9.41) to obtain the vertical line C' (that lies to the right of Re $\{l\} = -\frac{1}{2}$) and supplement it with two arcs of large radii in the first and fourth quadrants (Fig. 9.5). Because the partial amplitude decreases sufficiently fast as $|l| \to \infty$, the contributions of the integrals over the arcs can be neglected. However, since the scattering matrix is meromorphic in the complex moment plane for Re $\{l\} > -\frac{1}{2}$, we have to add to the integral over C' the sum of residues of the integrand of (9.41) associated with the scattering matrix poles in the plane of the complex angular momentum quantum number. As a result, the scattering amplitude can be written as

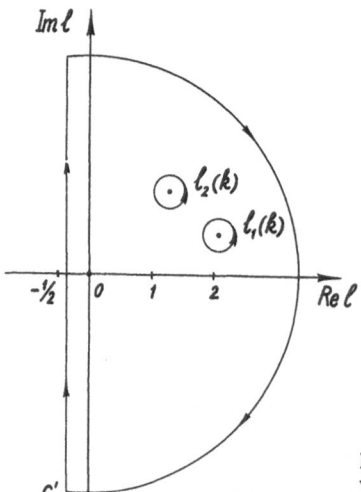

Fig. 9.5. Modification of the integration contour for the Watson–Sommerfeld integral

$$f(k,z) = \frac{1}{4k} \int_{C'} dl\, (2l+1) \left\{ S_l(k) - 1 \right\} P_l(-z) (\sin l\pi)^{-1}$$
$$+ \frac{i\pi}{2k} \sum \frac{2\,\alpha_r(k)+1}{\sin \alpha_r(k)\,\pi}\, b_r(k)\, P_{\alpha_r(k)}(-z) \quad , \tag{9.55}$$

where $\alpha_r(k)$ is the position and $b_r(k)$ is the residue of the scattering matrix pole.

The scattering amplitude represented by the integral over the contour C' can be analytically continued to the whole complex z plane with the exception of the cut along a part of the real axis. Indeed, the integrand in (9.55) is asymptotically limited by the quantity

$$N\,|l|^{1/2} \exp\left[-\mathrm{Re}\,\{\vartheta\}\,\mathrm{Im}\,\{l\}\right] \quad ,$$

hence, for $\mathrm{Re}\,\{\vartheta\} > 0$ the integral absolutely converges and, therefore, the scattering amplitude $f(k,z)$ is analytic in the entire complex z plane with a cut along the real axis from the point $z = 1$ to infinity. We have already shown that the amplitude is analytic within the Lehmann ellipse, so that actually the cut along the real axis is restricted to the interval from the point $z = 1 + m^2/2k^2$ to ∞. Note that the scattering amplitude written in the form of (9.55) can be analytically continued to z for which the convergence in (9.39, 41) is violated.

Thus we have shown that if the potential is a superposition of the Yukawa potentials, then the scattering amplitude $f(k,z)$ is analytic in the complex z plane with a cut along the real axis from $1 + m^2/2k^2$ to infinity.

Sometimes one intorduces a new variable

$$t = -q^2 \quad , \tag{9.56}$$

which is the negative of the square of the momentum transfer, in place of the scattering angle cosine z. Since the momentum transfer q is related to the scattering angle ϑ by (8.54), the quantities t and z must satisfy the linear relation

$$t = 2k^2(z-1) \quad . \tag{9.57}$$

Then it is convenient to introduce the variable

$$s = \frac{2\mu E}{\hbar^2} \tag{9.58}$$

in place of the energy E and to treat the scattering amplitude as a function of s and t. For the potentials described by superpositions of the Yukawa potentials, the scattering amplitude $f(s,t)$ which is a function of the complex variables s and t, is analytic in the complex s plane with a cut along the positive real semi-axis, and analytic in the complex t plane with a cut along the real axis from the point m^2 to infinity.

9.4 Asymptotic Behavior of the Scattering Amplitude for Large z

Having written the scattering amplitude in the form of (9.55), one can not only consider the analytic properties of the scattering amplitude in the complex z plane, one can also find the amplitude asymptotics in the nonphysical region where the cosine of the scattering angle becomes infinite, $z \to \infty$ [9.2]. It is not difficult to show by making use of (9.55) that the asymptotic behavior of the scattering amplitude $f(k, z)$ for $z \to \infty$ is determined by the singularities of the scattering matrix $S_l(k)$ in the complex l plane. Indeed, the asymptotic of the Legendre function $P_l(z)$ for large z is

$$P_l(z) \simeq z^l \quad , \qquad z \to \infty \quad , \tag{9.59}$$

therefore, the dominant contribution in (9.55) is due to the rightmost Regge pole, i.e, the one with the greatest real part $\alpha_r(k)$. If z becomes infinite, then the contributions of other poles and the rest of the integral over the vertical contour can be disregarded and thus the scattering amplitude asymptotic is given by

$$f(k, z) \simeq \frac{i\pi}{2k} \frac{2\,\alpha(k) + 1}{\sin\,\alpha(k)\pi}\, b(k)\,(-z)^{\alpha(k)} \quad . \tag{9.60}$$

This relation is of great importance, for reasons to be given below. In the relativistic theory, the amplitudes of crossing processes

1) $a + b = c + d \quad$ and
2) $a + \bar{c} = \bar{b} + d$

(the bars label the antiparticles) are related in such a way that large nonphysical values of the cosine of the scattering angle for process (1) correspond to large physical values of energy for reaction (2). Then the high energy behavior of the scattering amplitude for (2) is determined by the pole dislocation of the partial scattering amplitude of (1).

The variables s and t are the most suitable ones for the study of the crossing processes (1) and (2). By virtue of crossing symmetry, the amplitude of the process (2) is obtained from that of the process (1) by interchanging t and s, so that

$$f_1(s, t) = f_2(t, s) \quad . \tag{9.61}$$

Thus, the asymptotic of the amplitude of the process (1) with respect to the variable t determines the asymptotic of the amplitude of (2) with respect to s and vice versa.

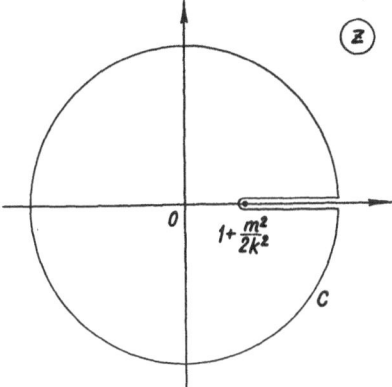

Fig. 9.6. Integration contour in the complex z plane used for the derivation of the momentum transfer dispersion relations

9.5 Momentum Transfer Dispersion Relations

The scattering amplitude corresponding to the potential given by (9.42) is analytic in the complex z plane with a cut along the real axis from the point $1 + m^2/2k^2$ to infinity. Therefore, we can derive the dispersion relations for the scattering amplitude $f(k, z)$ with respect to z for fixed k.

   We take the contour C in the complex z plane to encircle the cut clockwise, and supplement it with a circle of infinite radius with the center in the coordinate origin (Fig. 9.6). The direct derivation of the dispersion relation is complicated because the scattering amplitude $f(k, z)$ does not vanish as $|z| \to \infty$. Indeed, integrating the quantity $f(k, z')/(z' - z)$ (here z' is the integration variable and z is a fixed point within the contour), one fails to obtain the dispersion relation since the integral over the circle of infinite radius does not vanish. According to (9.60), however, the scattering amplitude grows for large $|z|$ not faster than z^N, where N is the smallest positive integer greater than $\text{Re}\{\alpha(k)\}$. Hence, if the scattering amplitude $f(k, z')$ is prior to integration, multiplied by

$$\prod_{i=1}^{N} \frac{(z - z_i)}{(z' - z_i)} \quad ,$$

where z_i are the coordinates of N arbitrary points which lie off the cut, then the integral over the circle of infinite radius vanishes. Thus, within the context of the Cauchy theorem we have

$$\frac{1}{2\pi i} \int_C dz' \, \frac{f(k, z')}{z' - z} \prod_{i=1}^{N} \frac{(z - z_i)}{(z' - z_i)}$$

$$= f(k, z) - \sum_{j=1}^{N} f(k, z_j) \prod_{i \neq j}^{N} \frac{(z - z_i)}{(z_j - z_i)} \quad . \tag{9.62}$$

When crossing the cut, the discontinuity of the scattering amplitude is given by

$$f(k, z + iO) - f(k, z - iO) \equiv 2i \operatorname{Im}_z \{f(k, z)\} \quad . \tag{9.63}$$

It is clear that for real k, $\operatorname{Im}_z \{f(k, z)\}$ coincides with the amplitude imaginary part $\operatorname{Im} \{f(k, z)\}$. Making use of the definition (9.63), we rewrite the dispersion relation (9.62) as

$$f(k, z) = \sum_{j=1}^{N} f(k, z_j) \prod_{i \neq j}^{N} \frac{(z - z_i)}{(z_j - z_i)}$$
$$+ \frac{1}{\pi} \int_{1+(m^2/2k^2)}^{\infty} dz' \, \frac{\operatorname{Im}_{z'} \{f(k, z')\}}{z' - z} \prod_{i=1}^{N} \frac{(z - z_i)}{(z' - z_i)} \quad . \tag{9.64}$$

Though z_i can be taken to be k-independent, in the general case the number N must depend on k since k enters $\operatorname{Re} \{\alpha(k)\}$. If $\operatorname{Re} \{\alpha(k)\}$ is limited as $k \to \infty$, then a finite number N exists which is greater than the maximum possible $\operatorname{Re} \{\alpha(k)\}$. Substituting this N into (9.64), we obtain a dispersion relation which holds for any value of k.

Introducing instead of z the negative of the square of the momentum transfer t, we rewrite the dispersion relation (9.64) as

$$f(s, t) = \sum_{j=1}^{N} f(s, t_j) \prod_{i \neq j}^{N} \frac{(t - t_i)}{(t_j - t_i)}$$
$$+ \frac{1}{\pi} \int_{m^2}^{\infty} dt' \, \frac{\operatorname{Im}_{t'} \{f(s, t')\}}{t' - t} \prod_{i=1}^{N} \frac{(t - t_i)}{(t' - t_i)} \quad , \tag{9.65}$$

where

$$\operatorname{Im}_t \{f(s, t)\} \equiv \frac{1}{2i} \{f(s, t + iO) - f(s, t - iO)\} \quad . \tag{9.66}$$

If we take all t_i to be zeros, than the dispersion relation (9.65) reduces to

$$f(s, t) = \sum_{j=1}^{N} t^j \, g_j(s) + \frac{t^N}{\pi} \int_{m^2}^{\infty} dt' \, \frac{\operatorname{Im}_{t'} \{f(s, t')\}}{t'^N (t' - t)} \quad , \tag{9.67}$$

where

$$g_j(s) = \frac{1}{j!} \frac{d^j}{dt^j} f(s, t) \bigg|_{t=0} \quad . \tag{9.68}$$

Note that the integral in the right-hand part of the momentum transfer dispersion relation contains only nonphysical values of the scattering amplitude.

Problems

9.1 Show that N_{l+1} (the number of bound states with quantum number $l + 1$) cannot exceed N_l (the number of bound states with quantum number l), i.e., that

$$N_{l+1} \leq N_l \quad . \tag{9.69}$$

Let k and l be real. Then the scattering matrix $S_l(k)$ can be expressed in terms of the real scattering phase shift $\delta_l(k)$, i.e.,

$$S_l(k) = \exp\left[2\mathrm{i}\,\delta_l(k)\right] \quad .$$

Making use of the radial equation (9.2) for the regular function $\Phi_l(k, r)$, we obtain

$$\frac{d}{dr}\, W\left(\Phi_l(k, r),\, \frac{\partial \Phi_l(k, r)}{\partial l}\right) = (2l + 1)\,\frac{\Phi_l(k, r)}{r^2} \quad .$$

We integrate this equation over r from 0 to ∞ taking into account (9.13, 19) to obtain

$$|f_l(k)|^2\left(\frac{\pi}{2} - \frac{\partial \delta_l(k)}{\partial l}\right) > 0 \quad ,$$

or

$$\frac{\partial \delta_l(k)}{\partial l} < \frac{\pi}{2} \quad . \tag{9.70}$$

This infers that the scattering phase shift $\delta_{l+1}(k)$ cannot exceed the scattering phase shift associated with the quantum number smaller by one $\delta_l(k)$, by more than $\pi/2$, i.e., that

$$\delta_{l+1}(k) < \delta_l(k) + \frac{\pi}{2} \quad . \tag{9.71}$$

Since the number of bound states with quantum number l can be expressed in terms of the scattering phase shift, $N_l = \delta_l(0)/\pi$, it follows from the inequality (9.71) that

$$N_{l+1} < N_l + \frac{1}{2} \quad . \tag{9.72}$$

And since both N_l and N_{l+1} are integers, we have

$$N_{l+1} \leq N_l \quad .$$

9.2 Consider the resonance behavior of the scattering matrix in the vicinity of a Regge pole making use of the representation (9.55) for the scattering amplitude $f(k, z)$.

The scattering matrix $S_l(k)$ can be expressed in terms of the scattering amplitude $f(k, z)$:

$$S_l(k) = 1 + ik \int_{-1}^{1} dz \, f(k, z) \, P_l(z) \quad . \tag{9.73}$$

Making use of the relation

$$\int_{-1}^{1} dz \, P_l(z) \, P_{\alpha_r}(-z) = \frac{2}{\pi} \frac{\sin \alpha_r \pi}{(l + \alpha_r + 1)(l - \alpha_r)} \quad ,$$

which holds for integer l, and substituting (9.55) for $f(k, z)$, we find the contribution of the r-th pole of the amplitude (9.55) to $S_l(k)$ to be

$$S_l^{(r)}(k) \simeq \frac{i}{k} \frac{b_r(k)}{l - \alpha_r(k)} \quad . \tag{9.74}$$

If α_r is close to an integer l, then the energy dependence of S_l is determined by (9.74). Expanding $\alpha_r(k)$ in a series about the value k_0 for which Re $\{\alpha_r(k_0)\} = l$, we obtain

$$S_l(k) \simeq -i \frac{\hbar^2}{\mu} \left[\frac{d\alpha(k_0)}{dk_0} \right]^{-1} \left[\frac{b_r(k_0)}{E - E_0 + \frac{i}{2} \Gamma_0} \right] \quad , \tag{9.75}$$

where E_0 and Γ_0 are given by (9.34). Thus, the energy dependence of the scattering matrix $S_l(k)$ near the Regge pole is of resonance type. Evidently, this resonance-like dependence can be manifested only if Γ_0 is small compared to the characteristic energies on which the scattering amplitude depends.

9.3 Consider the behavior of the Regge trajectories in the vicinity of the threshold $E = 0$ [9.3].

If the potential $V(r)$ satisfies the condition (7.31), then the Jost function $f(\lambda, k)$ is an analytic function of k in the region Im $\{k\} < m/2$ except for the point $k = 0$ which is a branch point for arbitrary λ and a multiple pole of the order $|\lambda| - 1/2$ for real half-integer λ. Thus, to define $f(\lambda, k)$ unambiguously, we have to cut the complex k plane along the imaginary axis from $k = 0$ to $k = im/2$. In the region Im $\{k\} < m/2$ and $k \neq i\kappa$ ($0 \leq \kappa \leq m/2$), the function $f(\lambda, k)$ is uniquely determined by the iteration expansion; its value is usually referred to as the principal value of $f(\lambda, k)$. The Jost function values on other sheets of the Riemann surface are defined by analytic continuation. Suppose $f(\lambda, k \exp(in\pi))$ is the value of the Jost function that is obtained by analytic continuation along the path that lies in the region Im $\{k\} < m/2$ and encircles the point $k = 0$ anticlockwise as the argument continuously varies from φ to $\varphi + n\pi$. By using (9.2, 6), we obtain the relation

$$f(\lambda, k \, e^{i\pi}) = f(\lambda, k \, e^{-i\pi}) + 2i \cos \lambda\pi \, f(\lambda, k) \quad , \tag{9.76}$$

which yields

$$S(\lambda, k\, e^{i\pi}) = 1 + e^{2i\lambda\pi} - e^{2i\lambda\pi}\, S^{-1}(\lambda, k) \quad . \tag{9.77}$$

It is not difficutl to verify that

$$Z(\lambda, k^2) \equiv i k^{2\lambda}\, \frac{S(\lambda, k) - e^{2i\lambda\pi}}{S(\lambda, k) - 1} \tag{9.78}$$

is a single-valued function of k^2 in the region $\mathrm{Im}\,\{k\} < m/2$. For half-integer λ, this function reduces to

$$Z(\lambda, k^2) = k^{2\lambda}\, \cot \delta(\lambda, k) \quad . \tag{9.79}$$

Within the context of the definition (9.78), the scattering matrix may be written as

$$S(\lambda, k) = \frac{Z(\lambda, k^2) - i k^{2\lambda}\, e^{2i\lambda\pi}}{Z(\lambda, k^2) - i k^{2\lambda}} \quad , \tag{9.80}$$

then its poles are determined by the condition

$$Z(\lambda, k^2) = i k^{2\lambda} \tag{9.81}$$

except for the integer values of λ for which we have to employ the limiting form

$$\frac{\partial Z(\lambda, k^2)}{\partial \lambda} = 2 i k^{2\lambda} \ln k \quad . \tag{9.82}$$

Assuming that the energy E is low, we expand $Z(\lambda, k^2)$ in a power series of k^2, so that

$$Z(\lambda, k^2) = Z_0(\lambda) + k^2\, Z_1(\lambda) + \dots \quad , \tag{9.83}$$

where $Z_0\left(\alpha(0) + \tfrac{1}{2}\right) = 0$. The function $Z_0(\lambda)$ is regular in the vicinity of $\lambda = \alpha(0) + \tfrac{1}{2}$.

If $\alpha(0) < \tfrac{1}{2}$, then the right-hand part of (9.81) is greater than the second term in (9.83) and thus

$$\alpha(k) - \alpha(0) = O\left(E^{\alpha(0) + (1/2)}\right) \quad . \tag{9.84}$$

If $\alpha(0) > \tfrac{1}{2}$, then the second term in (9.83) is greater than the right-hand side of (9.81) and we have

$$\alpha(k) - \alpha(0) = O(E) \quad . \tag{9.85}$$

If $\alpha(0) = \tfrac{1}{2}$, then (9.82) must be applied and we obtain

$$\alpha(k) - \alpha(0) = O(E \ln E) \quad . \tag{9.86}$$

The angle between the Regge trajectory and the real axis is determined by the relation

$$\arg\left[\alpha(k) - \alpha(0)\right] \underset{k \to 0}{\longrightarrow} \begin{cases} \left[\frac{1}{2} - \alpha(0)\right]\pi\,, & \alpha(0) \leq \frac{1}{2} \\ 0\,, & \alpha(0) \geq \frac{1}{2} \end{cases}, \qquad (9.87)$$

Therefore, for $E = 0$ the Regge trajectory abandons the real axis at finite angles to the right for waves with $l > 0$, at a right angle for waves with $l = 0$, and at finite angles to the left for waves with $l < 0$. The greater the threshold value $\alpha(0)$, the closer to the real axis the trajectory goes.

10. Double Dispersion Relations

10.1 Mandelstam Representation

The energy dispersion relations considered in Chap. 8 cannot be treated as equations for the scattering amplitude; they only reflect its analytic properties. In order to transform the dispersion relations into equations we must introduce an additional relation between the real and imaginary parts of the amplitude. When analyzing the Jost functions, we derived such an additional relation from the unitary condition for the partial amplitudes. In a similar manner, we can try to transform the energy dispersion relation into an equation using the unitarity condition (5.9) for the total scattering amplitude. According to (5.9), the relation between the amplitude and its imaginary part, with scattering angle fixed, depends on the values of the amplitude in the entire physical range of angles. At the same time, the dispersion relation (8.63) was derived only for a restricted range of angles. Therefore, it is impossible to combine the dispersion relation with the unitarity condition unless the former is extended to wider angle range. The momentum transfer dispersion relation for the scattering amplitude, derived in the previous chapter, also cannot be employed because the dispersion relation itself contains integration over the nonphysical region.

Thus, simple single-variable dispersion relations for the scattering amplitude and the unitarity condition are insufficient for deriving equations describing the dynamics of the physical system. Such equations, however, can be derived from the unitarity condition combined with the so-called double dispersion relation for the scattering amplitude, which was proposed by *Mandelstam* [10.1]. The double dispersion relation determines the analytic properties of the scattering amplitude with respect to both energy and momentum transfer, in the entire region of their variation.

The appropriate independent variables for the analysis of the double dispersion relation are s and t which are usually referred to as the Mandelstam variables. We remind the reader that

$$s = k^2 \quad , \qquad t = 2k^2(z - 1) \quad . \tag{10.1}$$

The energy dispersion relation (8.63) for the scattering amplitude can be rewritten in terms of s and t as

$$f(s,t) = f_B(t) + \sum_{l,n} \frac{R_{ln}(t)}{s - s_{ln}} + \frac{1}{\pi} \int_0^\infty ds' \frac{\mathrm{Im}_{s'}\{f(s',t)\}}{s' - s - \mathrm{i}O} \quad , \tag{10.2}$$

where

$$R_{ln}(t) \equiv (-1)^{l+1}(2l+1)\,\beta_{ln}^2\,P_l\!\left(1 + \frac{t}{2s_{ln}}\right) \quad,$$

(10.3)

$$\mathrm{Im}_s\{f(s,t)\} \equiv \frac{1}{2i}\left\{f(s+i0,t) - f(s-i0,t)\right\} \quad.$$

If the potential satisfies the condition (8.60), then the difference between the scattering amplitude and its Born approximation, $f(s,t) - f_B(t)$, is analytic in the whole complex t plane with a cut along the real axis from the point $4m^2$ to infinity. The function $\mathrm{Im}_s\{f(s,t)\}$ is also analytic in this region. The functions $R_{ln}(t)$ entering the sum in (10.2) are polynomials with respect to t. The degrees of these polynomials are determined by the bound state angular momenta which cannot exceed a certain value $\alpha(s)$. So, the dispersion relation (10.2), which was derived for the momentum transfer $t \geq -4m^2$, can be analytically continued to the entire complex t plane except the cut along the real axis from the point $4m^2$ to infinity. To carry out analytic continuation, we substitute for $\mathrm{Im}_s\{f(s,t)\}$ in (10.2) its representation which explicitly reflects analyticity with respect to the variable t; such a representation immediately follows from the dispersion relation (9.67), so that

$$\mathrm{Im}_s\{f(s,t)\} = \sum_{i=1}^{N} t^i G_i(s) + \frac{t^N}{\pi}$$
$$\times \int_{4m^2}^{\infty} dt' \, \frac{\mathrm{Im}_{t'}\left\{\mathrm{Im}_s\{f(s,t')\}\right\}}{t'^N(t'-t)} \quad,$$

(10.4)

where

$$G_i(s) = \frac{1}{i!}\frac{d^i}{dt^i}\,\mathrm{Im}_s\{f(s,t)\}\bigg|_{t=0} \quad.$$

(10.5)

The lower integration limit for t' in general depends on s, but cannot be smaller than $4m^2$.

Substituting (10.4) into (10.2), we obtain the so-called Mandelstam double spectral representation for the amplitude $f(s,t)$, i.e.,

$$f(s,t) = f_B(t) + \sum_{l,n} \frac{R_{ln}(t)}{s - s_{ln}} + \sum_i \frac{t^i}{\pi}\int_0^{\infty} ds' \, \frac{G_i(s')}{s'-s-i0}$$
$$+ \frac{t^N}{\pi^2}\int_0^{\infty} ds' \int_0^{\infty} dt' \, \frac{\varrho(s',t')}{t'^N(t'-t)(s'-s-i0)} \quad,$$

(10.6)

where $\varrho(s,t)$ is the spectral density determined by the amplitude discontinuities across the cuts,

$$\varrho(s,t) \equiv \mathrm{Im}_t\left\{\mathrm{Im}_s\{f(s,t)\}\right\} \quad.$$

(10.7)

If the potential is a superposition of the Yukawa potentials (9.42), then the Born amplitude is given by

$$f_B(t) = -\int_m^\infty d\lambda \, \frac{\sigma(\lambda)}{\lambda^2 - t} \quad . \tag{10.8}$$

In the simplest case when bound states do not occur and $\mathrm{Im}_s\{f(s,t)\} \to 0$ as $|t| \to \infty$, the Mandelstam double spectral representation takes the form

$$f(s,t) = f_B(t) + \frac{1}{\pi^2} \int_0^\infty ds' \int_0^\infty dt' \, \frac{\varrho(s',t')}{(t'-t)(s'-s-iO)} \quad . \tag{10.9}$$

The double dispersion relations (10.6, 9) determine the analytic properties of the scattering amplitude $f(s,t)$ which is regarded as a function of the two independent variables s and t. The spectral density $\varrho(s,t)$ entering the dispersion relations is expressed in terms of the amplitude discontinuities across the cuts as given by (10.7). Supplementing the dispersion relation with the unitarity condition, we obtain a closed scheme of the dynamical description of the system, equivalent to the Schrödinger equation [10.2].

10.2 Spectral Density and Unitarity Condition

Let us consider the simplest case when bound states do not occur and $\mathrm{Im}_s\{f(s,t)\}$ $\to 0$ as $|t| \to \infty$. By virtue of (5.9), the unitarity condition can be rewritten in terms of s and t, i.e.,

$$\mathrm{Im}_s\{f(s,t)\} = \frac{\sqrt{s}}{4\pi} \int do'' \, f^*(s, t_{k'-k''}) \, f(s, t_{k-k''}) \quad , \tag{10.10}$$

where $s = k^2 = k'^2$, $t = -(k-k')^2$, $t_{k-k''} = -(k-k'')^2$, and $t_{k'-k''} = -(k'-k'')^2$. Relation (10.10) holds only for the physical region $-4s \le t \le 0$, however, it can be analytically continued using (10.9).

Substituting (10.9) in the right-hand part of (10.10) yields integrals of the type

$$I = \int do'' \, \frac{1}{\left[t_1 + (k-k'')^2\right]\left[t_2 + (k'-k'')^2\right]}$$

$$= \frac{1}{4s^2} \int do'' \, \frac{1}{(\lambda_1 - nn'')(\lambda_2 - n'n'')} \quad , \tag{10.11}$$

where $\lambda_1 = 1 + t_1/2s$, $\lambda_2 = 1 + t_2/2s$, and so on. We employ the Feynmann transformation

$$\frac{1}{ab} \equiv 2 \int_{-1}^1 d\alpha \, \frac{1}{\left[(1+\alpha)a + (1-\alpha)b\right]^2} \quad , \tag{10.12}$$

then

$$I = \frac{1}{2s^2} \int_{-1}^{1} d\alpha \int do''$$

$$\times \frac{1}{\left\{ (1+\alpha)\lambda_1 + (1-\alpha)\lambda_2 - \left[(1+\alpha)n + (1-\alpha)n' \right] n'' \right\}^2} \quad . \tag{10.13}$$

The integration over angles results in

$$I = \frac{\pi}{s^2} \int_{-1}^{1} d\alpha \, \frac{1}{1-\alpha^2} \, \frac{1}{\zeta - \cos \vartheta} \quad , \tag{10.14}$$

where ϑ is the angle between the vectors n and n'; the quantity ζ is defined by

$$\zeta = \lambda_1 \lambda_2 + \frac{1}{2} \left[\frac{1+\alpha}{1-\alpha} \left(\lambda_1^2 - 1 \right) + \frac{1-\alpha}{1+\alpha} \left(\lambda_2^2 - 1 \right) \right] \quad . \tag{10.15}$$

We take ζ to be the integration variable in (10.14). Making use of the relation

$$\sqrt{\left(\zeta - \lambda_1 \lambda_2 \right)^2 - \left(\lambda_1^2 - 1 \right)\left(\lambda_2^2 - 1 \right)}$$

$$= \frac{1}{2} \left[\frac{1+\alpha}{1-\alpha} \left(\lambda_1^2 - 1 \right) - \frac{1-\alpha}{1+\alpha} \left(\lambda_2^2 - 1 \right) \right] \quad ,$$

we find that

$$\frac{d\alpha}{1-\alpha^2} = \frac{1}{2} \frac{d\zeta}{\left[\left(\zeta - \lambda_1 \lambda_2 \right)^2 - \left(\lambda_1^2 - 1 \right)\left(\lambda_2^2 - 1 \right) \right]^{1/2}} \tag{10.16}$$

and thus

$$I = \frac{\pi}{s^2} \int_{\lambda_1 \lambda_2 + \left[(\lambda_1^2 - 1)(\lambda_2^2 - 1) \right]^{1/2}}^{\infty} d\zeta \, \frac{1}{\zeta - \cos \vartheta}$$

$$\times \frac{1}{\left[\left(\zeta - \lambda_1 \lambda_2 \right)^2 - \left(\lambda_1^2 - 1 \right)\left(\lambda_2^2 - 1 \right) \right]^{1/2}} \quad . \tag{10.17}$$

It is convenient, in the expression obtained, to return to the variables t, t_1, and t_2. We note that $\cos \vartheta = 1 + t/2s$ and introduce the notation $\zeta = 1 + t'/2s$. Then the integral (10.11) reduces to

$$I = \frac{4}{\sqrt{s}} \int_{0}^{\infty} dt' \, \frac{1}{t' - t} \, K(s, t'; t_1, t_2) \quad , \tag{10.18}$$

where

$$K(s, t; t_1, t_2) = \frac{\pi}{2} \frac{\theta(t - t_0)}{\left[s(t - t_1 - t_2)^2 - (4s + t) t_1 t_2 \right]^{1/2}} \tag{10.19}$$

and

$$t_0\left(s; t_1, t_2\right) = t_1 + t_2 + \frac{t_1 t_2}{2s}$$
$$+ \frac{1}{2s} \left[\left(4s + t_1\right)\left(4s + t_2\right) t_1 t_2\right]^{1/2} . \tag{10.20}$$

Note that the denominator in (10.19) is symmetric with respect to the variables t, t_1, and t_2.

Within the context of (10.18), all the terms in the right-hand part of (10.10) can be continued to arbitrary values of t. We write the left-hand part of (10.10) as

$$\text{Im}_s\{f(s,t)\} = \frac{1}{\pi} \int_0^\infty dt' \, \frac{\varrho(s,t')}{t'-t}$$

and omit integration over t' in both parts. Then we obtain from (10.10) an equation for the spectral density $\varrho(s,t)$, i.e.,

$$\varrho(s,t) = \int_m^\infty d\mu_1 \, \sigma(\mu_1) \int_m^\infty d\mu_2 \, \sigma(\mu_2) \, K\left(s,t; \mu_1^2, \mu_2^2\right)$$
$$- \frac{2}{\pi^2} \int_0^\infty ds_1 \, \frac{1}{s_1 - s} \int_0^\infty dt_1 \, \varrho(s_1, t_1)$$
$$\times \int_m^\infty d\mu \, \sigma(\mu) \, K\left(s,t; t_1, \mu^2\right)$$
$$+ \frac{1}{\pi^4} \int_0^\infty ds_1 \int_0^\infty ds_2 \int_0^\infty dt_1$$
$$\times \int_0^\infty dt_2 \, \frac{\varrho(s_1, t_1) \, \varrho(s_2, t_2)}{(s_1 - s - iO)(s_2 - s - iO)} \, K\left(s,t; t_1, t_2\right) . \tag{10.21}$$

Knowledge of the spectral density is sufficient for calculating the scattering amplitude $f(s,t)$ after taking into account (10.9). Thus, the analyticity and unitarity relations for the scattering amplitude yield an equation which completely describes the dynamics of the system and therefore, is equivalent to the Schrödinger equation.

To solve (10.21) is not difficult. Note that the function $K(s,t; t_1, t_2)$ vanishes for moderate s and t and given t_1 and t_2. According to (10.19), $K(s,t; t_1, t_2) = 0$ for $t < t_0(s; t_1, t_2)$, where t_0 is a monotonic ascending function of t_1 and t_2. The function $K(s,t; t_1, t_2)$ with the lowest values of t_1 and t_2 (equal to m^2) enters the first term of the right-hand part of (10.21). Therefore, it immediately follows from (10.21) that $\varrho(s,t) = 0$ if

$$t < t_0\left(s; m^2, m^2\right) \equiv 4m^2 + \frac{m^4}{s} . \tag{10.22}$$

The spectral density $\varrho(s,t)$ is not equal to zero only in the region in the (s,t) plane above the boundary that is given by

$$t = t_0\left(s; m^2, m^2\right) \quad .$$
(10.23)

The second term in the right-hand part of (10.21) does not vanish for $t > t_0\left(s; 4m^2, m^2\right)$, and the third term is nonzero for $t > t_0\left(s; 4m^2, 4m^2\right)$. So, in the region

$$t_0\left(s; m^2, m^2\right) < t < t_0\left(s; 4m^2, m^2\right) \quad ,$$
(10.24)

the spectral density $\varrho(s,t)$ is completely determined by the first term in the right-hand part of (10.21), i.e.,

$$\varrho_2(s,t) = \int_m^\infty d\mu_1\, \sigma(\mu_1) \int_m^\infty d\mu_2\, \sigma(\mu_2)\, K\left(s,t; \mu_1^2, \mu_2^2\right) \quad .$$
(10.25)

The second Born approximation can easily be shown to yield the same expression for the spectral density $\varrho(s,t)$. It must be emphasized, however, that (10.25) is exact in the range given in (10.24). [In this region the second Born approximation results in the exact expression for $\varrho(s,t)$.]

The curve $t = t_0\left(s; 4m^2, m^2\right)$ separates the region in which $\varrho(s,t)$ differs from (10.25); its asymptotic is $t = 4m^2/s$ for small s and $t = 9m^2$ for large s. So if the condition

$$t_0\left(s; 4m^2, m^2\right) < t < t_0\left(s; 9m^2, m^2\right)$$
(10.26)

is satisfied, then (10.26) can be substituted for $\varrho(s,t)$ in the integrand in the second term of the right-hand part of (10.21) and we find $\varrho(s,t)$ in the region given in (10.26) to be

$$\varrho(s,t) = \varrho_2(s,t) - \frac{2}{\pi^2} \int_0^\infty ds_1\, \frac{1}{s_1 - s}$$
$$\times \int_{t_0(s_1;m^2,m^2)}^{t_0(s;4m^2,m^2)} dt_1\, \varrho_2\left(s_1, t_1\right)$$
$$\times \int_m^\infty d\mu\, \sigma(\mu)\, K\left(s,t; t_1, \mu^2\right) \quad .$$
(10.27)

This procedure can be continued: the plane (s,t) is divided into a sequence of domains; in each of these the density $\varrho(s,t)$ is directly expressed in terms of the function $\sigma(\mu)$ and reduced to a finite polynomial with respect to the interaction constant. After the resultant expressions are substituted into (10.9) and integration over the relevant domains is carried out, the scattering amplitude is determined by the limit of the polynomial sequence.

Problems

10.1   Calculate in the second Born approximation the scattering amplitude for the potential given by a superposition of the Yukawa potentials, i.e.,

$$V_r(r) = \int_m^\infty d\mu \, \sigma(\mu) \, \frac{e^{-\mu r}}{r} \quad . \tag{10.28}$$

Making use of the general definition of the scattering amplitude (3.15) and the Lippmann–Schwinger equation (3.12), we can write the second Born approximation correction to the scattering amplitude as

$$f^{(2)}(s,t) = -\frac{\mu}{2\pi\hbar^2} \int d\mathbf{r} \int d\mathbf{r}' \, e^{-i\mathbf{k}'\mathbf{r}} \, V(r)$$
$$\times G_0^{(+)}(s;\mathbf{r},\mathbf{r}') \, V(r') \, e^{i\mathbf{k}\mathbf{r}'} \quad , \tag{10.29}$$

$$V(r) \equiv \frac{\hbar^2}{2\mu} \, V_r(r) \quad ,$$

where $G_0^{(+)}(s;\mathbf{r},\mathbf{r}')$ is the Green function given by (3.9, 11). First of all we calculate the scattering amplitude for the simple Yukawa potential

$$V_r(r) = V_{r0} \, \frac{e^{-\mu_0 r}}{r} \quad . \tag{10.30}$$

We substitute (10.30) for the potential and (3.11) for the Green function in (10.29), then the addition to the amplitude $f^{(2)}(s,t)$ may be written as

$$f^{(2)}(s,t) = -\frac{V_{r0}^2}{2\pi^2} \int d\mathbf{q} \, \frac{1}{s - q^2 + i0} \, \frac{1}{\mu_0^2 + (\mathbf{k} - \mathbf{q})^2}$$
$$\times \frac{1}{\mu_0^2 + (\mathbf{k}' - \mathbf{q})^2} \quad , \tag{10.31}$$

$$s = k^2 \, , \, t = -(\mathbf{k} - \mathbf{k}')^2 \quad .$$

Integration over angles can be done by using the Feynman transformation (10.12). Extending the integration to cover the range of negative q, we thus obtain

$$f^{(2)}(s,t) = -\frac{V_{r0}^2}{4\pi} \int_{-1}^{1} d\alpha \, \frac{1}{P(\alpha)} \int dq \, q \, \frac{1}{s - q^2 + i0}$$
$$\times \frac{1}{\mu_0^2 - s + q^2 - 2q \, P(\alpha)} \quad , \tag{10.32}$$

where

$$P(\alpha) = \sqrt{s + \tfrac{1}{4}(1 - \alpha^2)t} \quad . \tag{10.33}$$

We supplement the q integral over the real axis by the integral over a large radius semicircle in the upper half-plane. The integrand of (10.32) has poles within the contour under consideration at the points

$$q = \sqrt{s} + iO \qquad \text{and} \qquad q = P(\alpha) + i\sqrt{\mu_0^2 - \tfrac{1}{4}(1 - \alpha^2)t} \quad ,$$

and thus the residue theorem yields

$$f^{(2)}(s, t) = \frac{V_{r0}^2}{4} \int_{-1}^{1} d\alpha \; \frac{1}{\lambda(\alpha)\left[\mu_0^2 - 2i\sqrt{s}\,\lambda(\alpha)\right]} \quad , \tag{10.34}$$

where

$$\lambda(\alpha) = \sqrt{\mu_0^2 - \tfrac{1}{4}(1 - \alpha^2)t} \quad . \tag{10.35}$$

The function (10.34) is analytic in the upper half-plane of k for $t < 4\mu_0^2$ and vanishes at infinity. Hence, the second Born approximation correction to the amplitude (10.34) can be written as

$$f^{(2)}(s, t) = \frac{1}{\pi} \int_0^{\infty} ds' \; \frac{w(s', t)}{s' - s - iO} \quad , \tag{10.36}$$

if the notation

$$w(s, t) \equiv \mathrm{Im}_s\left\{f^{(2)}(s, t)\right\}$$
$$= \frac{1}{2} V_{r0}^2 \sqrt{s} \int_{-1}^{1} d\alpha \; \frac{1}{\mu_0^4 + s\left[4\mu_0^2 - (1 - \alpha^2)t\right]} \tag{10.37}$$

is introduced. We employ the identical transformation

$$\frac{1}{\mu_0^4 + s\left[4\mu_0^2 - (1 - \alpha^2)t\right]}$$
$$\equiv \frac{1}{(1 - \alpha^2)s} \int_0^{\infty} dt' \; \frac{1}{t' - t} \, \delta\!\left[t' - \left(4\mu_0^2 + \frac{\mu_0^4}{s}\right)\frac{1}{1 - \alpha^2}\right]$$

and carry out integration over the parameter α in (10.37). Then

$$w(s, t) = \frac{1}{2} V_{r0}^2 \int_0^{\infty} dt' \; \frac{1}{t' - t} \frac{1}{\sqrt{st'}} \frac{\theta\!\left(t' - 4\mu_0^2 - \mu_0^4/s\right)}{\left(t' - 4\mu_0^2 - \mu_0^4/s\right)^{1/2}} \quad . \tag{10.38}$$

This relation determines the analytic properties of $w(s, t)$ which is a function of the complex variable t. According to (10.38), $w(s, t)$ is analytic in the complex t plane with a cut along the real axis from the point $t = 4\mu_0^2 + \mu_0^4/s$ to infinity. Making use of (10.38), we rewrite the spectral representation of the amplitude $f^{(2)}(s, t)$ as a function of two complex variables s and t, i.e.,

$$f^{(2)}(s, t) = \frac{1}{\pi^2} \int_0^{\infty} ds' \int_0^{\infty} dt' \; \frac{\varrho^{(2)}(s', t')}{(s' - s - iO)(t' - t)} \quad , \tag{10.39}$$

where

$$\varrho^{(2)}(s,t) = \frac{\pi}{2}\,\frac{V_{r0}^2}{\sqrt{st}}\,\frac{\theta\left(t - 4\mu_0^2 - \mu_0^4/s\right)}{\left(t - 4\mu_0^2 - \mu_0^4/s\right)^{1/2}}\,. \tag{10.40}$$

The region in the (s,t) plane where the spectral density is nonzero is determined by the θ function. This region is hatched in Fig. 10.1. Note that the limiting value $t = 4\mu_0^2$ is attained only for infinite energy $s \to \infty$.

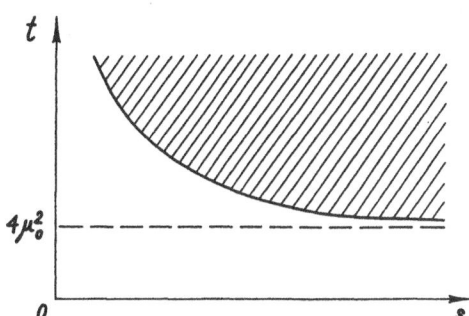

Fig. 10.1. The region in the (s,t) plane in which the spectral density $\varrho^{(2)}(s,t)$ is nonzero

The subsequent terms of the Born expansion of the amplitude,

$$f(s,t) = \sum_{n=1}^{\infty} f^{(n)}(s,t)\ , \tag{10.41}$$

can be found in the same manner. The spectral representation of $f^{(n)}(s,t)$ can be shown to be similar to that of $f^{(2)}(s,t)$. However, the region in which the spectral density $\varrho^{(n)}(s,t)$ does not vanish is smaller. The boundary of the n-th Born approximation domain for infinitely high energy is $n^2\mu_0^2$.

For a superposition of the Yukawa potentials (10.28), the second Born approximation correction to the scattering amplitude is given by (10.39) as before, but the spectral density $\varrho^{(2)}(s,t)$ is described by the double integral

$$\varrho^{(2)}(s,t) = \frac{\pi}{2} \int_m^\infty d\mu_1\, \sigma(\mu_1) \int_m^\infty d\mu_2\, \sigma(\mu_2)$$

$$\times \frac{\theta\left[t - t_0\left(s; \mu_1^2, \mu_2^2\right)\right]}{\left[s\left(t - \mu_1^2 - \mu_2^2\right)^2 - (4s + t)\,\mu_1^2\mu_2^2\right]^{1/2}}\,, \tag{10.42}$$

where

$$t_0\left(s; t_1, t_2\right) = t_1 + t_2 + \frac{t_1 t_2}{2s} + \frac{1}{2s}\left[(4s + t_1)(4s + t_2)\,t_1 t_2\right]^{1/2}$$

The expression (10.42) reproduces (10.25).

10.2   Consider the analytic properties of the partial scattering amplitudes in the complex s plane for the potential (10.28).

In order to set the partial amplitudes $f_l(s)$ in the complex s plane, we expand the Mandelstam representation (10.9) of the total amplitude $f(s,t)$ in terms of partial components. We multiply (10.9) termwise by $P_l(z)/2$ and integrate over z from -1 to 1. The procedure produces integrals of the type

$$\int_{-1}^{1} dz \, \frac{P_l(z)}{t' - 2s(1+z)} = \frac{1}{s} Q_l\left(1 + \frac{t}{2s}\right) \quad , \tag{10.43}$$

where $Q_l(z)$ is a Legendre function of the second kind; it is analytic in the complex z plane with a cut along the segment $-1 \leq z \leq 1$. Substituting the explicit expression for the Born amplitude (10.8), we find the partial scattering amplitude to be

$$f_l(s) = -\frac{1}{2s} \int_{m}^{\infty} d\mu \, \sigma(\mu) \, Q_l\left(1 + \frac{\mu^2}{2s}\right)$$
$$+ \frac{1}{2\pi^2 s} \int_{0}^{\infty} ds' \int_{0}^{\infty} dt' \, \frac{\varrho(s',t')}{s' - s - iO} \, Q_l\left(1 + \frac{t'}{2s}\right) \quad . \tag{10.44}$$

The partial amplitude (10.44) is analytic in the complex s plane with the cuts $(-\infty, -m^2/4)$ and $(0, \infty)$. The first Born term in (10.44) is responsible for the left-hand cut $(-\infty, -m^2/4)$, the second one gives rise to the cuts $(-\infty, -m^2)$ and $(0, \infty)$ because the spectral function $\varrho(s,t)$ vanishes for $t < 4m^2$. For $s \to 0$, $f_l(s) \sim s^l$.

10.3   Expand the function $\cot \delta_l(s)$, where $\delta_l(s)$ is the scattering phase shift, in power series of s [10.3].

We write the partial scattering amplitude $f_l(s)$ as a ratio of two functions,

$$f_l(s) = \frac{N_l(s)}{D_l(s)} \quad , \tag{10.45}$$

where $N_l(s)$ is an analytic function in the s plane with the left-hand cut $(-\infty, -m^2/4)$, which is real for real positive s; $D_l(s)$ is analytic in the whole s plane except the right-hand cut $(0, \infty)$. [$D_l(s)$ coincides with the function (8.26).] Moreover, we assume that $D_l(s) \to 1$ and $N_l(s) \sim s^l$ as $s \to 0$.

As follows from the unitarity condition (5.10) for the partial scattering amplitude, for s real and positive we have

$$\text{Im}\left\{(-1)^l D_l(s)\right\} = -\sqrt{s} \, N_l(s) \quad . \tag{10.46}$$

[The simplest way to derive this equation is to employ (8.29) and (8.31).] We introduce the notation

$$\mathcal{D}_l(s) \equiv (-1)^l D_l(s) + i\sqrt{s} \, N_l(s) \quad , \tag{10.47}$$

then (10.46) reduces to

$$\text{Im} \left\{ \mathcal{D}_l(s) \right\} = 0 \quad , \qquad s > 0 \quad . \tag{10.48}$$

In (10.46, 47), we consider the branch of the function $s^{1/2}$ for which $\left(s + i0\right)^{1/2} > 0$ for $s > 0$. Equation (10.48) implies that the analytic function (10.47) has no right-hand cut on the real positive semi-axis. According to the definition of $\mathcal{D}_l(s)$ and $N_l(s)$, (10.47) has, however, the left-hand cut $(-\infty, -m^2/4)$. Therefore, (10.47) can be expanded in a power series within the circle with center at $s = 0$ and radius of convergence $m^2/4$. It is not difficult to verify that

$$\text{Re} \left\{ \mathcal{D}_l(s) \right\} = s^{1/2} N_l(s) \cot \delta_l(s) \quad , \tag{10.49}$$

where $\delta_l(s)$ is the scattering phase shift. Having expanded the real part of the function $\mathcal{D}_l(s)$ in a power series, we obtain

$$s^{1/2} N_l(s) \cot \delta_l(s) = \sum_{n=0}^{\infty} a_n^l \, s^n \quad . \tag{10.50}$$

Because $N_l(s)$ is analytic within the same circle and proportional to s^l as $s \to 0$, for small s we have

$$s^{l+1/2} \cot \delta_l(s) = \sum_{n=0}^{\infty} b_n^l \, s^n \quad . \tag{10.51}$$

Retaining several leading terms of the expansion in the right-hand part, we obtain the formula for the approximation of the effective range. The convergence domain of the series in the right-hand part is determined by the potential range $\sim m^{-1}$. It is clear that the wider the convergence domain of the series (10.51), the less the number of expansion terms required for satisfactory description of the low energy partial scattering amplitude.

11. The Inverse Problem
   of Scattering Theory

11.1 Integral Representation of the Solutions
   of the Scattering Problem

In the previous chapters we considered in detail the properties of the scattering matrix $S_l(k)$ associated with the potentials which satisfy the conditions (7.2, 3). Particle scattering by a potential field was shown to be completely determined by the wave function asymptotics at infinity. At the same time, it is a matter of great practical interest to consider the inverse problem, namely, to reproduce the potential from the scattering data, that is, the wave function asymptotics at infinity. In principle, the inverse scattering problem can be solved if the scattering phase shift associated with a fixed angular momentum is given for all energies, or if all scattering phase shifts corresponding to different angular momenta are given for a fixed energy. We shall consider in detail the first formulation of the problem. If the system possesses bound states, knowledge of the scattering phase shifts for all energies is insufficient for the calculation of the potential. In order to find it unambiguously, we have to set, along with the scattering phase shifts for all energies, the eigenvalues of the bound state energies and the normalization constants of the bound state wave functions [11.1].

It is convenient to treat the inverse problem in terms of integral representations of the various solutions of (7.1) associated with the scattering problem. We write the Jost solution f(k, r) as

$$f(k, r) = e^{-ikr} + \int_r^\infty dr' \, K(r, r') e^{-ik'r} \quad , \qquad \text{Im}\,\{k\} \leq 0 \quad , \qquad (11.1)$$

where $K(r, r')$ is some function which is nonzero for $r' > r$. (For the sake of simplicity we consider only the case $l = 0$.) We substitute (11.1) for the Jost function in (7.24) and change the sequence of integration in the double integral. The Fourier transformation reduces the relation obtained to the integral equation for the function $K(r, r')$, so that

$$K(r,r') = \frac{1}{2} \int_{(r+r')/2}^{\infty} dr'' \, V_r(r'')$$

$$+ \frac{1}{2} \int_{r}^{(r+r')/2} dr'' \, V_r(r'') \int_{-r''+r'+r}^{r''+r'-r} dr''' \, K(r'',r''')$$

$$+ \frac{1}{2} \int_{(r+r')/2}^{\infty} dr'' \, V_r(r'') \int_{r''}^{r''+r'-r} dr''' \, K(r'',r''') \ . \tag{11.2}$$

The solution of this equation can be found by successive approximations. It is not difficult to show that if the conditions (7.2, 3) are satisfied, then the successive approximation series uniformly converges in the range $0 < r \le r'$.

Formally, (11.1) expresses the solution of the scattering equation with the potential $V_r(r)$ in terms of the solution of the corresponding equation of free motion, the function $K(r,r')$ being the kernel of the integral transformation. A peculiar feature of this transformation is that $K(r,r')$ is energy-independent.

As follows from (11.2), the derivatives of $K(r,r')$ with respect to each argument are continuous. If $V_r(r)$ is differentiable, then $K(r,r')$ also has second derivatives. Differentiating (11.2) yields a differential equation for $K(r,r')$,

$$\frac{\partial^2}{\partial r^2} K(r,r') - \frac{\partial^2}{\partial r'^2} K(r,r') = V_r(r) K(r,r') \ . \tag{11.3}$$

Moreover, (11.2) immediately yields the relation

$$K(r,r) = \frac{1}{2} \int_{r}^{\infty} dr' \, V_r(r') \ , \tag{11.4}$$

which leads to

$$V_r(r) = -2 \frac{d}{dr} K(r,r) \ . \tag{11.5}$$

Integral representations of the form (11.1) can also be derived for the solutions of (7.1) associated with other boundary conditions, in particular, for the solution $u(k,r)$ that is regular at the point $r = 0$.

Let us consider a complete set of functions describing particle motion in the field of the potential $V_r(r)$. This set includes both the functions of the continuum spectrum of the system,

$$u(k,r) = \frac{i}{(2\pi)^{1/2}} \left\{ f(k,r) - \exp\left[2i\,\delta(k)\right] f(-k,r) \right\} \ , \tag{11.6}$$

$$E = \frac{\hbar^2 k^2}{2\mu} > 0 \ ,$$

and the bound state functions associated with the sequence of discrete levels $E_n = -\hbar^2 \kappa_n^2 / 2\mu < 0 \ (n = 1, 2, \ldots)$,

$$u_n(r) = \beta_n \, f(-i\kappa_n, r) \ . \tag{11.7}$$

At large distances, the functions $u(k, r)$ and $u_n(r)$ reduce to the asymptotics

$$u(k, r) \rightarrow \bar{u}(k, r) \equiv \sqrt{\frac{2}{\pi}} \exp [i \delta(k)] \sin [kr + \delta(k)] \quad , \qquad (11.8)$$

$$u_n(r) \rightarrow \bar{u}_n(r) \equiv \beta_n \exp (-\kappa_n r) \quad . \qquad (11.9)$$

Normalization of the functions (11.6) is determined by the requirement that the completeness condition for the set of functions must take the form

$$\int_0^\infty dk \, u(k, r) \, u^*(k, r') + \sum_n u_n(r) \, u_n(r') = \delta(r - r') \quad . \qquad (11.10)$$

Making use of the relation between $u(k, r)$ and $f(k, r)$ and the representation (11.1), we find the integral representation for the functions $u(k, r)$ to be

$$u(k, r) = \bar{u}(k, r) + \int_0^\infty dr' \, K(r, r') \, \bar{u}(k, r') \quad . \qquad (11.11)$$

Since the transformation kernel $K(r, r')$ is energy-independent, an analogous relation occurs for the bound state functions, i.e.,

$$u_n(r) = \bar{u}_n(r) + \int_r^\infty dr' \, K(r, r') \, \bar{u}_n(r') \quad . \qquad (11.12)$$

Inverting (11.11), we obtain

$$\bar{u}(k, r) = u(k, r) - \int_r^\infty dr' \, \overline{K}(r, r') \, u(k, r') \quad , \qquad (11.13)$$

where $\overline{K}(r, r')$ is a function which is nonzero for $r' > r$. It is not difficult to derive for $\overline{K}(r, r')$ an equation analogous to (11.2).

11.2 Reproducing the Potential by the Scattering Phase Shifts

Let us show that the function $K(r, r')$ satisfies a linear integral equation with the kernel depending on the scattering phase shift $\delta(k)$, the eigenvalues of the bound state energies E_n, and the corresponding bound state normalization constants β_n.

We multiply (11.11) by $\bar{u}^*(k, r')$ and (11.12) by $\bar{u}_n(r')$ and carry out summation over energy in the obtained relations. We then have

$$\int_0^\infty dk \, u(k, r) \, \bar{u}^*(k, r') + \sum_n u_n(r) \, \bar{u}_n(r')$$

$$= \int_0^\infty dk \, \bar{u}(k, r) \, \bar{u}^*(k, r') + \sum_n \bar{u}_n(r) \, \bar{u}_n(r') \qquad (11.14)$$

$$+ \int_r^\infty dr'' \, K(r, r'') \left[\int_0^\infty dk \, \bar{u}(k, r'') \, \bar{u}^*(k, r') + \sum_n \bar{u}_n(r'') \, \bar{u}_n(r') \right] .$$

We introduce the notation

$$F(r, r') \equiv \int_0^\infty dk \left[\overline{u}(k, r) \, \overline{u}^*(k, r') - \frac{2}{\pi} \sin kr \sin kr' \right]$$
$$+ \sum_n \overline{u}_n(r) \, \overline{u}_n(r') \quad , \tag{11.15}$$

then, since

$$\frac{2}{\pi} \int_0^\infty dk \, \sin kr \, \sin kr' = \delta(r - r') \quad ,$$

we find from (11.14) that for $r < r'$

$$\int_0^\infty dk \, u(k, r) \, \overline{u}^*(k, r') + \sum_n u_n(r) \, \overline{u}_n(r')$$
$$= F(r, r') + \delta(r - r') + K(r, r') \tag{11.16}$$
$$+ \int_r^\infty dr'' \, K(r, r'') \, F(r'', r') \quad .$$

Taking $r' \to \infty$ in (11.10), we have

$$\int_0^\infty dk \, u(k, r) \, \overline{u}^*(k, r') + \sum_n u_n(r) \, \overline{u}_n(r') = \delta(r - r') \quad . \tag{11.17}$$

Comparing (11.16) to (11.17), we obtain a linear integral equation for the function $K(r, r')$:

$$K(r, r') + F(r, r') + \int_r^\infty dr'' \, K(r, r'') \, F(r'', r') = 0 \quad (r' > r) \quad . \tag{11.18}$$

By virtue of (11.8, 9, 15) the kernel of this equation is a function of the scattering phase shift, bound state energies, and normalization constants, i.e.,

$$F(r, r') = \frac{1}{2\pi} \int_{-\infty}^\infty dk \left[1 - \exp\left(2\mathrm{i}\,\delta(k)\right) \right] \exp\left[\mathrm{i}\,k(r + r')\right]$$
$$+ \sum_n \beta_n^2 \exp\left[-\kappa_n(r + r')\right] \quad . \tag{11.19}$$

Equation (11.18) has a unique solution for $r \geq 0$. Indeed, were the solutions of (11.18) not unique, the homogeneous equation would have a nontrivial solution. We assume this solution to be real and denote it (for fixed r) as $\chi(r')$, so that

$$\chi(r') + \int_r^\infty dr'' \, \chi(r'') \, F(r'', r') = 0 \quad . \tag{11.20}$$

We multiply (11.20) by $\chi(r')$ and integrate over r'. Then

$$\int_r^\infty dr' \, \chi^2(r') + \int_r^\infty dr' \int_r^\infty dr'' \, \chi(r') \, \chi(r'') \, F(r'', r') = 0 \quad .$$

Making use of (11.19), we have

$$\int_r^\infty dr' \, \chi^2(r') + \frac{1}{2\pi} \int_{-\infty}^\infty dk \left(1 - e^{2i\,\delta(k)}\right) \chi^2(-k)$$
$$+ \sum_n \beta_n^2 \, \chi^2(-i\kappa_n) = 0 \quad, \tag{11.21}$$

where

$$\chi(k) \equiv \int_r^\infty dr' \, e^{-ikr'} \chi(r') \quad . \tag{11.22}$$

The function $\chi(k)$ is analytic and bounded in the lower half-plane of k. Since $\chi(r') = 0$ for $r' < r$, we have

$$\frac{1}{2\pi} \int_{-\infty}^\infty dk \, \chi^2(-k) = \int_r^\infty dr' \, \chi(r') \chi(-r') = 0 \quad,$$

$$\frac{1}{2\pi} \int_{-\infty}^\infty dk \, |\chi(-k)|^2 = \int_r^\infty dr' \, \chi^2(r') \quad,$$

and hence (11.21) can be rewritten as

$$\frac{1}{2\pi} \int_{-\infty}^\infty dk \left\{ |\chi(-k)|^2 - \chi^2(-k) \, e^{2i\,\delta(k)} \right\}$$
$$+ \sum_n \beta_n^2 \, \chi^2(-i\kappa_n) = 0 \quad. \tag{11.23}$$

Since $\chi(r')$ is real, it follows from (11.22) that $\chi(-i\kappa_n)$ is real and $\chi^2(-i\kappa_n) \geq 0$. Therefore, (11.23) is satisfied if

$$\chi(-i\kappa_n) = 0 \quad\text{and}\quad \chi^2(k) = |\chi(k)|^2 \, e^{2i\,\delta(k)} \quad .$$

Hence, the argument of $\chi^2(k)$ on the real axis is twice the scattering phase shift, $2\delta(k)$, and the zeros of the function $\chi(k)$ coincide with those of the Jost function $f(k)$. Therefore, the function $\chi^2(k)/f^2(k)$ is analytic in the lower half-plane and real on the real axis. It can be analytically continued to the whole plane, the result being bounded, i.e., constant, because $f(k) \to 1$ as $|k| \to \infty$. This constant must be equal to zero since

$$\int_{-\infty}^\infty dk \, |\chi(k)|^2 < \infty \quad .$$

Therefore, $\chi(r') = 0$ and thus (11.18) has a unique solution.

Thus, once we know the scattering phase shifts for all energies, the bound state energies $E_n = -\hbar^2\kappa_n^2/2\mu$, and the bound state normalization constants β_n, we can find the kernel $F(r,r')$ of the linear integral equation (11.18). Then we solve (11.18) for all $r \geq 0$ and find the function $K(r,r')$ which, according to (11.5), determines the potential [11.1, 2]. Note that if $f(k)$ is defined not only in the lower but also in the upper half-plane, then the normalization constants β_n

can be expressed in terms of the Jost function by means of (7.74). This assertion concerns the finite-range potentials. According to (8.14), the Jost function itself is unambiguously determined by the scattering phase shift $\delta(k)$ and the bound state energies E_n. Therefore, in this case the normalization constants cannot be regarded as independent parameters of the problem.

Problems

11.1 Find the potential from the observed energy dependence of the intensity of particle scattering at a given angle. Apply the Born approximation.

The Born amplitude of elastic scattering in the centrally symmetric field is given by

$$f_B(k, \vartheta) \equiv f_B(q) = -\int_0^\infty dr \, r \, \frac{\sin qr}{q} \, V_r(r) \quad , \tag{11.24}$$

where

$$q = 2k \, \sin \frac{\vartheta}{2} \quad .$$

The quantity $f_B(q)$ can be found from the measured data. For k fixed, we find the potential from (11.24) to be

$$V_r(r) = -\frac{2}{\pi} \int_0^\infty dq \, q \, f_B(q) \, \frac{\sin qr}{q} \quad . \tag{11.25}$$

11.2 Reproduce the potential by the given scattering phase shift $\delta(k)$ under the assumptions that the perturbation theory is applicable and no bound states occur in the system.

If bound states do not occur, then the function (11.19), which determines the kernel of the integral (11.18), is given by

$$F(r, r') = \frac{1}{2\pi} \int_{-\infty}^\infty dk \, \left(1 - e^{2i\,\delta(k)}\right) \exp\left[ik(r + r')\right] \quad . \tag{11.26}$$

Having substituted this expression into (11.18) and solved the linear integral equation (11.18) by successive approximations, we find that

$$K(r, r') = \sum_{n=1}^\infty K^{(n)}(r, r') \quad , \tag{11.27}$$

where

$$K^{(n)}(r, r') = (-1)^n \int_r^\infty dr_1 \int_r^\infty dr_2 \ldots \int_r^\infty dr_{n-1} \, F(r, r_1)$$
$$\times F(r_1, r_2) \ldots F(r_{n-1}, r') \quad . \tag{11.28}$$

Within the context of (11.5), the potential $V_r(r)$ may be written as a series

$$V_r(r) = \sum_n V_r^{(n)}(r) \quad , \tag{11.29}$$

where

$$V_r^{(n)}(r) = -2 \frac{d}{dr} K^{(n)}(r, r) \quad . \tag{11.30}$$

The first- and second-order terms are

$$V_r^{(1)}(r) = -\frac{4}{\pi} \int_0^\infty dk \, k \left\{ \sin 2kr - \sin 2(kr + \delta(k)) \right\} \quad , \tag{11.31}$$

$$V_r^{(2)}(r) = \frac{4}{\pi^2} \left\{ \int_0^\infty dk \, [\cos 2kr - \cos 2(kr + \delta(k))] \right\}^2 \quad . \tag{11.32}$$

11.3 Reproduce the potential by the scattering phase shift $\delta(b)$ (b is the impact parameter) in the high energy approximation.

In the high energy approximation, $\delta(b)$ and the potential satisfy the integral relation

$$\delta(b) = -\frac{1}{4k} \int_{-\infty}^\infty dz \, V_r\left(\sqrt{b^2 + z^2}\right) \quad . \tag{11.33}$$

We introduce a new integration variable, $r = (b^2 + z^2)^{1/2}$, instead of z. Then

$$\delta(b) = -\frac{1}{2k} \int_b^\infty dr \, r \, \frac{V_r(r)}{(r^2 - b^2)^{1/2}} \quad . \tag{11.34}$$

This is the Abel integral equation. The inverse transformation yields

$$V_r(r) = \frac{4k}{\pi} \frac{1}{r} \frac{d}{dr} \int_r^\infty db \, b \, \frac{\delta(b)}{(b^2 - r^2)^{1/2}} \quad . \tag{11.35}$$

Thus, in the high energy approximation the potential $V_r(r)$ is completely determined by the given impact-parameter-dependence of the scattering phase shift $\delta(b)$.

12. Separable Representation of the Scattering Amplitude

12.1 The Scattering Amplitude off the Energy Shell

Let us calculate the scattering amplitude off the energy shell by using the general relation between the scattering amplitude and the transition matrix derived in Chap. 3. We remind the reader that the transition operator t is governed by the Lippmann–Schwinger equation (2.55). In the momentum representation, this equation may be written as ($\hbar = 1$)

$$\langle k'|t(z)|k\rangle = \langle k'|V|k\rangle + \int \frac{dq}{(2\pi)^3} \frac{\langle k'|V|q\rangle \langle q|t(z)|k\rangle}{z - (q^2/2\mu)} \quad , \tag{12.1}$$

$$z = E + iO \quad ,$$

where E is the energy of two-particle relative motion which is conserved under elastic scattering, μ is the reduced mass. Equation (12.1) determines the t matrix both on the energy shell, when the initial and final momenta are related to energy by

$$\frac{k^2}{2\mu} = \frac{k'^2}{2\mu} = E \quad , \tag{12.2}$$

and off the energy shell, when

$$\frac{k^2}{2\mu} \neq \frac{k'^2}{2\mu} \neq E \quad . \tag{12.3}$$

According to (3.73), the elastic scattering amplitude $f(k, k')$ can be expressed in terms of the t matrix on the energy shell, so that

$$f(k, k') = -\frac{\mu}{2} \langle k'|t(E + iO)|k\rangle \quad . \tag{12.4}$$

The scattering amplitude off the energy shell, $f(k, k'; z)$, can be found under the assumption that (12.4) holds when the condition (12.2) is violated.

Both the scattering amplitude $f(k, k'; z)$ and the transition matrix $\langle k'|t(z)|k\rangle$ possess singularities in the plane of complex energy z. These are poles associated with the discrete spectrum and the cut along the positive real semi-axis given rise to by the continuum spectrum of the system. The singularities can be explicitly derived from the so-called spectral representation of the t matrix,

$$\langle \boldsymbol{k}'|t(z)|\boldsymbol{k}\rangle = \langle \boldsymbol{k}'|V|\boldsymbol{k}\rangle + \sum_N \frac{g_N(\boldsymbol{k}')\, g_N(\boldsymbol{k})}{z + (\kappa_N^2/2\mu)}$$

$$+ \int \frac{d\boldsymbol{q}}{(2\pi)^3}\, \frac{\langle \boldsymbol{k}'|t\left[(q^2/2\mu)+\mathrm{i}O\right]|\boldsymbol{q}\rangle\, \langle \boldsymbol{q}|t\left[(q^2/2\mu)-\mathrm{i}O\right]|\boldsymbol{k}\rangle}{z - (q^2/2\mu)}$$

(12.5)

where $g_N(\boldsymbol{k}) = \left[(k^2 + \kappa_N^2)/2\right]\varphi_N(\boldsymbol{k})$, $N = \{n,l,m\}$, $\kappa_N^2/2\mu$ is the binding energy, $\varphi_N(\boldsymbol{k})$ is the bound state wave function in the momentum representation.

If interaction is centrosymmetric, then the partial component expansion of the potential is given by

$$\langle \boldsymbol{k}'|V|\boldsymbol{k}\rangle = \sum_l (2l+1)\, V_l(k,k')\, P_l(\cos\vartheta)$$

$$= 4\pi \sum_{lm} V_l(k',k)\, Y_{lm}^*(\boldsymbol{n}')\, Y_{lm}(\boldsymbol{n}) \quad, \tag{12.6}$$

where $Y_{lm}(\boldsymbol{n})$ and $Y_{lm}(\boldsymbol{n}')$ are the spherical functions of the angles describing the directions of the vectors $\boldsymbol{k}$ and $\boldsymbol{k}'$; ϑ is the angle between $\boldsymbol{k}$ and $\boldsymbol{k}'$. Each term in (12.6) describes the interaction in the state with the angular momentum quantum number l. The partial component of the potential, $V_l(k',k)$, is given by

$$V_l(k',k) = 4\pi \int_0^\infty dr\, r^2\, j_l(k',r)\, V(r)\, j_l(kr) \quad. \tag{12.7}$$

The t matrix and the scattering amplitude can be expanded in terms of partial components, too, so that

$$\langle \boldsymbol{k}'|t(z)|\boldsymbol{k}\rangle = \sum_l (2l+1)\, t_l(k',k;z)\, P_l(\cos\vartheta) \quad, \tag{12.8}$$

$$f(\boldsymbol{k},\boldsymbol{k}';z) = \sum_l (2l+1)\, f_l(k,k';z)\, P_l(\cos\vartheta) \quad. \tag{12.9}$$

Here $t_l(k',k;z)$ is the partial transition matrix and $f_l(k,k';z)$ is the partial scattering amplitude off the energy shell. The equation for the partial transition matrix readily follows from (12.1), i.e.,

$$t_l(z) = V_l + V_l\, \frac{1}{z - H_0}\, t_l(z) \quad, \tag{12.10}$$

or

$$t_l(k',k;z) = V_l(k',k)$$

$$+ \frac{1}{2\pi^2} \int_0^\infty dq\, q^2\, V_l(k',q)\, \frac{1}{z - (q^2/2\mu)}\, t_l(q,k;z) \quad. \tag{12.11}$$

The partial scattering amplitude and the partial transition matrix satisfy a relation analogous to (12.4):

$$f_l(k, k'; z) = -\frac{\mu}{2\pi} t_l(k', k; z) \quad . \tag{12.12}$$

The partial scattering amplitude on the energy shell is a function of the scattering phase shift $\delta_l(k)$, i.e.,

$$f_l\left(k, k; \frac{k^2}{2\mu}\right) \equiv f_l(k) = \frac{1}{k} \exp\left(i\,\delta_l\right) \sin \delta_l \quad . \tag{12.13}$$

Two-particle scattering is completely described by the given scattering amplitude on the energy shell because, according to (2.70), the scattering cross section is directly expressed in terms of the phase shifts. In the case of three-particle scattering, knowledge of the two-particle amplitudes off the energy shell is also required. The three-particle problem becomes much simpler if the two-particle amplitudes are separable, i.e., when they can be written as products of factors depending on different variables k, k', and z.

It is not difficult to show that the scattering amplitude is separable if the potential is nonlocal and its arguments are separated,

$$V_l(k', k) = -\frac{\lambda}{2\mu} g_l(k') g_l(k) \quad . \tag{12.14}$$

Although realistic physical potentials are local and, to be rigorous, cannot be described by the separable formula (12.14), this expression is nevertheless a good approximation for short-range square-integrable potentials. Thus it makes it possible to derive a separable representation for the two-particle scattering amplitude.

12.2 Hilbert–Schmidt Expansion of the Scattering Amplitude

Let us discuss the factorization approach to the analysis of potentials and two-particle scattering amplitudes based on the Hilbert–Schmidt theorem for symmetric integral equations [12.1, 2].

The similitude transformation reduces the Lippmann–Schwinger equation (12.10) to the symmetric form,

$$\bar{t}_l(z) = \bar{V}_l(z) + \bar{V}_l(z)\,\bar{t}_l(z) \quad , \tag{12.15}$$

where

$$\bar{V}_l(z) = -\left(H_0 - z\right)^{-1/2} V_l \left(H_0 - z\right)^{-1/2} \quad ,$$

$$\bar{t}_l(z) = -\left(H_0 - z\right)^{-1/2} t_l(z) \left(H_0 - z\right)^{-1/2} \quad .$$

The idea of the method is to employ the eigenfunctions of the kernel of the integral equation (12.15), i.e., the solutions of the equation

$$\bar{V}_l(z)\,\bar{g}_{nl}(z) = \eta_{nl}(z)\,\bar{g}_{nl}(z) \tag{12.16}$$

Both eigenvalues $\eta_{nl}(z)$ (n are the quantum numbers which label the eigenvalues in descending order of their absolute values) and eigenfunctions $\overline{g}_{nl}(z)$ depend on z as a parameter. The eigenfunctions $\overline{g}_{nl}(z)$ belong to the Hilbert space, i.e., their norms are assumed to be finite, so that

$$\left(\overline{g}_{nl}(z), \overline{g}_{nl}(z)\right) = \delta_{nn'} \quad . \tag{12.17}$$

At the same time, the eigenfunctions $g_{nl}(z)$ of the nonsymmetrized integral Lippmann–Schwinger equation (12.10), i.e., of the operator $V_l G_0(z)$, given by

$$V_l G_0(z) g_{nl}(z) = \eta_{nl}(z) g_{nl}(z) \quad , \tag{12.18}$$

are more suitable for practical calculations. Since the operators $V_l G_0(z)$ and $\overline{V}_l(z)$ are related by the similitude transformation, their eigenvalues coincide, whereas $g_{ln}(z)$ and $\overline{g}_{nl}(z)$ satisfy the relation

$$g_{nl}(z) = \left(H_0 - z\right)^{1/2} \overline{g}_{nl}(z) \quad . \tag{12.19}$$

In the momentum representation, (12.18) may be written as

$$\frac{1}{2\pi^2} \int_0^\infty dk' \, k'^2 \, V_l(k, k') \left(z - \frac{k'^2}{2\mu}\right)^{-1} g_{nl}(k', z)$$
$$= \eta_{nl}(z) g_{nl}(k, z) \quad . \tag{12.20}$$

The orthonormalization condition for the eigenfunctions $g_{nl}(k, z)$ is

$$\frac{1}{2\pi^2} \int_0^\infty dk \, k^2 \left(z - \frac{k^2}{2\mu}\right) g_{n'l}^*(k, z^*) \, g_{nl}(k, z) = -\delta_{nn'} \quad . \tag{12.21}$$

Though the set of eigenfunctions of the kernel of the integral Lippmann–Schwinger equation (12.10) is incomplete, both the solution $t_l(z)$ and the free term V_l can be expanded in terms of these functions.

We write the solution of (12.10) in the momentum representation as a series

$$t_l(k', k; z) = \sum_n a_{nl}(k', z) g_{nl}(k, z) \quad . \tag{12.22}$$

To find the coefficients a_{nl}, we substitute (12.22) into (12.10) and use (12.20) and the orthonormalization condition (12.21). We then obtain the separable representation of the transition matrix:

$$t_l(k', k; z) = -\sum_n \frac{\eta_{nl}(z)}{1 - \eta_{nl}(z)} g_{nl}^*(k', z^*) g_{nl}(k, z) \quad . \tag{12.23}$$

The separable representation of the interaction potential,

$$V_l(k', k) = -\sum_n \eta_{nl}(z) g_{nl}^*(k', z^*) g_{nl}(k, z) \quad , \tag{12.24}$$

can be obtained in the same manner.

As n increases, the rate of decrease of the expansion terms in (12.23, 24) is determined by that of the eigenvalues $\eta_{nl}(z)$ (the eigenfunctions $g_{ln}(z)$ are bounded functions of n). Comparing (12.23) to (12.24), we can separate the interaction potential in (12.23) and thus find the transition matrix :

$$t_l(k', k; z) = V_l(k', k) - \sum_n \frac{\eta_{nl}^2(z)}{1 - \eta_{nl}(z)} \, g_{nl}^*(k', z^*) \, g_{nl}(k, z) \quad . \tag{12.25}$$

The series in (12.25) converges faster than (12.23) because it contains squares of η_{nl} which decrease as n grows. However, (12.25) is not separable; for this reason we shall study three-particle systems in terms of the t matrix expansion (12.23) in which we retain several first terms.

12.3 Properties of Eigenvalues and Eigenfunctions of the Kernel of the Lippmann–Schwinger Equation

Let us consider the main properties of the eigenvalues and eigenfunctions which occur in the Hilbert–Schmidt method. First we note that the kernel of the symmetrized equation (12.15) is Hermitian only for real negative values of z. In this case the Hermitian property of the kernel $\overline{V}_l(z)$ means that the eigenvalues $\eta_{nl}(z)$ are always real for $z < 0$. In other cases, the operator $\overline{V}_l(z)$ is not Hermitian (though symmetric) and its eigenvalues are complex.

It is convenient to introduce instead of $g(z)$ the functions

$$\psi(z) = G_0(z) \, g(z) \quad , \tag{12.26}$$

which satisfy the equation

$$G_0(z) \, V \, \psi(z) = \eta(z) \, \psi(z) \quad . \tag{12.27}$$

[The operators $G_0(z)V$ and $VG_0(z)$ are related by the similitude transformation.] The solutions of (12.27) must be square integrable. The eigenvalues $\eta(z)$ can be found from the requirement that such solutions must exist for arbitrary z (except for the values lying on the positive real semi-axis).

Equation (12.27) can be reduced to the Schrödinger equation

$$\left\{ H_0 + \frac{V}{\eta(z)} - z \right\} \psi(z) = 0 \tag{12.28}$$

with a z-dependent generalized potential $V/\eta(z)$. If z is real and negative, then this potential is Hermitian and the solution $\psi(z)$ of (12.28) can be regarded as the bound state wave function associated with the binding energy $-z$. Therefore, for $z < 0$ the eigenvalue $\eta(z)$ determines the number by which the potential must be divided in order for the system to possess a bound state with binding energy $-z$. For $\eta(z) = 1$, (12.28) reduces to the usual Schrödinger equation which has solutions decreasing at infinity only for negative energies $z = -\varepsilon$. These values are associated with bound states and can be found from the condition

$$\eta(-\varepsilon) = 1 \quad . \tag{12.29}$$

The bound state wave function in the momentum representation can be written as

$$\varphi_{nl}(k) = N_{nl} \frac{g_{nl}(k, -\varepsilon_{nl})}{(k^2/2\mu) + \varepsilon_{nl}} \quad , \tag{12.30}$$

where N_{nl} is the normalization constant and $g_{nl}(k, -\varepsilon_{nl})$ is the eigenfunction of (12.18). Thus for $z < 0$, of the whole set of $\psi_{nl}(z)$, only those with n and $z = -\varepsilon_{nl}$ such that $\eta_{nl}(z) = 1$, are physical wave functions associated with the bound states of the system.

As follows from (12.27),

$$\eta_{nl}(z) = -\frac{2}{\pi} \frac{\int_0^\infty dk \, k^2 \left[(k^2/2\mu) - z\right]^{-1} \left| \int_0^\infty dr \, r^2 \, j_l(kr) \, V(r) \, \psi_{nl}(r, z) \right|^2}{\int_0^\infty dr \, r^2 \, V(r) \left| \psi_{nl}(r, z) \right|^2} \quad . \tag{12.31}$$

Therefore, for real negative z the eigenvalues $\eta(z)$ have fixed signs, namely, the attractive potentials $(V < 0)$ produce positive eigenvalues $\eta(z) > 0$; the repulsive potentials $(V > 0)$ give rise to negative eigenvalues $\eta(z) < 0$.

The diagonal matrix element $\langle \psi(z) | (H_0 - z) \eta(z) + V | \psi(z) \rangle$ is identically equal to zero. We differentiate it with respect to z to obtain

$$\frac{1}{\eta(z)} \frac{d\eta(z)}{dz} = \frac{\langle \psi(z) | \psi(z) \rangle}{\langle \psi(z) | H_0 - z | \psi(z) \rangle} \quad . \tag{12.32}$$

Since the operator $H_0 - z$ is positive definite for $z < 0$, (12.32) yields the inequality

$$\frac{1}{|z|} > \frac{1}{\eta(z)} \frac{d\eta(z)}{dz} > 0 \tag{12.33}$$

which is valid for real negative energies z. Thus, for potentials with fixed signs and z negative, the eigenvalues $\eta(z)$ are either positive ascending (for attractive potentials) or negative descending (for repulsive potentials) functions of z. According to (12.29), this implies that a bound state with negative energy $z = -\varepsilon_{nl}$ can occur only if

$$\eta_{nl}(0) \geq 1 \quad . \tag{12.34}$$

If z is complex, then (12.27) has solutions only if $\eta(z)$ is complex. If for complex energy z_0 with a small positive imaginary part the eigenvalue $\eta(z)$ is such that

$$\text{Re}\left\{\eta(z_0)\right\} \simeq 1 \quad \text{and} \quad \text{Im}\left\{\eta(z_0)\right\} \ll 1 \quad, \tag{12.35}$$

then the corresponding wave function describes either a quasidiscrete resonance or a virtual state of the system.

In the coordinate representation, calculating the eigenfunctions $\psi_{nl}(r,z) \equiv u_{nl}(r,z)/r$ and eigenvalues $\eta_{nl}(z)$ reduces to solving the differential equation

$$\left\{-\frac{1}{2\mu}\frac{d^2}{dr^2} + \frac{l(l+1)}{2\mu r^2} + \frac{V(r)}{\eta_{nl}(z)} - z\right\} u_{nl}(r,z) = 0 \tag{12.36}$$

with the boundary conditions

$$u_{nl}(r,z) \sim r^{l+1} \quad, \qquad r \to 0 \quad, \tag{12.37}$$

$$u_{nl}(r,z) \sim e^{iqr} \quad, \qquad r \to \infty \quad, \tag{12.38}$$

where

$$q = (2\mu z)^{1/2} \quad, \qquad \text{Im}\left\{q\right\} > 0 \quad.$$

Let us consider the Schrödinger equation with the potential $g\,V(r)$; the binding constant g is separated out for the sake of convenience. For the energy $z = q^2/2\mu$, the solution $\Phi_l(q,r;g)$ of this equation which satisfies the boundary condition (12.37), may be written as

$$\Phi_l(q,r;g) \sim \left\{f_l(-q;g)\,f_l(q,r;g) - f_l(q;g)\,f_l(-q,r;g)\right\} \quad, \tag{12.39}$$

where $f_l(\pm q;g)$ is the Jost function, $f_l(\pm q,r;g)$ are the solutions of the Schrödinger equation with the boundary condition at infinity

$$\lim_{r\to\infty} e^{\pm iqr} f_l(\pm q,r;g) = 1 \quad.$$

It is clear that the solution of (12.36) can be written in the same form as (12.39), where g must be regarded as the generalized binding constant $g/\eta(z)$. Since the proper solution of (12.36), in contrast to (12.39), must contain only a diverging wave at infinity, the z dependence of η can be found from the condition that the coefficient before the converging wave must vanish, i.e.,

$$f_l\left(-q; \frac{g}{\eta}\right) = 0 \quad. \tag{12.40}$$

The roots of this equation determine the eigenvalues $\eta_{nl}(z)$.

The Jost function $f_l(-q;g)$ coincides with the Fredholm determinant $D_l(z;g)$ of the integral equation (12.15). Note that $D_l(z;g)$ can be written as an infinite series with respect to the equation kernel $\overline{V}_l(z)$, i.e.,

$$D_l(z,g) \equiv 1 + \sum_{m=1}^{\infty} \frac{(-1)^m}{m!} \int_0^\infty dk_1 \dots \int_0^\infty dk_m \tag{12.41}$$

$$\times \begin{vmatrix} \overline{V}_l(k_1,k_1;z) & \overline{V}_l(k_1,k_2;z) & \dots & \overline{V}_l(k_1,k_m;z) \\ \overline{V}_l(k_2,k_1;z) & \overline{V}_l(k_2,k_2;z) & \dots & \overline{V}_l(k_2,k_m;z) \\ \vdots & \vdots & \ddots & \vdots \\ \overline{V}_l(k_m,k_1,;z) & \overline{V}_l(k_m,k_2;z) & \dots & \overline{V}_l(k_m,k_m;z) \end{vmatrix} \quad.$$

So, making use of the expansion (12.24), we obtain

$$D_l(z; g) = \prod_n \left[1 - \eta_{nl}(z) \right] \quad . \tag{12.42}$$

Thus, the Jost function $f_l(-q; g)$ is determined by the eigenvalues $\eta_{nl}(z)$ [12.3]:

$$f_l(-q; g) = \prod_n \left[1 - \eta_{nl}(z) \right] \quad . \tag{12.43}$$

Relation (12.24) is linear with respect to g and $\eta_{nl}(z)$. It is not difficult to verify that

$$f_l\left(-q; \frac{g}{\eta}\right) = \prod_n \left[1 - \frac{\eta_{nl}(z)}{\eta} \right] \quad , \tag{12.44}$$

where η is an arbitrary parameter. Equating the right-hand part of (12.44) to zero because of (12.40), we indeed obtain an equation for the eigenvalues $\eta_{nl}(z)$.

Equation (12.43) makes it possible to find the relation between the two-particle scattering phase shift $\delta_l(q)$ which enters the on-energy-shell component of the two-particle t matrix, and the eigenvalues $\eta_{nl}(E+iO)$ for $E = q^2/2\mu > 0$. Indeed, the scattering phase shift is the argument of the Jost function $f_l(q; g)$. Therefore, bearing in mind that $\delta_l(q)$ is an odd function of q and using (12.43), we find that

$$\delta_l(q) = - \sum_n \arg \left[1 - \eta_{nl}(E+iO) \right] \quad , \tag{12.45}$$

or

$$\delta_l(q) = \sum_n \arctan \frac{\mathrm{Im}\left\{ \eta_{nl}(E+iO) \right\}}{1 - \mathrm{Re}\left\{ \eta_{nl}(E+iO) \right\}} \tag{12.46}$$

If g is complex, then the Jost function $f_l(-q; g)$ is an entire function of g for arbitrary q and z. Hence, for z fixed, the eigenvalues $\eta_{nl}(z)$, determined by the zeros of the Jost function (12.40), form a discrete set. Only a finite number of eigenvalues $\eta_{nl}(z)$ lie outside the circle of finite radius. Since $\eta_{nl}(z)$ are discrete, they can be labeled with integers n (in the order of descending absolute values), which was implied before.

The analytic properties of the Jost function $f_l(-q; g)$ in the complex energy plane are determined by the behavior of the potential $V(r)$. We assume that the singularity of the potential at the point $r = 0$ is weaker than r^{-2} and that the decrease of the potential at infinity is sufficiently fast (say, faster than r^{-3}). In this case the Jost function $f_l(-q; g)$ is analytic and has no singularities on the physical sheet of the Riemann surface of complex z, i.e., in the whole complex z plane except the cut along the real positive semi-axis. The analytic properties of the eigenvalues $\eta_{nl}(z)$ follow from those of the Jost function $f_l(-q; g)$. Since the Jost function $f_l(-q; g)$ is analytic with respect to both variables and in the general case $\partial f_l / \partial g \neq 0$, (12.40) has a unique solution $\eta_{nl}(z)$ which is analytic with respect to z in the analyticity region of $f_l(-q; g)$. Thus, the eigenvalues $\eta_{nl}(z)$ of the Hilbert–Schmidt operator, similarly to the Jost function $f_l(-q; g)$, are analytic functions without singularities in the whole complex z plane with the exception of the cut along the real positive semi-axis.

Since $\eta_{nl}(z)$ are real for $z < 0$, the Schwartz reflection principle holds for real z, i.e.,

$$\eta_{nl}^*(z) = \eta_{nl}(z^*) \quad . \tag{12.47}$$

Note that for real negative z we can choose the phase factor in such a way that $g_{nl}(k, z)$ will be real. In this case the reflection principle holds for the eigenfunctions, too:

$$g_{nl}^*(k, z) = g_{nl}(k, z^*) \quad . \tag{12.48}$$

We have already mentioned that the eigenvalues $\eta_{nl}(z)$ are real only on the real negative semi-axis. Therefore, as long as the sign of $\text{Im}\left\{\eta_{nl}(z)\right\}$ does not change in both the upper and the lower half-planes of z, we find from the inequality (12.33) that in the upper half-plane

$$\text{Im}\left\{\eta_{nl}(z)\right\} > 0 \quad \text{for} \quad V < 0$$

and

$$\text{Im}\left\{\eta_{nl}(z)\right\} < 0 \quad \text{for} \quad V > 0 \quad . \tag{12.49}$$

In the lower half-plane, the imaginary part of $\eta_{nl}(z)$ has the opposite sign.

The eigenvalues $\eta_{nl}(z)$ vanish as $|z| \to \infty$, hence we can employ the Cauchy theorem and write the dispersion relation for the analytic function $\eta_{nl}(z)$:

$$\eta_{nl}(E) = \frac{1}{\pi} \int_0^\infty dE' \frac{\text{Im}\left\{\eta_{nl}(E')\right\}}{E' - E - iO} \quad . \tag{12.50}$$

Making use of (12.32), it is not difficult to show that for low energies E, the imaginary part of $\dot\eta_{nl}$ is given by

$$\text{Im}\left\{\eta_{nl}(E + iO)\right\} = C_{nl} E^{l+1/2} \quad , \tag{12.51}$$

where C_{nl} is a real constant. Substituting (12.51) into (12.50), we obtain the expansion of the eigenvalue η_{nl} for real positive energies,

$$\eta_{nl}(E + iO) \simeq \eta_{nl}(0) + B_{nl} E + iC_{nl} E^{l+1/2} \quad , \tag{12.52}$$

where B_{nl} is a real constant.

The operator trace is independent of the representation, hence

$$\sum_{n=1}^\infty \left[\eta_{nl}(z)\right]^p = \text{Tr}\left\{G_{0l}(z) V\right\}^p \quad , \qquad p = 1, 2, \ldots \quad . \tag{12.53}$$

For $p = 1$, we have

$$\sum_{n=1}^{\infty} \eta_{nl}(z) = -2i\mu q \int_0^{\infty} dr \, r^2 \, j_l(qr) \, V(r) \, h_l^{(+)}(qr) \quad , \tag{12.54}$$

$$q = (2\mu z)^{1/2} \quad .$$

The integral in the right-hand part of (12.54) is convergent if the range of the potential is finite and the singularity at the point $r = 0$ is weaker than r^{-2}. For the series in the left-hand part of (12.54) to be convergent, the eigenvalues $\eta_{nl}(z)$ must decrease with growing n as $n^{-\gamma}$, where $\gamma > 1$. For $z < 0$, the n dependence of $\eta_{n0}(z)$ for large n can easily be found in the quasiclassical approximation. We apply the Bohr-Sommerfeld quantization rule,

$$(2\mu)^{1/2} \int_0^R dr \, \sqrt{-\frac{V(r)}{\eta_{n0}(z)} + z} = n\pi \quad , \tag{12.55}$$

where R must be found from the condition $V(R) = z\,\eta_{n0}(z)$. If n is sufficiently large, then

$$|z|\,\eta_{n0}(z) \ll V_0 \tag{12.56}$$

(V_0 is the potential depth) and hence the integration range in (12.55) can be extended to infinity; the second term in the square root can be disregarded. We then obtain

$$\eta_{n0}(z) = \frac{2\mu \, V_0 \tilde{R}^2}{\pi^2 n^2} \quad , \tag{12.57}$$

where $\tilde{R} = \int_0^{\infty} dr \, [-V(r)/V_0]^{1/2}$ is the effective range of the potential. Making use of (12.57), we rewrite the condition (12.56) as

$$(-2\mu z)^{1/2} \tilde{R} \ll n\pi \quad . \tag{12.58}$$

Since the eigenfunctions (12.18) are bounded functions of n, the terms of the transition matrix expansion (12.23) decrease with increasing n with a rate determined by the rate of decrease of the eigenvalues $\eta_{nl}(z)$. According to (12.57), convergence of (12.23) for $z \leq 0$ is quite good. It is not difficult to show that the weaker the potential singularity and the shorter the potential range, the better the convergence.

The eigenvalues of the Hulthen and square well potentials for $z = 0$ are given, respectively, by

$$\eta_{n0}(0) = \frac{g}{n^2} \quad \text{and} \quad \eta_{n0}(0) = \frac{4g}{(2n - 1)^2 \, \pi^2} \tag{12.59}$$

($g = 2\mu \, V_0 R^2$, V_0 is the potential depth, R is the potential range). The ratios of the eigenvalues $\eta_{nl}(0)$ for $n = 1, 2, 3,$ and 4 for these potentials are

$$1 : \frac{1}{4} : \frac{1}{9} : \frac{1}{16} \quad \text{and} \quad 1 : \frac{1}{9} : \frac{1}{25} : \frac{1}{49} \quad .$$

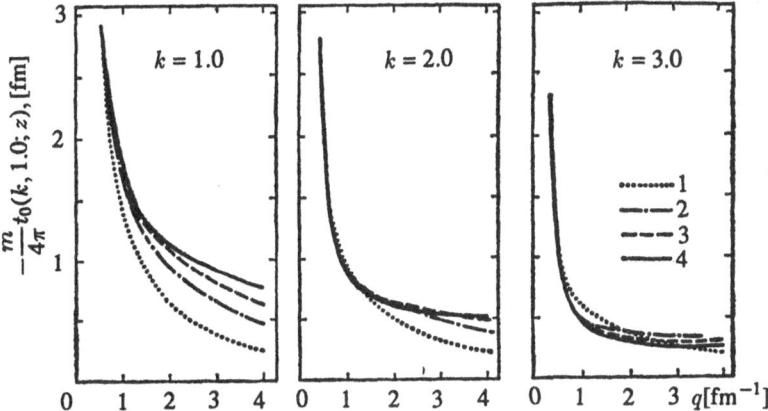

Fig. 12.1. Partial two-particle t matrix ($l = 0$) as a function of the parameter $q \equiv (m|z|)^{1/2}$ ($z < 0$) for three values of the momentum k in the case of Hulthen potential. The number of each curve indicates the number of terms retained in the separable expansion

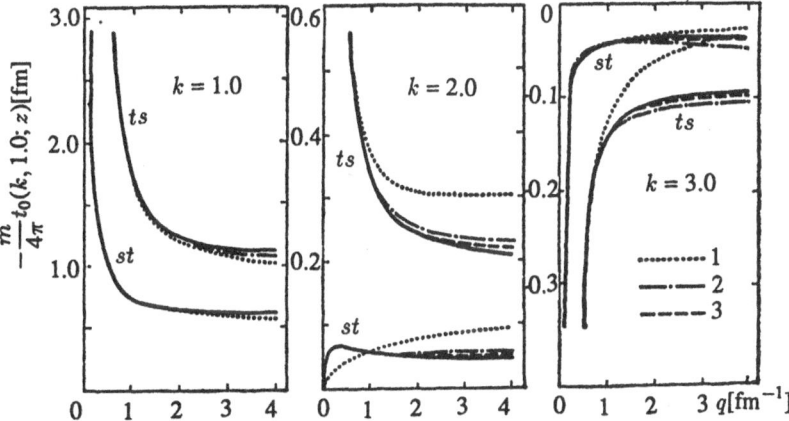

Fig. 12.2. The partial two-particle t matrix ($l = 0$) as a function of the parameter $q \equiv (m|z|)^{1/2}$ ($z < 0$) in the case of a square well potential. The curves are numbered as in Fig. 12.1

Convergence of the transition matrix expansion (12.23) for different momenta k, negative z $\left(q = \sqrt{2\mu|z|}\right)$, and $l = 0$ is shown in Fig. 12.1 for the Hulthen potential [12.4] and in Fig. 12.2 for the square well potential [12.5]. As follows from Fig. 12.2, if the transition matrix corresponding to the square well potential is calculated within the context of (12.23) taking into account only the two first terms, the result is very close to the exact value. [The number of each curve shows the order of the approximation, that is, the number of terms retained in the expansion (12.23). The subscripts ts and st label the potential parameters associated with the two-nucleon interaction in the triplet-singlet and singlet-triplet spin-isospin states, respectively.]

Problems

12.1 Find the scattering amplitude off the energy shell assuming that the particle interaction is described by the Yamaguchi potential

$$\langle \boldsymbol{k}'|V|\boldsymbol{k}\rangle = -\frac{\lambda}{2\mu}\, g(k)\, g(k') \quad, \tag{12.60}$$

where λ is the binding constant ($\lambda > 0$ corresponds to attractive forces) and $g(k)$ is an arbitrary real function of the momentum $\boldsymbol{k}$.

In the coordinate representation, the Yamaguchi potential preserves its factorized form:

$$\langle \boldsymbol{r}'|V|\boldsymbol{r}\rangle = -\frac{\lambda}{2\mu}\, g(r)\, g(r') \quad, \tag{12.61}$$

where

$$g(r) = \int \frac{d\boldsymbol{k}}{(2\pi)^3}\, g(k)\, e^{i\boldsymbol{k}\boldsymbol{r}} \quad.$$

Note that the potential (12.60) describes only S-state interactions.

In the momentum representation, the bound state wave function for the potential (12.60) is given by

$$\varphi_0(k) = N\,\frac{g(k)}{\alpha^2 + k^2} \quad, \tag{12.62}$$

where $E_0 = -\alpha^2/2\mu$ is the binding energy; the constant N is determined by the normalization condition, so that

$$N^{-2} = \int \frac{d\boldsymbol{k}}{(2\pi)^3}\,\frac{g^2(k)}{\left(\alpha^2 + k^2\right)^2} \quad.$$

The binding energy is governed by the condition

$$\lambda^{-1} = \int \frac{d\boldsymbol{k}}{(2\pi)^3}\,\frac{g^2(k)}{\alpha^2 + k^2} \quad. \tag{12.63}$$

The exact solution of the Lippmann–Schwinger equation with the potential (12.60) can be found analytically. Substituting (12.60) into (12.1), we find that

$$t(\boldsymbol{k}', \boldsymbol{k}; z) = -\frac{\lambda}{2\mu}\,\frac{g(k)\, g(k')}{1 + (\lambda/2\mu)\int \left[d\boldsymbol{q}/(2\pi)^3\right]\left[g^2(q)/(z - (q^2/2\mu))\right]} \quad, \tag{12.64}$$

$$z = E + i0$$

As should be anticipated, the t matrix has a pole associated with the energy of the bound state of the system. The elastic scattering amplitude and the scattering phase shift for the Yamaguchi potential are given by

$$f(k) = \frac{\lambda}{4\pi} \frac{g^2(k)}{1 + (\lambda/2\pi^2) \int_0^\infty dq\, q^2 \left[g^2(q)/(k^2 - q^2 + i0)\right]} \quad , \tag{12.65}$$

$$k \cot \delta = \frac{4\pi}{\lambda\, g^2(k)} \left\{ 1 + \frac{\lambda}{2\pi^2} \int_0^\infty dq\, q^2 \frac{g^2(q)}{k^2 - q^2} \right\} \tag{12.66}$$

The wave function describing scattering from the Yamaguchi potential can also be found explicitly. We obtain

$$\psi_k(q) = (2\pi)^3\, \delta(q - k) \tag{12.67}$$

$$+ \frac{\lambda\, g(k)\, g(q)}{(q^2 - k^2 - i0) \left\{ 1 + (\lambda/2\pi^2) \int_0^\infty dq'\, q'^2 \left[g^2(q')/(k^2 - q'^2 + i0)\right] \right\}} .$$

As an example, we consider the potential (12.60) with the function $g(k)$ of the form

$$g(k) = \frac{(2\pi)^{3/2}}{k^2 + \beta^2} \quad . \tag{12.68}$$

In the coordinate representation, we have

$$g(r) = \sqrt{\frac{\pi}{2}} \frac{e^{-\beta r}}{r} \quad . \tag{12.69}$$

The binding energy and the binding constant satisfy the relation

$$\lambda^{-1} = \frac{\pi^2}{(\alpha + \beta)^2\, \beta} \quad .$$

The normalization constant entering (12.62) is given by

$$N = \frac{1}{\pi} \sqrt{\alpha(\alpha + \beta)^2\, \beta} \quad .$$

The bound state wave function in the coordinate representation is

$$\varphi_0(r) = \sqrt{\frac{\alpha(\alpha + \beta)\beta}{2\pi(\beta - \alpha)^2}} \frac{e^{-\alpha r} - e^{-\beta r}}{r} \quad . \tag{12.70}$$

Lastly, the scattering wave function in the coordinate representation may be written as

$$\psi_k(r) = e^{ikr} + f(k) \frac{e^{ikr} - e^{-\beta r}}{r} \quad , \tag{12.71}$$

where the scattering amplitude is

$$f(k) = -\left[\beta - \frac{\beta^2 + k^2}{2\beta} - \frac{(\beta^2 + k^2)^2}{2(\alpha + \beta)^2 \beta} + ik\right]^{-1} \quad . \tag{12.72}$$

12.2   Find the eigenfunctions of the kernel of the Lippmann–Schwinger equation with the Hulthen potential [12.1].

The radial dependence of the Hulthen potential is given by

$$V(r) = -\frac{V_0}{\exp(r/R) - 1} \quad , \qquad V_0 \equiv \frac{g}{2\mu R^2} \quad . \tag{12.73}$$

At short distances ($r \ll R$), the Hulthen potential is singular as r^{-1}, at long distances ($r \gg R$) it decreases exponentially.

We consider only spherically symmetric states ($l = 0$, the subscript l will be omitted). The Jost function for the Hulthen potential is

$$f(-q; g) = \prod_{n=1}^{\infty} \left[1 - \frac{g}{n(n - 2iqR)}\right] \quad . \tag{12.74}$$

The zeros of the Jost function (12.74) determine the bound state energies

$$E_n = -\frac{\kappa_n^2}{2\mu} \quad , \qquad \kappa_n = \frac{1}{2R}\left(\frac{g}{n} - n\right) \quad , \qquad n = 1, 2, \dots \quad . \tag{12.75}$$

The bound state functions are described by the expressions

$$\varphi_n(k) = \frac{N_n}{k^2 + \kappa_n^2} \sum_{\nu=1}^{n} a_{n\nu} \frac{\nu(\nu + 2\kappa_n R)}{(kR)^2 + (\nu + \kappa_n R)^2} \quad , \tag{12.76}$$

where

$$a_{n\nu} = (-1)^{\nu+1} \frac{n!}{(n - \nu)! \, \nu!} \frac{\Gamma[(g/n) + \nu] \, \Gamma[(g/n) - n + 1]}{\Gamma(g/n) \, \Gamma[(g/n) - n + \nu + 1]} \quad .$$

Once the Jost function is known, it is not difficult to find from (12.40) the eigenfunctions of the kernel of the Lippmann–Schwinger equation to be

$$\eta_n(z) = \frac{g}{n(n - 2i\sqrt{2\mu z}\,R)} \quad , \qquad \mathrm{Im}\{z^{1/2}\} \geq 0 \quad . \tag{12.77}$$

The corresponding eigenfunctions, that is, the solutions of (12.18), are

$$g_n(k, z) = C_n(z) \sum_{\nu=1}^{n} A_{n\nu}(z) \frac{g}{\eta_\nu(z)} \frac{1}{k^2 R^2 - 2\mu z R^2 + (g/\eta_\nu(z))} \quad , \tag{12.78}$$

$$A_{n\nu}(z) = (-1)^{\nu+1} \sum_{\sigma=1}^{n} \frac{n - \sigma + 1}{n + \sigma - 1} \frac{\eta_\sigma(z)}{\eta_{n+\sigma-1}(z)} \quad ,$$

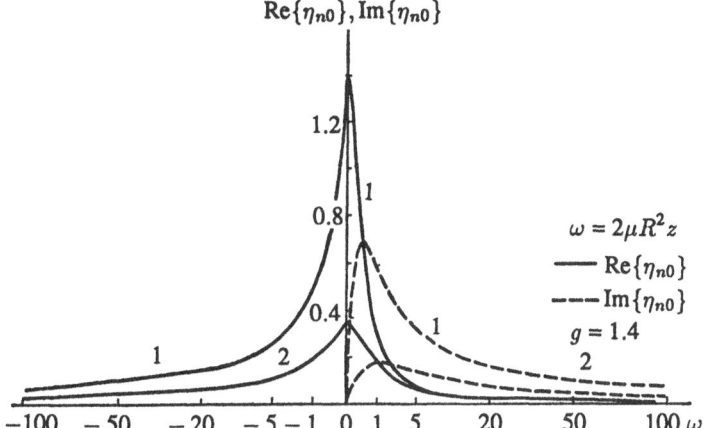

Fig. 12.3. Re $\{\eta_{n0}\}$ (*solid lines*) and Im $\{\eta_{n0}\}$ (*dashed lines*) as functions of the parameter $\omega = 2\mu z R^2$ for the Hulthen potential ($g = 1.403$). The numerical labels on the curves are equal to n

$$C_n^2(z) = \frac{n\pi\, g^{2n-1} R}{\mu(n!)^4} \left[\eta_{2n}(z) \prod_{\nu=1}^{n-1} \eta_\nu^2(z) \right]^{-1}$$

and are normalized according to (12.21).

As the parameter z varies from $-\infty$ to 0, the eigenvalues $\eta_n(z)$ are real and vary from 0 to g/n^2. The bound states occur for z such that $\eta_n(z) = 1$. With further variation of z from 0 to $\infty + iO$ (along the upper edge of the cut), the eigenvalues $\eta_n(z)$ become complex and move in the complex plane along the semicircles of radii $g/2n^2$ with the centers at the points $g/2n^2$ on the real axis. The expansion of $\eta_{n0}(E+iO)$ for low real energies E is determined by the constants B_{n0} and C_{n0},

$$B_{n0} = -\frac{8\mu g R^2}{n^4} \quad , \quad C_{n0} = \frac{2\sqrt{2\mu}\, g R}{n^3} \quad . \tag{12.79}$$

Therefore, if g/n^2 is somewhat greater than one, then a resonance state with energy E_n can occur, It corresponds to η_n with Re $\{\eta_n(E_n + iO)\} \simeq 1$ and Im $\{\eta_n(E_n + iO)\} \ll 1$. Figure 12.3 shows Re $\{\eta_n(z)\}$ and Im $\{\eta_n(z)\}$ as functions of the energy parameter z which varies from $-\infty$ to $\infty + iO$; $g = 1.4$ (the two-nucleon interaction in the triplet spin state) [12.1].

12.3 Find the eigenvalues and eigenfunctions of the kernel of the Lippmann–Schwinger equation with the square well potential

$$V(r) = \begin{cases} -V_0 & , & r < R \ ; \\ 0 & , & r > R \ ; \end{cases} \qquad V_0 \equiv \frac{g}{2\mu R^2} \ . \tag{12.80}$$

The Jost function associated with the square well potential is

$$f(-q; g) = e^{iqR} \left\{ \cos \left(g + q^2 R^2 \right)^{1/2} \right.$$

$$\left. - i \frac{qR}{(g + q^2 R^2)^{1/2}} \sin \left(g + q^2 R^2 \right)^{1/2} \right\} \quad . \tag{12.81}$$

The bound state energies are determined by the roots of the transcendental equation

$$\sqrt{g - \kappa^2 R^2} \cot \sqrt{g - \kappa^2 R^2} = -\kappa R \quad , \tag{12.82}$$

and the bound state wave functions are

$$\varphi_n(k) = \frac{N_n}{k^2 + \kappa_n^2} \frac{\cos kR + (\kappa_n/R) \sin kR}{g - (\kappa_n^2 + k^2) R^2} \quad . \tag{12.83}$$

The eigenfunctions $\eta_n(z)$ of the square well potential can be found by solving the transcendental equation

$$\sqrt{\frac{g}{\eta} + q^2 R^2} \cot \sqrt{\frac{g}{\eta} + q^2 R^2} = iqR \quad . \tag{12.84}$$

For small z, one can find the explicit expression for the eigenvalues $\eta_n(z)$ to be

$$\eta_n(z) \simeq \frac{4g}{(2n-1)^2 \pi^2} \left\{ 1 + \frac{8i}{(2n-1)^2 \pi^2} qR \right.$$

$$\left. + \frac{4}{(2n-1)^2 \pi^2} \left(1 - \frac{20}{(2n-1)^2 \pi^2} \right) q^2 R^2 \right\} \quad , \tag{12.85}$$

$$|q|R \ll 1 \quad .$$

The eigenfunctions $g_n(k, z)$ of the square well potential are given by

$$g_n(k, z) = C_n(z) \frac{g}{\eta_n(z)} \frac{\cos kR - (iq/k) \sin kR}{[g/\eta_n(z)] + (q^2 - k^2) R^2} \quad , \tag{12.86}$$

$$C_n^2(z) = \frac{4\pi R}{\mu} \frac{1 + q^2 R^2 [\eta_n(z)/g]}{[g/\eta_n(z)] - iqR} \quad .$$

The real and imaginary parts of the eigenvalues $\eta_n(z)$ associated with the square well potential ($g = 3.6$) are given in Fig. 12.4 as functions of the parameter z which varies along the real axis from $-\infty$ to $\infty + iO$ [12.6]. The coefficients of the low energy expansion of $\eta_n(E + iO)$ in the case of the square well potential are

$$B_{n0} = \frac{32 \mu g R^2}{(2n-1)^4 \pi^4} \left[1 - \frac{20}{(2n-1)^2 \pi^2} \right] \quad ,$$

$$C_{n0} = \frac{32 (2\mu)^{1/2} g R}{(2n-1)^4 \pi^4} \quad . \tag{12.87}$$

For $n \geq 2$, the coefficient B_{n0} is positive, $B_{n0} > 0$. Therefore, for $n \geq 2$ the resonance states of the system occur if g/n^2 is somewhat smaller than one; they are associated with the energies E_n for which $\mathrm{Re}\left\{\eta_n(E_n + iO)\right\} \simeq 1$ and $\mathrm{Im}\left\{\eta_n(E_n + iO)\right\} \ll 1$.

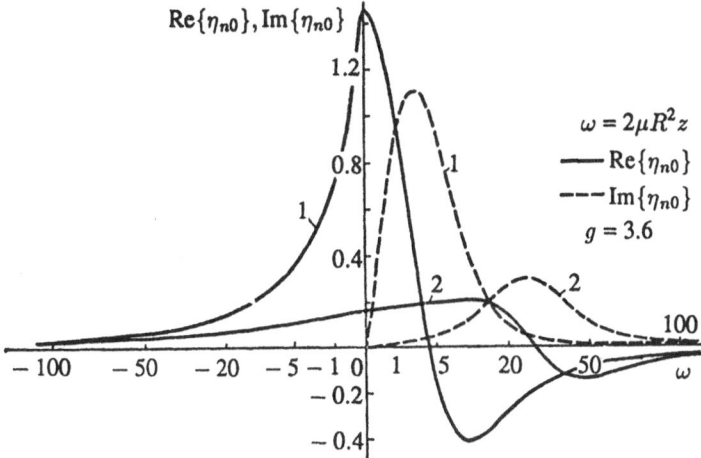

Fig. 12.4. Re $\{\eta_{n0}\}$ (*solid lines*) and Im $\{\eta_{n0}\}$ (*dashed lines*) as functions of the parameter $\omega = 2\mu z R^2$ for the square well potential ($g = 3.608$). The numerical labels on the curves are equal to n

13. Three-Particle Scattering

13.1 The Faddeev Equations

Let us consider scattering in a three-particle system. The approach that directly employs the Lippmann–Schwinger equation is inapplicable in this case. With the center-of-mass motion being taken into account, the Lippmann–Schwinger equation has ambiguous solutions even in the case of a two-particle system (Chap. 3). A similar ambiguity occurs for systems of three and more particles. Since the subsystems contained in the many-particle system can possess bound states just as the entire system, the solutions of the inhomogeneous Lippmann–Schwinger equations are always ambiguous, and only the homogeneous equation for the bound state of the whole system in the c.m.s. has single-valued solution. Ambiguity can be eliminated by appropriate rearrangement of the Lippmann–Schwinger equations. The resultant equations are usually referred to as the Faddeev equations [13.1 – 3].

Let us derive the integral Faddeev equations for a system of three nonrelativistic spinless particles. The total Hamiltonian of the system may be written as

$$H = H_0 + V \quad , \tag{13.1}$$

where H_0 is the particle kinetic energy operator, and V is the interaction potential. Taking into account only two-particle forces, we can write V as a sum of three terms

$$V + V_{12} + V_{23} + V_{31} \quad , \tag{13.2}$$

where V_{ij} describes interaction between the particles i and j which decreases with increasing distance between the particles.

The transition operator T is determined by the Lippmann–Schwinger equation

$$T(Z) = V + V G_0(Z) T(Z) \quad , \tag{13.3}$$

where

$$G_0(Z) = \left(Z - H_0\right)^{-1} \quad , \qquad Z = E + iO \quad . \tag{13.4}$$

Note that the kernel of the integral equation (13.3) is singular by virtue of delta functions which appear due to momentum conservation for the particle that is not involved in interaction with the other particles forming the pair.

In the case of the two-particle interaction given by (13.2), the transition operator T can also be written as a sum

$$T(Z) = T^{(1)}(Z) + T^{(2)}(Z) + T^{(3)}(Z) \quad , \tag{13.5}$$

the individual terms of which are defined by

$$T^{(k)}(Z) = V_{ij} + V_{ij} \, G_0(Z) \, T(Z) \quad , \qquad ijk = 123, \, 231, \, 312 \quad . \tag{13.6}$$

Having substituted (13.5) for $T(Z)$ in the right-hand parts of (13.6), we can treat these relations as a system of interconnected operator equations for the individual terms of (13.5). Note that the iteration series $T^{(k)}(Z)$ in (13.6) contains both singular terms (of the type $V_{23} \, G_0(Z) \, V_{23}$, $V_{23} \, G_0(Z) \, V_{23} \, G_0(Z) \, V_{23}$, and so on) and the terms in which the delta functions are eliminated by the intermediate integration (of the type $V_{23} \, G_0(Z) V_{31}$ etc.). The resultant system (13.6) is equivalent to (13.3) and therefore its solutions are ambiguous, similarly to those of the Lippmann–Schwinger equation.

In order to eliminate ambiguity, we rearrange the system (13.6) regarding the two-particle transition operators T_{ij} as being known. Suppose the operators T_{ij} are determined by

$$T_{ij}(Z) = V_{ij} + V_{ij} \, G_0(Z) \, T_{ij}(Z) \quad , \tag{13.7}$$

which follows from (13.3) if interaction between the particles i, j and the third particle is neglected. Note that the right-hand part of (13.7) contains all the singularities of (13.6) and hence can serve to exclude them. Separating in (13.6) the diagonal part

$$\left[1 - V_{ij} \, G_0(Z) \right] T^{(k)}(Z) = V_{ij} + V_{ij} \, G_0(Z) \left[T^{(i)}(Z) + T^{(j)}(Z) \right] \quad ,$$
$$(ijk = 123, \, 231, \, 312)$$

inverting the two-particle operator $\left[1 - V_{ij} \, G_0(Z) \right]$, and then making use of (13.7), we rewrite the system of interconnected equations for individual terms of the transition operator as

$$T^{(k)}(Z) = T_{ij}(Z) + T_{ij}(Z) \, G_0(Z) \left[T^{(i)}(Z) + T^{(j)}(Z) \right] \quad , \tag{13.8}$$
$$(ijk = 123, \, 231, \, 312) \quad .$$

In contrast to (13.6), this system of integral equations has a single-valued solution. The iteration series of the second term in the right-hand part of (13.8) contains no singularities. Therefore, the system of integral equations (13.8) can be solved by the Fredholm methods.

By using (3.79) for the relation between the transition operator T and the Green function G, we obtain from (13.8) a system of equations for G. Within the context of (13.5), we have

$$G(Z) = G_0(Z) + G^{(1)}(Z) + G^{(2)}(Z) + G^{(3)}(Z) \quad , \tag{13.9}$$

$$G^{(i)}(Z) = G_0(Z)\, T^{(i)}(Z)\, G_0(Z) \quad , \qquad i = 1,2,3 \ .$$

The functions $G^{(i)}$ satisfy the equations

$$G^{(k)}(Z) = G_{ij}(Z) - G_0(Z) + G_0(Z)\, T_{ij}(Z) \left[G^{(i)}(Z) + G^{(j)}(Z) \right] \quad ,$$
$$ijk = 123,\ 231,\ 312 \quad , \tag{13.10}$$

where

$$G_{ij}(Z) = G_0(Z) + G_0(Z)\, T_{ij}(Z)\, G_0(Z) \quad .$$

The equations for the corresponding wave function of the system can be derived from (13.10) by using (3.60).

In a three-particle system, the range of motion of all three particles as well as of one particle relative to the bound state of the other two can be infinite. We denote the asymptotic wave functions as Φ_{123} and Φ_i, where $i = 1, 2,$ or 3. (The subscripts label unbound particles. All three particles are assumed to be different.)

We apply (13.10), multiplied by $i\varepsilon$, to the function Φ_{123} and use (3.60). Then we obtain an equation for the wave function of the system of three unbound particles Ψ_{123}, i.e.,

$$\Psi_{123} = \Phi_{123} + \Psi_{123}^{(1)} + \Psi_{123}^{(2)} + \Psi_{123}^{(3)} \quad ,$$

$$\Psi_{123}^{(1)} + \Phi_{1(23)} - \Phi_{123} + G_0(Z)\, T_{23}(Z) \left[\Psi_{123}^{(2)} + \Psi_{123}^{(3)} \right] \quad ,$$

$$\Psi_{123}^{(2)} = \Phi_{2(31)} - \Phi_{123} + G_0(Z)\, T_{31}(Z) \left[\Psi_{123}^{(3)} + \Psi_{123}^{(1)} \right] \quad , \tag{13.11}$$

$$\Psi_{123}^{(3)} = \Phi_{3(12)} - \Phi_{123} + G_0(Z)\, T_{12}(Z) \left[\Psi_{123}^{(1)} + \Psi_{123}^{(2)} \right] \quad ,$$

$$Z = E + iO \quad ,$$

where

$$\Phi_{1(23)} = \lim_{\varepsilon \to 0} i\varepsilon\, G_{23}(E + i\varepsilon)\, \Phi_{123} , \ \dots \quad .$$

The function $\Phi_{1(23)}$, in contrast to Φ_{123}, takes into account interaction between the particles 2 and 3. It is not difficult to verify that at large distances the difference $\Phi_{1(23)} - \Phi_{123}$ is a divergent wave with respect to the distance between particles 2 and 3.

If particle 1 is scattered by a bound state of particles 2 and 3, then the wave fucntion Ψ_1 is determined by the equations

$$\Psi_1 = \Psi_1^{(1)} + \Psi_1^{(2)} + \Psi_1^{(3)} \quad ,$$

$$\Psi_1^{(1)} = \Phi_1 + G_0(Z)\, T_{23}(Z) \left[\Psi_1^{(2)} + \Psi_1^{(3)} \right] \quad ,$$

$$\Psi_1^{(2)} = G_0(Z)\, T_{31}(Z) \left[\Psi_1^{(3)} + \Psi_1^{(1)} \right] \quad , \tag{13.12}$$

$$\Psi_1^{(3)} = G_0(Z) \, T_{12}(Z) \left[\Psi_1^{(1)} + \Psi_1^{(2)} \right] \quad ,$$

$$Z = E + iO \quad .$$

The functions Ψ_2 and Ψ_3 are governed by analogous systems of equations.

The system of three integral equations (13.12) can be reduced to a system of two integral equations for the functions $\Psi^{(I)}$ and $\Psi^{(II)}$, which are defined as

$$\Psi_1^{(I)} = \Psi_1^{(1)} \quad , \qquad \Psi_1^{(II)} = \Psi_1^{(2)} + \Psi_1^{(3)} \quad ,$$

and can be easily shown to satisfy the equations

$$\Psi_1^{(I)} = \Phi_1 + G_0(Z) \, T_{23}(Z) \, \Psi_1^{(II)} \quad , \qquad \Psi_1^{(II)} = G_0 \, T_1(Z) \, \Psi_1^{(I)} \quad , \tag{13.13}$$

where T_1 is the operator describing the scattering of an individual particle by the system of other two noninteracting particles, i.e.,

$$T_1(Z) = V_1 + V_1 \, G_0(Z) \, T_1(Z) \quad , \qquad V_1 = V_{12} + V_{13} \quad . \tag{13.14}$$

Note that both infinite motion of all three particles and scattering of one particle by the bound state of the other two are described by nonuniform systems of integral equations which have unique solutions.

The wave function Ψ_0 of the bound state of the whole system is determined by a uniform set of integral equations which can be derived in a manner similar to the previous analysis. We obtain

$$\Psi_0 = \Psi_0^{(1)} + \Psi_0^{(2)} + \Psi_0^{(3)} \quad ,$$

$$\Psi_0^{(1)} = G_0(Z) \, T_{23}(Z) \left[\Psi_0^{(2)} + \Psi_0^{(3)} \right] \quad ,$$

$$\Psi_0^{(2)} = G_0(Z) \, T_{31}(Z) \left[\Psi_0^{(3)} + \Psi_0^{(1)} \right] \quad , \tag{13.15}$$

$$\Psi_0^{(3)} = G_0(Z) \, T_{12}(Z) \left[\Psi_0^{(1)} + \Psi_0^{(2)} \right] \quad .$$

The uniform set of (13.15) has solutions only for energies associated with bound states of the system. In the c.m.s., these energies are negative ($E < 0$). Note that the set (13.15) can be formally derived from (13.12) if we put $\Phi_1 = 0$.

13.2 Coordinates and Momenta in the Three–Particle System

We consider three particles with masses m_1, m_2, and m_3, coordinates r_1, r_2, and r_3, and momenta k_1, k_2, and k_3. To describe three-particle relative motion, we introduce the Jacobi coordinates

$$x_1 = r_1 - \frac{m_2 r_2 + m_3 r_3}{m_2 + m_3} \quad ,$$

$$r_{23} = r_2 - r_3 \quad , \tag{13.16}$$

$$R = \frac{1}{M}\left(m_1 r_1 + m_2 r_2 + m_3 r_3\right) \quad ,$$

where $M = m_1 + m_2 + m_3$ is the total mass of the system. Instead of relative coordinates x_1 and r_{23}, the pairs x_2 and r_{31} or x_3 and r_{12} can be used:

$$x_2 = -\frac{m_1}{m_1 + m_3}\, x_1 + \frac{m_3 M}{(m_1 + m_3)(m_2 + m_3)}\, r_{23} \quad ,$$

$$r_{31} = -x_1 - \frac{m_2}{m_2 + m_3}\, r_{23} \quad ; \tag{13.17}$$

or

$$x_3 = -\frac{m_1}{m_1 + m_2}\, x_1 - \frac{m_2 M}{(m_1 + m_2)(m_2 + m_3)}\, r_{23} \quad ,$$

$$r_{12} = x_1 - \frac{m_3}{m_2 + m_3}\, r_{23} \quad . \tag{13.18}$$

The Jacobi momenta are defined by

$$p_1 = \frac{1}{M}\left[(m_2 + m_3)\, k_1 - m_1\left(k_2 + k_3\right)\right] \quad ,$$

$$k_{23} = \frac{m_3 k_2 - m_2 k_3}{m_2 + m_3} \quad , \tag{13.19}$$

$$K = k_1 + k_2 + k_3 \quad .$$

The equivalent relative momenta are p_2 and k_{31}, or p_3 and k_{12}:

$$p_2 = -\frac{m_2}{m_2 + m_3}\, p_1 + k_{23} \quad ,$$

$$k_{31} = -\frac{m_3 M}{(m_1 + m_3)(m_2 + m_3)}\, p_1 - \frac{m_1}{m_1 + m_3}\, k_{23} \quad , \tag{13.20}$$

or

$$p_3 = -\frac{m_3}{m_2 + m_3}\, p_1 - k_{23} \quad ,$$

$$k_{12} = \frac{m_2 M}{(m_1 + m_2)(m_2 + m_3)}\, p_1 - \frac{m_1}{m_1 + m_2}\, k_{23} \quad . \tag{13.21}$$

The kinetic energy operator of the system can be written in terms of the Jacobi coordinates x_1, r_{23}, and R, so that

$$H_0 = -\frac{1}{2\mu_1}\, \Delta_1 - \frac{1}{2\mu_{23}}\, \Delta_{23} - \frac{1}{2M}\, \Delta_R \quad , \tag{13.22}$$

with the reduced masses being

$$\mu_1 = \frac{m_1\left(m_2 + m_3\right)}{M} \quad , \qquad \mu_{23} = \frac{m_2 m_3}{m_2 + m_3} \quad .$$

In the momentum representation, H_0 takes the form

$$H_0 = \frac{p_1^2}{2\mu_1} + \frac{k_{23}^2}{2\mu_{23}} + \frac{K^2}{2M} \quad . \tag{13.23}$$

Now let us consider explicit expressions for the asymptotic functions. We label the initial momenta by the superscript zero. The free motion of all particles is described by the asymptotic function Φ_{123}

$$\Phi_{123} = \exp\left[i\, p_1^0 x_1 + i\, k_{23}^0 r_{23} + i\, K^0 R\right] \quad , \tag{13.24}$$

$$E_{123} = \frac{p_1^{0^2}}{2\mu_1} + \frac{k_{23}^{0^2}}{2\mu_{23}} + \frac{K^{0^2}}{2M} \quad .$$

Note that Φ_{123} is invariant with respect to interchanges of the Jacobi coordinates.
The asymptotic function Φ_1 is given by

$$\Phi_1 = \exp\left[i\, p_1^0 x_1 + i\, K^0 R\right] \varphi_{\kappa_{23}}(r_{23}) \quad , \tag{13.25}$$

$$E_1 = \frac{p_1^{0^2}}{2\mu_1} - \frac{\kappa_{23}^2}{2\mu_{23}} + \frac{K^{0^2}}{2M} \quad ,$$

where $\varphi_{\kappa_{23}}(r_{23})$ is the solution of the equation

$$\left(-\frac{1}{2\mu_{23}} \Delta_{23} + V_{23} - E_{23}\right) \varphi_{\kappa_{23}}(r_{23}) = 0 \tag{13.26}$$

for negative energies of relative motion $E_{23} = -\kappa_{23}^2/2\mu_{23} < 0$, i.e., the bound state wave function of the particles 2 and 3.
The asymptotic function $\Phi_{1(23)}$ can be written as

$$\Phi_{1(23)} = \exp\left[i\, p_1^0 x_1 + i\, K^0 R\right] \varphi_{k_{23}^0}(r_{23}) \quad , \qquad E_{1(23)} = E_{123} \quad , \tag{13.27}$$

where $\varphi_{k_{23}^0}(r_{23})$ is the solution of (13.26) with $E_{23} = \left(k_{23}^0\right)^2/2\mu_{23} > 0$, which reduces at infinity to the sum of a plane wave and a diverging spherical wave.

13.3 Momentum Representation

The Faddeev equations are simplest in the momentum representation. Each component $\Psi^{(i)}$ of the total wave function can be written as a function of the relevant set of coordinates, i.e.,

$$\Psi^{(i)} \equiv \Psi^{(i)}\left(r_{jk}, x_i, R\right) \quad , \qquad ijk = 123, 231, 312 \quad . \tag{13.28}$$

In the momentum representation, $\Psi^{(i)}$ is defined by

$$\Psi^{(i)}\left(k_{jk}, p_i, K\right) = \int dR \, dx_i \, dr_{jk} \, \exp\left[-i \, k_{jk} \, r_{jk} - i \, p_i \, x_i - i \, K R\right]$$
$$\times \Psi^{(i)}\left(r_{jk}, x_i, R\right) \quad . \tag{13.29}$$

The kinetic energy operator in the momentum representation reduces to the multiplication operator, hence the Green function $G_0(Z)$ is diagonal:

$$\langle k'_{jk} \, p'_i \, K' | G_0(Z) | K p_i k_{jk} \rangle = (2\pi)^9 \left(Z - \frac{k_{jk}^2}{2\mu_{jk}} - \frac{p_i^2}{2\mu_i} - \frac{K^2}{2M} \right)^{-1} \tag{13.30}$$
$$\times \delta\left(k_{jk} - k'_{jk}\right) \delta\left(p_i - p'_i\right) \delta\left(K - K'\right) \quad .$$

The two-particle scattering operator $T_{jk}(Z)$ is diagonal in the representation of the total momentum of the system and the free particle momentum p_i, i.e.,

$$\langle k'_{jk} \, p'_i \, K' \, | T_{jk}(Z) | \, K p_i k_{jk} \rangle \tag{13.31}$$

$$= (2\pi)^6 \left\langle k'_{jk} \left| t_{jk}\left(Z - \frac{p_i^2}{2\mu_i} - \frac{K^2}{2M} \right) \right| k_{jk} \right\rangle \delta\left(p_i - p'_i\right) \delta(K - K') \, ,$$

where $t_{jk}(Z)$ is the two-particle t matrix determined by the integral equation (12.1).

Since all the operators entering the Faddeev equations are diagonal in the K-representation, the wave function Ψ must contain the factor $\delta\left(K - K^0\right)$ which reflects conservation of the total momentum of the system. Therefore, all K-dependences can be excluded from consideration as soon as we convert to the center-of-mass system.

For example, we write in the momentum representation the system of integral Faddeev equations (13.12) for the case of particle scattering by the bound state of the other two particles:

$$\Psi_1 = \Psi_1^{(1)}\left(k_{23}, p_1\right) + \Psi_1^{(2)}\left(k_{31}, p_2\right) + \Psi_1^{(3)}\left(k_{12}, p_3\right) \quad , \tag{13.32}$$

$$\Phi_1^{(1)}\left(\boldsymbol{k}_{23},\boldsymbol{p}_1\right) = \Phi_1\left(\boldsymbol{k}_{23},\boldsymbol{p}_1\right)$$

$$+ \left(Z - \frac{k_{23}^2}{2\mu_{23}} - \frac{p_1^2}{2\mu_1}\right)^{-1}$$

$$\times \int \frac{d\boldsymbol{k}_{23}'}{(2\pi)^3} \left\langle \boldsymbol{k}_{23} \left| t_{23}\left(Z - \frac{p_1^2}{2\mu_1}\right)\right| \boldsymbol{k}_{23}' \right\rangle$$

$$\times \left\{ \Psi_1^{(2)}\left(\boldsymbol{k}_{31}',\boldsymbol{p}_2'\right) + \Psi_1^{(3)}\left(\boldsymbol{k}_{12}',\boldsymbol{p}_3'\right) \right\} \quad,$$

$$\Psi_1^{(2)}\left(\boldsymbol{k}_{31},\boldsymbol{p}_2\right) = \left(Z - \frac{k_{31}^2}{2\mu_{31}} - \frac{p_2^2}{2\mu_2}\right)^{-1}$$

$$\times \int \frac{d\boldsymbol{k}_{31}'}{(2\pi)^3} \left\langle \boldsymbol{k}_{31} \left| t_{31}\left(Z - \frac{p_2^2}{2\mu_2}\right)\right| \boldsymbol{k}_{31}' \right\rangle \qquad (13.33)$$

$$\times \left\{ \Psi_1^{(3)}\left(\boldsymbol{k}_{12}',\boldsymbol{p}_3'\right) + \Psi_1^{(1)}\left(\boldsymbol{k}_{23}',\boldsymbol{p}_1'\right) \right\} \quad,$$

$$\Psi_1^{(3)}\left(\boldsymbol{k}_{12},\boldsymbol{p}_3\right) = \left(Z - \frac{k_{12}^2}{2\mu_{12}} - \frac{p_3^2}{2\mu_3}\right)^{-1}$$

$$\times \int \frac{d\boldsymbol{k}_{12}'}{(2\pi)^3} \left\langle \boldsymbol{k}_{12} \left| t_{12}\left(Z - \frac{p_3^2}{2\mu_3}\right)\right| \boldsymbol{k}_{12}' \right\rangle$$

$$\times \left\{ \Psi_1^{(1)}\left(\boldsymbol{k}_{23}',\boldsymbol{p}_1'\right) + \Psi_1^{(2)}\left(\boldsymbol{k}_{31}',\boldsymbol{p}_2'\right) \right\} \quad,$$

$$Z \equiv \frac{p_1^{0^2}}{2\mu_1} - \frac{\kappa_{23}^2}{2\mu_{23}} + iO \quad.$$

The quantities $\boldsymbol{k}_{31}'$, $\boldsymbol{p}_2'$ and $\boldsymbol{k}_{12}'$, $\boldsymbol{p}_3'$ entering the integrand in the first equation must be expressed in terms of $\boldsymbol{p}_1$ and $\boldsymbol{k}_{23}'$ according to (13.20, 21) etc. The function $\Phi_1\left(\boldsymbol{k}_{23},\boldsymbol{p}_1\right)$, by virtue of (13.25), is given by

$$\Phi_1\left(\boldsymbol{k}_{23},\boldsymbol{p}_1\right) = (2\pi)^3\,\delta\left(\boldsymbol{p}-\boldsymbol{p}_1^0\right)\varphi_{\kappa_{23}}(\boldsymbol{k}_{23}) \quad, \qquad (13.34)$$

where $\varphi_{\kappa_{23}}(\boldsymbol{k}_{23})$ is the wave function of the two-particle bound state with the binding energy $\kappa_{23}^2/2\mu_{23}$; $\boldsymbol{p}_1^0$ is the initial momentum of relative motion of the system.

If all three particles are identical and have zero spin and isotopic spin, then the total wave function of the system Ψ_1 must be symmetric with respect to permutations of any pairs of particles. In this case we have

$$\Psi_1^{(1)}(\boldsymbol{k},\boldsymbol{p}) = \Psi_1^{(2)}(\boldsymbol{k},\boldsymbol{p}) = \Psi_1^{(3)}(\boldsymbol{k},\boldsymbol{p}) \equiv \psi(\boldsymbol{k},\boldsymbol{p}) \qquad (13.35)$$

and the wave function can be written as

$$\Psi_1 = \psi\left(k_{23}, p_1\right) + \psi\left(k_{31}, p_2\right) + \psi\left(k_{12}, p_3\right) \quad . \tag{13.36}$$

[The relative momenta k_{ij} and p_k (where $ijk = 123, 231,$ and 312) are given by (13.19) with $m_1 = m_2 = m_3 = m$.] Then, instead of the system of three equations (13.33), we obtain a single integral equation for the function $\psi(k, p)$:

$$\psi(k, p) = \varphi(k, p) + \left(Z - \frac{k^2}{2m} - \frac{3}{4}\frac{p^2}{m}\right)^{-1}$$
$$\times \int \frac{dp'}{(2\pi)^3} \left\{ \left\langle k \left| t\left(Z - \frac{3}{4}\frac{p^2}{m}\right) \right| \frac{p}{2} + p' \right\rangle \right. \tag{13.37}$$
$$\left. + \left\langle k \left| t\left(Z - \frac{3}{4}\frac{p^2}{m}\right) \right| -\frac{p}{2} - p' \right\rangle \right\} \psi\left(p + \frac{p'}{2}, p'\right) \quad ,$$

where $\varphi(k, p)$ is a function whose symmetrization yields the initial wave function of the system

$$\Phi_1 = \varphi\left(k_{23}, p_1\right) + \varphi\left(k_{31}, p_2\right) + \varphi\left(k_{12}, p_3\right) \quad . \tag{13.38}$$

For a three-particle bound state we have $\varphi = 0$, $Z = E_0 < 0$. In the case of particle scattering by a two-particle bound state,

$$\varphi(k, p) = (2\pi)^3 \, \delta\left(p - p^0\right) \varphi_{N_0}(k) \quad , \tag{13.39}$$

$$N_0 = \{n_0, l_0, m_0\} \quad , \qquad Z = \frac{3}{4}\frac{p^{0^2}}{m} - \frac{\kappa_{N_0}^2}{m} + iO \quad .$$

Note that both initial and final states of the matrix elements $\langle k | t(Z - \frac{3}{4}(p^2/m)) | k' \rangle$ entering (13.37) lie off the energy shell, i.e., $k^2/m \neq k'^2/m \neq Z - \frac{3}{4}(p^2/m)$.

13.4 Partial Wave Expansion

In the general case the wave function of the system described by the integral equation (13.37) depends on six variables (two relative vectors). For a centrosymmetric interaction potential, the three-dimensional integral equation (13.37) can be reduced to a system of two-dimensional integral equations by expanding the wave function in terms of angular functions and separating the angle variables [13.4].

Let us consider a system of three identical particles. Suppose l is the angular momentum of the two-particle relative motion and λ is the analogous but associated with the relative motion of the third particle with respect to the center-of-mass of the other two particles. The total angular momentum L of the system is equal to the vector sum of l and λ, i.e.,

$$L = l + \lambda \quad . \tag{13.40}$$

The wave function of the state with total angular momentum L may be taken in the form

$$\mathcal{Y}_{l\lambda LM}(n_k, n_p) = \sum_{m\mu} (lm\lambda\mu \,|\, LM)\, Y_{lm}(n_k)\, Y_{\lambda\mu}(n_p) \quad . \tag{13.41}$$

The functions $\mathcal{Y}_{l\lambda LM}$ are orthonormalized and form a complete set.

The three-particle wave function $\varphi(k, p)$ depends, along with k and p, on the initial relative momentum of the system p_0. We assume the angular momentum of the two-particle bound state, which is the scatterer of the third particle, to be zero $(l_0 = 0)$. Then, since the wave function $\psi(k, p)$ is scalar, its expansion in terms of functions (13.41) is given by

$$\psi(k, p) \equiv \psi(k, p; p_0)$$
$$= \sum_{l\lambda LM} \psi_{l\lambda L}(k, p; p_0)\, \mathcal{Y}_{l\lambda LM}(n_k, n_p)\, Y^*_{LM}(n_{p_0}) \quad . \tag{13.42}$$

Substituting (13.42) and (12.8) into (13.37) and bearing in mind that the functions (13.41) are orthonormalized, we obtain a system of integral equations for the expansion coefficients $\psi_{l\lambda L}$, i.e.,

$$\psi_{l\lambda L}(k, p; p_0) = (2\pi)^3 \, \varphi_{10}(k) \, \frac{\delta(p - p_0)}{p^2} \, \delta_{l0} \, \delta_{\lambda L}$$
$$+ 2\Delta_l \left(z_p - \frac{k^2}{m} \right)^{-1} \frac{1}{2\pi^2} \int dp' \int do_p$$
$$\times \mathcal{Y}^*_{l\lambda L0}(n_q, n_p)\, t_l(q, k; z_p)$$
$$\times \sum_{l'\lambda'} \psi_{l'\lambda'L}(q', p'; p_0)\, \mathcal{Y}_{l'\lambda'L0}(n_{q'}, n_{p'}) \quad , \tag{13.43}$$

$$z_p \equiv Z - \frac{3}{4} \frac{p^2}{m} \quad ,$$

where $\Delta_l = \frac{1}{2}[1 + (-1)^l]$, $q = p/2 + p'$, and $q' = p + p'/2$. Having introduced in the integrand in the right-hand part of (13.43) a delta-function and additional integration $[dk'^2 \, \delta(k'^2 - q'^2)]$, we obtain

$$\psi_{l\lambda L}(k, p; p_0) = (2\pi)^3 \, \varphi_{10}(k) \, \frac{\delta(p - p_0)}{p^2} \, \delta_{l0} \, \delta_{\lambda L}$$
$$+ 2\Delta_l \left(z_p - \frac{k^2}{m} \right)^{-1} \sum_{l'\lambda'} \frac{1}{2\pi^2} \int_0^\infty dp' \int_{|p-p'/2|}^{p+p'/2} dk' \, \frac{k'p'}{p}$$
$$\times t_l(q, k; z_p) K^{(L)}_{l\lambda,l'\lambda'}(p, p'; k')\, \psi_{l'\lambda'L}(k', p'; p_0) \quad , \tag{13.44}$$

where $q^2 \equiv k'^2 + \frac{3}{4}p'^2 - \frac{3}{4}p^2$ and

$$K^{(L)}_{l\lambda,l'\lambda'}(p, p'; k)$$
$$\equiv 2 \int do_p \, do_{p'} \, \mathcal{Y}^*_{l\lambda L0}(n_q, n_p) \, \delta\left(\cos\theta - \frac{k'^2 - p^2 - p'^2/4}{pp'} \right)$$

$$\times \, \mathcal{Y}_{l' \lambda' L0}\left(\boldsymbol{n}_{q'}, \boldsymbol{n}_{p'}\right) \quad . \tag{13.45}$$

Thus, the integral equation (13.37) for the function $\psi(\boldsymbol{k}, \boldsymbol{p})$ is reduced to an infinite set of two-dimensional integral equations for the expansion coefficients $\psi_{l\lambda L}(k, p)$. The set (13.44) is finite if particle interaction is described by a finite-range potential.

After explicit expressions are substituted for the functions $\mathcal{Y}_{l\lambda L0}$ and angle-integration is carried out, one finds the kernels of the integral equations to be

$$
\begin{aligned}
K_{l\lambda, l'\lambda'}^{(L)}(p, p'; k') &= (4\pi)^{3/2}(2\lambda' + 1)^{-1/2} \sum_{mm'} (-1)^{l+l'-m-m'} \\
&\quad \times \left(lmL - m \mid \lambda 0\right)\left(l'm'L - m \mid \lambda'm' - m\right) \\
&\quad \times Y_{lm}^*(\vartheta, 0)\, Y_{l'm'}(\vartheta', 0)\, Y_{\lambda'm-m'}(\theta, 0) \quad ,
\end{aligned}
\tag{13.46}
$$

where the angles θ, ϑ, and ϑ' are given by

$$\cos\theta = \frac{k'^2 - p^2 - p'^2/4}{pp'}$$

$$\cos\vartheta = \frac{k'^2 - p^2/2 - p'^2/4}{p\left(k'^2 + \frac{3}{4}p'^2 - \frac{3}{4}p^2\right)^{1/2}} \quad , \tag{13.47}$$

$$\cos\vartheta' = \frac{k'^2 + p^2 - p'^2/4}{2pk'} \quad .$$

Note that

$$K_{00,00}^{(0)} = 1 \quad . \tag{13.48}$$

The three-particle bound state wave function can also be expanded in terms of the angular functions (13.41). For the bound state with total angular momentum L and projection M, this expansion is given by

$$\psi_{LM}(\boldsymbol{k}, \boldsymbol{p}) = \sum_{l\lambda} \psi_{l\lambda L}(k, p)\, \mathcal{Y}_{l\lambda LM}\left(\boldsymbol{n}_k, \boldsymbol{n}_p\right) \quad . \tag{13.49}$$

We substitute (13.49) into (13.37) and note that the function φ entering (13.37) is equal to zero for the bound state of the system; thus we obtain a uniform set of two-dimensional integral equations for the expansion coefficients $\psi_{l\lambda L}(k, p)$,

$$
\begin{aligned}
\psi_{l\lambda L}(k, p) &= 2\Delta_l\left(z_p - \frac{k^2}{m}\right)^{-1} \sum_{l'\lambda'} \frac{1}{2\pi^2} \int_0^\infty dp' \int_{|p-p'/2|}^{p+p'/2} dk'\, \frac{k'p'}{p} \\
&\quad \times t_l(q, k; z_p)\, K_{l\lambda, l'\lambda'}^{(L)}(p, p'; k')\, \psi_{l'\lambda' L}(k', p') \quad ,
\end{aligned}
\tag{13.50}
$$

where $K_{l\lambda, l'\lambda'}^{(L)}$ is given by (13.46) as before. This system immediately follows from (13.44) if we set the inhomogeneous terms equal to zero.

Because of the factor Δ_l in (13.44, 50), the components $\psi_{l\lambda L}$ with odd l vanish due to the symmetry of the wave function Ψ with respect to permutations of any pairs of particles.

In the case of short-range nuclear potentials, the two-particle t matrix elements $t_l(k', k; z)$ decrease rapidly as l increases: for small k or k', the elements $t_l(k', k; z)$ are proportional to $(kk')^l$; for large k, the components t_l are small for all l and, in addition, the contribution of large k in the equations is suppressed by the factor $\left(z_p - k^2/m\right)^{-1}$. Therefore, in the equations obtained, summation over l (and hence over λ for given L) can be restricted to a finite number of terms and the systems of integral equations become finite.

The sets of (13.44, 50) are appreciably simplified if the two-particle t matrix is separable. In this case, these sets can be reduced to sets of one-dimensional integral equations which can be solved numerically.

13.5 Separable Expansion of the Two-Particle t Matrix and One-Dimensional Form of the Faddeev Equations

Let us apply the separable representation of the two-particle scattering amplitude introduced in Chap. 12 to the study of the three-particle problem. Namely, we shall reduce the systems of two-dimensional integral equations (13.44, 50) to systems of one-dimensional integral equations [13.4].

Substituting the separable expansion (12.23) for the two-particle t matrix and (12.30) for the two-particle bound state wave function into (13.44), we find the function $\psi_{l\lambda L}$ to be

$$\psi_{l\lambda L}(k, p; p_0) = (2\pi)^3 \, N_{10} \sum_n \frac{g_{nl}(k, z_p)}{(k^2/m) - z_p} \tag{13.51}$$

$$\cdot \times \left\{ \delta_{n1} \, \delta_{l0} \, \delta_{\lambda L} \frac{\delta(p - p_0)}{p^2} + \tau_{nl}(z_p) \, a_{nl\lambda L}(p, p_0) \right\} \ ,$$

where

$$\tau_{nl}(z_p) \equiv \left\{ 1 + \frac{1}{2\pi^2} \int_0^\infty dq \, q^2 \, \frac{g_{nl}^2(q, z_p)}{z_p - (q^2/2)} \right\}^{-1} \ ,$$

$$a_{nl\lambda L}(p, p_0) \equiv \frac{1}{8\pi^5} \frac{\Delta_l}{N_{10}} \sum_{l'\lambda'} \int_0^\infty dp' \int_{|p-p'/2|}^{p+p'/2} dk' \, \frac{k'p'}{p} \tag{13.52}$$

$$\times K_{l\lambda, l'\lambda'}^{(L)}(p, p'; k') \, g_{nl}(q, z_p) \, \psi_{l'\lambda' L}(k', p'; p_0) \ .$$

The factor Δ_l vanishes for odd l, hence all components $a_{nl\lambda L}$ with odd l vanish, too. This property is directly associated with the assumption of identical particles which leads to the wave function Ψ being symmetric with respect to permutation of any pair of particles. Substituting (13.51) into (13.52), we find that the functions $a_{nl\lambda L}(p, p_0)$ are determined by the following system of integral equations:

$$
\begin{aligned}
a_{nl\lambda L}(p, p_0) = {} & U_{nl\lambda L, 10Ll}(p, p_0; Z) \\
& + \sum_{n'l'\lambda'} \int_0^\infty dp' \, p'^2 \, U_{nl\lambda L, n'l'\lambda'L}(p, p'; Z) \\
& \times \tau_{n'l'}(z_{p'}) \, a_{n'l'\lambda'L}(p', p_0) \quad,
\end{aligned}
\tag{13.53}
$$

where

$$
U_{nl\lambda L, n'l'\lambda'L}(p, p'; Z) \equiv \frac{\Delta_l \Delta_{l'}}{\pi^2} \int_{|p-p'|/2}^{p+p'/2} dp' \, \frac{k'}{pp'} \, K_{l\lambda, l'\lambda'}^{(L)}(p, p'; k')
$$
$$
\times \frac{g_{nl}(q, z_p) \, g_{n'l'}(q', z_{p'})}{(k'^2/m) - z_{p'}} \quad.
\tag{13.54}
$$

(In deriving these formulas, we use $\Delta_l^2 = \Delta_l$.) We introduce in (13.54) the new integration variable

$$
y = \frac{k'^2 - p^2 - p'^2/4}{pp'} \quad.
$$

Then the integration limits become independent of p and p', so that

$$
U_{nl\lambda L, n'l'\lambda'L}(p, p'; Z) = \frac{\Delta_l \Delta_{l'}}{2\pi^2} \int_{-1}^1 dy \, K_{l\lambda, l'\lambda'}^{(L)}(p, p'; y)
$$
$$
\times \frac{g_{nl}(q, z_p) \, g_{n'l'}(q', z_{p'})}{(p^2 + p'^2 + pp'y)/m - Z} \quad,
\tag{13.55}
$$

$$
q = \left[\tfrac{1}{4} p^2 + p'^2 + pp'y \right]^{1/2} \quad, \qquad q' = \left[p^2 + \tfrac{1}{4} p'^2 + pp'y \right]^{1/2} \quad.
$$

The function $K_{l\lambda, l'\lambda'}^{(L)}(p, p'; y)$ in (13.55) is determined by (13.46) in which the angles θ, ϑ, and ϑ' must be expressed in terms of the variable y:

$$
\cos \theta = y \quad,
$$

$$
\cos \vartheta = \frac{p/2 + p'y}{\left[\tfrac{1}{4} p^2 + p'^2 + pp'y \right]^{1/2}} \quad,
\tag{13.56}
$$

$$
\cos \vartheta' = \frac{p + p'y/2}{\left[p^2 + \tfrac{1}{4} p'^2 + pp'y \right]^{1/2}} \quad.
$$

The components $U_{nl\lambda L, n'l'\lambda'L}$ with odd l or l' vanish by virtue of the factor $\Delta_l \Delta_{l'}$ that enters (13.55).

The amplitude of elastic scattering of one particle by a bound state of the other two particles is given by

$$f(\mathbf{p}_0, \mathbf{p}) = -\frac{m}{3\pi} \int d\mathbf{x} \, d\mathbf{r} \, e^{-i\mathbf{p}\mathbf{x}} \, \varphi(\mathbf{r})$$

$$\times \left[V\left(\mathbf{x} + \tfrac{1}{2}\mathbf{r}\right) + V\left(\mathbf{x} - \tfrac{1}{2}\mathbf{r}\right) \right] \Psi_1(\mathbf{r}, \mathbf{x}; \mathbf{p}_0)$$

$$= \left[\frac{(p^2 - p_0^2)}{4\pi} \int \frac{d\mathbf{k}}{(2\pi)^3} \, \varphi(\mathbf{k}) \, \Psi_1(\mathbf{k}, \mathbf{p}; \mathbf{p}_0) \right]_{p=p_0} . \tag{13.57}$$

Substituting the expansion

$$f(\mathbf{p}_0, \mathbf{p}) = 4\pi \sum_{LM} f_L(p_0, p_0) \, Y_{LM}^*(\mathbf{n}_p) \, Y_{LM}(\mathbf{n}_{p_0}) \tag{13.58}$$

and using (13.36, 42, 51), one finds the partial amplitude $f_L(p_0, p_0)$ to be a function of a_{10LL} for $p = p_0$, i.e.,

$$f_L(p_0, p_0) = \left[\frac{(p^2 - p_0^2)}{4(2\pi)^5} \int_0^\infty dk \, k^2 \, \varphi_{10}(k) \, \psi_{0LL}(k, p; p_0) \right]_{p=p_0}$$

$$= \frac{2\pi}{3} \, m \left(\frac{d\eta_{10}(Z)}{dZ} \right)_{z=-\varepsilon_{10}}^{-1} a_{10LL}(p_0, p_0) . \tag{13.59}$$

Thus, the three-particle wave function $\psi_{l\lambda L}$ is expressed in terms of the functions $a_{nl\lambda L}(p, p_0)$ which, as a matter of fact, are the amplitudes of the off-energy-shell scattering of a particle by a two-particle bound state with quantum numbers $n_0 = 1$, $l_0 = 0$ (the final state is described by the quantum numbers n, l). The functions $U_{nl\lambda L, n'l'\lambda'L}$ serve as matrix elements of the potential responsible for the interaction between the particle and the two-particle bound state, and the functions $\tau_{nl}(z_p)$ are the propagators of the free particle and the bound state of the pair of interacting particles with quantum numbers n, l. The effective potential depends on energy and is nonlocal.

If the incident particle energy is equal to zero $(p_0 = 0)$, then, according to (13.46, 55), the free term of (13.53) reduces to

$$U_{nl\lambda L, 10LL}(p, 0; -\varepsilon_{10}) = \delta_{l\lambda} \, \delta_{L0} \, U_{nl0,1000}(p, 0; -\varepsilon_{10}) \tag{13.60}$$

and hence all amplitudes with nonzero total angular momenta L vanish. The amplitudes $a_{nl\lambda L}(p, 0)$ with $L = 0$ and $l = \lambda$ are given by

$$a_{nl\lambda L}(p, 0) = \delta_{l\lambda} \, \delta_{L0} \, a_{nll0}(p, 0) . \tag{13.61}$$

The functions $a_{nll0}(p, 0) \equiv a_{nl}(p, 0)$ are determined by the integral equations

$$a_{nl}(p, 0) = U_{nl,10}(p, 0; -\varepsilon_{10})$$

$$+ \sum_{n'l'} \int_0^\infty dp' \, p'^2 \, U_{nl,n'l'}(p, p'; -\varepsilon_{10})$$

$$\times \tau_{n'l'}\left(-\varepsilon_{10} - \frac{3}{4} \frac{p'^2}{m} \right) a_{n'l'}(p', 0) , \tag{13.62}$$

where

$$U_{nl,n'l'}(p,p';Z) \equiv U_{nl0,n'l'l'0}(p,p';Z)$$

$$= (-1)^{l+l'} \, \Delta_l \Delta_{l'} \, (2l+1)^{1/2}(2l'+1)^{1/2} \frac{1}{2\pi^2}$$

$$\times \int_{-1}^{1} dy \, P_l(\cos\vartheta) \frac{g_{nl}(q,z_p) \, g_{n'l'}(q',z_{p'})}{(p^2+p'^2+pp'y)/m - Z}$$

$$\times P_{l'}(\cos(\theta-\vartheta')) \quad . \tag{13.63}$$

For the scattering length that is minus the zero-energy elastic scattering amplitude we obtain the expression

$$A = -\frac{2\pi}{3} \, m \left[\frac{d\eta_{10}(z)}{dz} \right]_{z=-\varepsilon_{10}}^{-1} a_{10}(0,0) \quad . \tag{13.64}$$

Let us derive the integral equations describing a three-particle bound state with total angular momentum quantum number L and projection M. Substituting the separable expansion (12.23) for the two-particle t matrix in (13.50), we find

$$\psi_{l\lambda L}(k,p) = \sum_n \frac{g_{nl}(k,z_p)}{(k^2/m) - z_p} \, \tau_{nl}(z_p) \, a_{nl\lambda L}(p) \quad , \tag{13.65}$$

and obtain a uniform set of one-dimensional integral equations for the partial amplitudes $a_{nl\lambda L}(p)$, i.e.,

$$a_{nl\lambda L}(p) = \sum_{n'l'\lambda'} \int_0^\infty dp' \, p'^2 \, U_{nl\lambda L, n'l'\lambda'L}(p,p';Z)$$

$$\times \tau_{n'l'}(z_{p'}) \, a_{n'l'\lambda'L}(p') \quad . \tag{13.66}$$

If the total angular momentum of the system is equal to zero, $L = 0$, then

$$U_{nl\lambda 0, n'l'\lambda'0}(p,p';Z) = \delta_{l\lambda} \, \delta_{l'\lambda'} \, U_{nl,n'l'}(p,p'Z) \quad ,$$

$$a_{nl\lambda 0}(p) = \delta_{l\lambda} \, a_{nl}(p) \quad , \tag{13.67}$$

and (13.66) reduces to

$$a_{nl}(p) = \sum_{n'l'} \int_0^\infty dp' \, p'^2 \, U_{nl,n'l'}(p,p';Z) \, \tau_{n'l'}(z_{p'}) \, a_{n'l'}(p') \quad . \tag{13.68}$$

The system of one-dimensional integral equations (13.62, 68) can be solved numerically. The results are shown in Figs. 13.1–3.

Equations (13.62, 68) were employed to calculate the binding energy E_0 of a system of three identical particles in the ground state with $L = 0$ and the length A of particle scattering by a two-particle bound state [13.5, 6]. Two cases were considered, where the particle interaction was described by the Hulthen potential (12.73) and the square well potential (12.80).

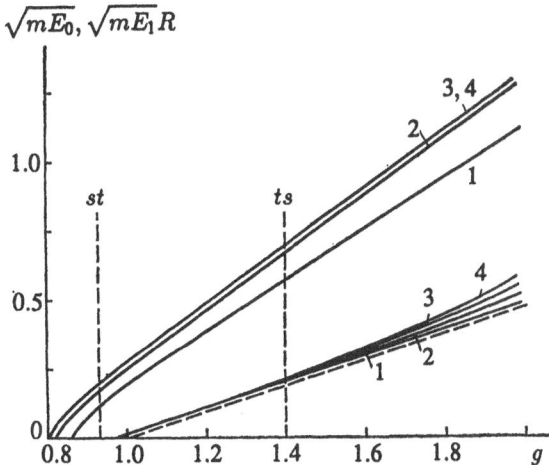

Fig. 13.1. The quantities $(mE_0)^{1/2} R$ and $(mE_1)^{1/2} R$ (E_0 and E_1 are the binding energies of the system of three identical spinless particles in the ground and first excited states with $L = 0$) as functions of the effective depth g of the two-particle interaction described by the Hulthen potential. The numerals on the curves indicate the order of the approximation. The *dashed line* corresponds to the two-particle system $[(m\epsilon)^{1/2} R$ versus g, ϵ is the two-particle ground state binding energy for $l = 0]$

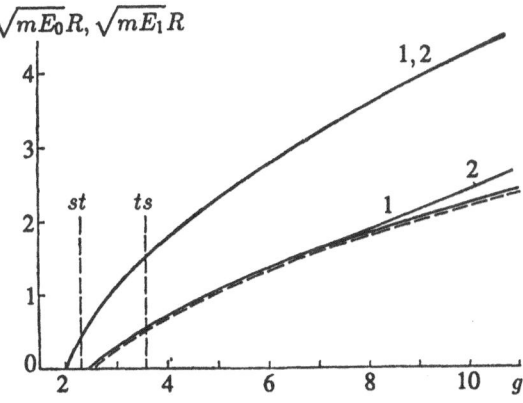

Fig. 13.2. The quantities $(mE_0)^{1/2} R$ and $(mE_1)^{1/2} R$ as functions of the effective depth g of a two-particle square well potential (see Fig. 13.1)

Figures 13.1,2 give the quantities $(mE_0)^{1/2} R$ and $(mE_1)^{1/2} R$ as functions of the effective length g of the two-particle interaction described by the Hulthen and square well potentials, respectivley. The g-dependence of the scattering length A for the Hulthen potential is shown in Fig. 13.3. The results shown in Figs. 13.1–3 were calculated by taking into account only the S-state interaction between pairs of particles ($l = 0$). Different curves correspond to different numbers of terms retained in the separable expansion (12.23) (the numeral at each curve shows the order of the approximation, i.e., the number of terms of (12.23) retained).

In the case of the Hulthen potential, curves 3 and 4 for the three-particle ground state binding energy actually coincide. Therefore, we need to account for only the first three terms in the separable expansion (12.23). If the effective

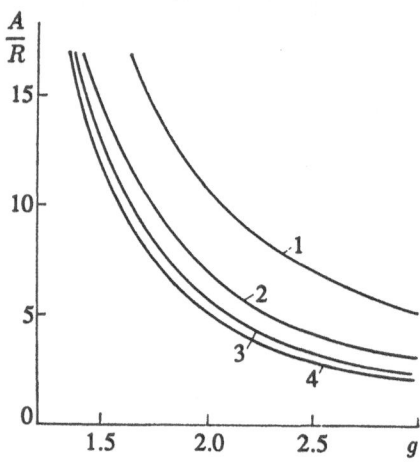

Fig. 13.3. The length A of a particle scattering by a bound state of two particles as a function of g for a system of three identical spinless particles interacting in a Hulthen potential. The numerical labels on the curves indicate the order of the approximation

depth g is large, then curve 4 asymptotically approaches a straight line, i.e.,

$$\sqrt{mE_0}\, R \simeq Cg - \tfrac{3}{4}\, C^{-1} \quad , \qquad C \simeq 1.03 \quad . \tag{13.69}$$

In the case of a square well potential, convergence of the solution is even better than for a Hulthen potential. As is seen in Fig. 13.2, calculations including one or two terms of the separable expansion (12.23) yield very similar three-particle binding energies E_0.

The results shown in the figures clearly indicate that the properties of the three-particle system are very sensitive to the form of the two-particle interaction even for short-range potentials, in contrast to two-particle systems which behave almost independently of the potential form. Thus, we see that having fitted the parameters of the Hulthen and square well potentials in such a way that the two-particle binding energy, scattering length, and effective interaction range coincide, one obtains considerably differing results for the three-particle binding energy and length of particle scattering by a two-particle bound state.

Problems

13.1   Consider the motion of three identical particles under the assumption that their interaction occurs only in the S-state and is described by a nonlocal potential with separated variables (Yamaguchi potential) given by

$$\langle \mathbf{k}'|V|\mathbf{k}\rangle = -\frac{\lambda}{2\mu}\, g(k)\, g(k') \quad . \tag{13.70}$$

The wave function of the system of three identical particles is symmetric with respect to permutations of any pairs of particles, hence, in the c.m.s. it can be written in the form of (13.36), with the function $\psi(\mathbf{k}, \mathbf{p})$ determined by the integral equation (13.37). For the Yamaguchi potential (13.70), the two-particle

t matrix is factorized. By assuming that the free term in (13.37) is equal to zero and using (12.64), we find the three-particle bound state wave function to be [13.7]

$$\psi_0(\boldsymbol{k},\boldsymbol{p}) = \frac{a(\boldsymbol{p})\,g(\boldsymbol{k})}{k^2 + \frac{3}{4}\,p^2 + \kappa^2} \quad , \qquad E = -\frac{\kappa^2}{m} \quad , \tag{13.71}$$

and the function $a(\boldsymbol{p})$ to be determned by the integral equation

$$a(\boldsymbol{p}) = 2\lambda\,\tau\left(-\frac{\kappa^2}{m} - \frac{3}{4}\frac{p^2}{m}\right)$$
$$\times \int \frac{d\boldsymbol{p}'}{(2\pi)^3} \frac{g[(\boldsymbol{p}/2)+\boldsymbol{p}']\,g[\boldsymbol{p}+\boldsymbol{p}'/2)]}{\kappa^2 + p^2 + p'^2 + \boldsymbol{p}\boldsymbol{p}'}\,a(\boldsymbol{p}') \quad , \tag{13.72}$$

$$\tau(z) = \left[1 + \frac{\lambda}{m}\int \frac{d\boldsymbol{q}}{(2\pi)^3}\frac{g^2(q)}{z - (q^2/m)}\right]^{-1} \quad .$$

If the total angular momentum of the three-particle bound state is equal to zero, $L = 0$, then the function $a(\boldsymbol{p})$ depends only on the modulus of $\boldsymbol{p}$, and the integral equation (13.72) reduces to the one-dimensional integral equation

$$a(p) = 4\pi\lambda\tau\left(-\frac{\kappa^2}{m} - \frac{3}{4}\frac{p^2}{m}\right)\int_0^\infty dp'\,p'^2\,I(p,p';\kappa^2)\,a(p') \quad , \tag{13.73}$$

where

$$I(p,p';\kappa^2)$$
$$= \frac{1}{(2\pi)^3}\int_{-1}^{1} dy\,\frac{g\left(\sqrt{\frac{1}{4}\,p^2 + p'^2 + pp'y}\right)\,g\left(\sqrt{p^2 + \frac{1}{4}\,p'^2 + pp'y}\right)}{\kappa^2 + p^2 + p'^2 + pp'y} \quad .$$

Substituting (12.68) for $g(k)$, we obtain the following expressions for τ and I:

$$\tau\left(-\frac{\kappa^2}{m} - \frac{3}{4}\frac{p^2}{m}\right) = \left\{1 - \frac{(\alpha+\beta)^2}{\left(\beta + \sqrt{\kappa^2 + \frac{3}{4}\,p^2}\right)^2}\right\}^{-1} \quad , \tag{13.74}$$

$$I(p,p';\kappa^2) = \frac{1}{\left(\beta^2 - \kappa^2 - \frac{3}{4}\,p^2\right)\left(\beta^2 - \kappa^2 - \frac{3}{4}\,p'^2\right)pp'}$$
$$\times \left\{\ln\frac{\kappa^2 + p^2 + p'^2 + pp'}{\kappa^2 + p^2 + p'^2 - pp'}\right.$$
$$+ \frac{4}{3}\frac{\beta^2 - \kappa^2 - \frac{3}{4}\,p^2}{p^2 - p'^2}\,\ln\frac{\beta^2 + p^2 + \frac{1}{4}\,p'^2 + pp'}{\beta^2 + p^2 + \frac{1}{4}\,p'^2 - pp'}$$
$$\left.+ \frac{4}{3}\frac{\beta^2 - \kappa^2 - \frac{3}{4}\,p'^2}{p^2 p'^2}\,\ln\frac{\beta^2 + \frac{1}{4}\,p^2 + p'^2 + pp'}{\beta^2 + \frac{1}{4}\,p^2 + p'^2 - pp'}\right\} \quad . \tag{13.75}$$

If the parameters λ and β are given, then the one-dimensional integral equation (13.73) can be solved numerically. This enables one to find the energy and the wave function of the three-particle bound state.

Considering particle scattering by a two-particle bound state, we must take the free term in the integral equation (13.37) in the form of (13.39), then the function $\psi(k,p)$ may be written as

$$\psi(k,p) = (2\pi)^3 \, \frac{N \, g(k)}{k^2 + \frac{3}{4}p^2 - mZ} \left\{ \delta(p - p_0) + \frac{1}{2\pi^2} \frac{F(p_0,p)}{p^2 - p_0^2 - iO} \right\} \ ,$$

$$Z = \frac{3}{4} \frac{p_0^2}{m} - \frac{\alpha^2}{m} + iO \quad . \tag{13.76}$$

The function $F(p_0,p)$ satisfies the integral equation

$$
\begin{aligned}
F(p_0,p) = \frac{2\lambda}{m} & \left(p^2 - p_0^2 \right) \tau\left(Z - \frac{3}{4}\frac{p^2}{m} \right) \\
& \times \left\{ \frac{1}{4\pi} \frac{g(p+p_0/2) \, g(p/2+p_0)}{(1/m)(p^2 + p_0^2 + pp_0) - Z} \right. \\
& \left. + \int \frac{dp'}{(2\pi)^3} \frac{g(p+p'/2) \, g(p/2+p') \, F(p_0,p')}{(p'^2 - p_0^2 - iO)\left[(1/m)(p^2 + p'^2 + pp') - Z \right]} \right\} .
\end{aligned}
\tag{13.77}
$$

The elastic scattering amplitude $f(p_0,p)$ is directly determined by the value of the function $F(p_0,p)$ on the energy shell, i.e.,

$$f(p_0,p) = F(p_0,p)_{p^2 = p_0^2} \quad . \tag{13.78}$$

In the case of elastic scattering of a zero-energy particle by a two-particle bound state, (13.77) for the amplitude $F(p) \equiv F(0,p)$ reduces to the one-dimensional integral equation

$$
\begin{aligned}
F(p) = 4\pi\lambda p^2 \tau & \left(-\frac{\alpha^2}{m} - \frac{3}{4}\frac{p^2}{m} \right) \\
& \times \left\{ \frac{1}{8\pi^2} \frac{g(p) \, g(p/2)}{\alpha^2 + p^2} + \int_0^\infty dp' \, I(p,p';\alpha^2) \, F(p') \right\} \quad .
\end{aligned}
\tag{13.79}
$$

Substituting (12.68) for the function $g(k)$ and allowing the parameter β to extend to infinity (this corresponds to the potential range tending to zero), we obtain from (13.79) the Ter–Martirosian–Skorniakov equation [13.8], i.e.,

$$
\begin{aligned}
F(p) = \frac{8}{3} & \left(\alpha + \sqrt{\alpha^2 + \frac{3}{4}p^2} \right) \left\{ \frac{1}{\alpha^2 + p^2} \right. \\
& \left. + \frac{1}{\pi} \int_0^\infty dp' \, \frac{1}{pp'} \, \ln \frac{\alpha^2 + p^2 + p'^2 + pp'}{\alpha^2 + p^2 + p'^2 - pp'} \, F(p') \right\} \quad .
\end{aligned}
\tag{13.80}
$$

Solving (13.79) numerically, one obtains the zero-energy elastic scattering amplitude $F(0)$ or the scattering length $A = -F(0)$.

13.2   Find the wave function asymptotics in the coordinate representation for particle scattering by a two-particle bound state [13.9].

If the two-particle t matrix associated with interaction of an individual particle pair possesses only simple poles, then the singularities of the three-particle wave function can easily be separated explicitly. The three-particle wave function describing particle scattering by a two-particle bound state is determined by the relations (13.32, 33).

For the sake of simplicity we assume that each pair of particles has a single bound state and write the t matrix as

$$\langle k'|t(z)|k \rangle = \frac{g(k)\, g(k')}{z + \kappa^2/2\mu} + \langle k'|\tilde{t}(z)|k \rangle \quad , \tag{13.81}$$

where $\tilde{t}(z)$ is a smooth function of energy z, $g(k)$ is the vertex function of the bound state of the pair $\varphi(k)$. Substituting (13.81) into (13.32) and (13.33), we find the three-particle function to be

$$\Psi_1\left(p_1, k_{23}; p_1^0\right) = \Phi_1\left(p_1, k_{23}; p_1^0\right) + \left(Z - \frac{p_1^2}{2\mu_1} - \frac{k_{23}^2}{2\mu_{23}}\right)^{-1}$$
$$\times B_1\left(p_1, k_{23}; p_1^0, Z\right) \quad , \tag{13.82}$$

$$Z = E + iO \quad , \qquad E = \frac{p_1^{0^2}}{2\mu_1} - \frac{k_{23}^2}{2\mu_{23}} \quad ,$$

where

$$B_1\left(p_1, k_{23}; p_1^0, Z\right) = \sum_i K_{i1}\left(p_i, k_{jk}; p_1^0, Z\right) \quad , \tag{13.83}$$

$$K_{i1}\left(p_i, k_{jk}; p_1^0, Z\right) = P_{i1}\left(p_i, k_{jk}; p_1^0, Z\right)$$
$$+ \frac{g_{jk}\left(k_{jk}\right)}{Z - \left(p_i^2/2\mu_i\right) + \left(\kappa_{jk}^2/2\mu_{jk}\right)}$$
$$\times Q_{i1}\left(p_i; p_1^0, Z\right) \quad . \tag{13.84}$$

The quantities P and Q are defined by the relations

$$P_{i1}\left(p_i, k_{jk}; p_1^0, Z\right) \equiv \int \frac{dk'_{jk}}{(2\pi)^3} \left\langle k_{jk}\left|\tilde{t}_{jk}\left(Z - \frac{p_i^2}{2\mu_i}\right)\right|k'_{jk}\right\rangle$$
$$\times \left\{\Psi_1^{(j)}\left(p'_j, k'_{ki}; p_1^0\right) + \Psi_1^{(k)}\left(p'_k, k'_{ij}; p_1^0\right)\right\} \quad ,$$

$$\tag{13.85}$$

$$Q_{i1}\left(p_i; p_1^0, Z\right) \equiv \int \frac{dk'_{jk}}{(2\pi)^3}\, g_{jk}\left(k'_{jk}\right)$$
$$\times \left\{ \Psi_1^{(j)}\left(p'_j, k'_{ki}; p_1^0\right) + \Psi_1^{(k)}\left(p'_k, k'_{ij}; p_1^0\right)\right\} \quad .$$

According to (13.82), the three-particle wave function describing scattering of particle 1 by the bound state of particles 2 and 3, has poles at the points

$$Z = \frac{p_1^2}{2\mu_1} + \frac{k_{23}^2}{2\mu_{23}} \qquad \text{and} \qquad Z = \frac{p_i^2}{2\mu_i} - \frac{\kappa_{jk}^2}{2\mu_{jk}} \quad .$$

The quantities B_1 and Q_{i1} determine the wave function residues at the corresponding points. We assume that functions P_{i1} and Q_{i1} have no singularities. It is easy to verify that

$$Q_{i1}\left(p_i; p_1^0, E\right) = T_{1\to i} \equiv \left(\Phi_i, (V_{ij} + V_{ik})\, \Psi_1\right) \quad . \tag{13.86}$$

Therefore, $Q_{11}\left(p_1; p_1^0, E\right)$ directly determines the amplitude of elastic particle scattering by the two-particle bound state, whereas $Q_{21}\left(p_2; p_1^0, E\right)$ and $Q_{31}\left(p_3; p_1^0, E\right)$ are the amplitudes of particle exchange scattering. Respectively, the cross sections of elastic and particle exchange scattering are

$$d\sigma_{1\to 1} = \frac{\mu_1^2}{4\pi^2}\left|Q_{11}\left(p_1; p_1^0, E\right)\right|^2 do_1 \quad , \tag{13.87}$$

$$d\sigma_{1\to i} = \frac{\mu_1 \mu_i}{4\pi^2}\, \frac{p_i}{p_1^0}\left|Q_{i1}\left(p_i; p_1^0, E\right)\right|^2 do_i \quad . \tag{13.88}$$

The validity of

$$B_1\left(p_1, k_{23}; p_1^0, E\right) = T_{1\to 123} \equiv \left(\Phi_{123}, (V_{12} + V_{23} + V_{31})\, \Psi_1\right) \tag{13.89}$$

can be verified in a similar manner. (Φ_{123} is the wave function of the system of three unbound particles with momenta p_1 and k_{23}.) Therefore, $B_1\left(p_1, k_{23}; p_1^0, Z\right)$ is the amplitude of scattering accompanied by disintegration of the scatterer. The disintegration cross section is given by

$$d\sigma_{1\to 123} = \frac{\mu_1^2}{4\pi^2}\, \frac{p_1}{p_1^0}\left|B_1\left(p_1, k_{23}; p_1^0, E\right)\right|^2 \frac{dk_{23}}{(2\pi)^3}\, do_1 \quad . \tag{13.90}$$

The wave function of the system in the coordinate representation can be derived from (13.83) by the Fourier transformation; the result is

$$\Psi_1\left(x_1, r_{23}; p_1^0\right) = \int \frac{dp_1}{(2\pi)^3} \int \frac{dk_{23}}{(2\pi)^3}\, \exp\left[ip_1 x_1 + ik_{23} r_{23}\right]$$
$$\times \Psi_1\left(p_1, k_{23}; p_1^0\right) \quad . \tag{13.91}$$

Let us consider the contribution of the pole terms of (13.83) in the wave function asymptotics. First we consider the terms which contain singularities of the form

$$\left(Z - \frac{p_i^2}{2\mu_i} - \frac{k_{jk}^2}{2\mu_{jk}}\right)^{-1} \left(Z - \frac{p_i^2}{2\mu_i} + \frac{\kappa_{jk}^2}{2\mu_{jk}}\right)^{-1} \quad .$$

We introduce the notation

$$\psi^{(i)}(\boldsymbol{x}, \boldsymbol{r}) \equiv \int \frac{d\boldsymbol{p}}{(2\pi)^3} \int \frac{d\boldsymbol{k}}{(2\pi)^3} \exp\left[i\boldsymbol{p}\boldsymbol{x} + i\boldsymbol{k}\boldsymbol{r}\right]$$
$$\times \frac{R_i(\boldsymbol{p}, \boldsymbol{k})}{Z - (p^2/2\mu_i) + (\kappa_{jk}^2/2\mu_{jk})} \quad , \tag{13.92}$$

$$R_i(\boldsymbol{p}, \boldsymbol{k}) \equiv \frac{g_{jk}(\boldsymbol{k}) \, Q_{i1}\left(\boldsymbol{p}, \boldsymbol{p}_1^0, Z\right)}{Z - (p^2/2\mu_i) - (k^2/2\mu_{jk})} \quad . \tag{13.93}$$

It is not difficult to show that

$$\psi^{(i)}(\boldsymbol{x}, \boldsymbol{r}) = -\frac{\mu_i}{2\pi} \int \frac{d\boldsymbol{k}}{(2\pi)^3} e^{i\boldsymbol{k}\boldsymbol{r}} \int d\boldsymbol{x}'$$
$$\times \frac{\exp\left[i\sqrt{2\mu_i\left(E + \kappa_{jk}^2/2\mu_{jk}\right)} \, |\boldsymbol{x} - \boldsymbol{x}'|\right]}{|\boldsymbol{x} - \boldsymbol{x}'|} R_i(\boldsymbol{x}', \boldsymbol{k}) \quad ,$$

$$\tag{13.94}$$

$$R_i(\boldsymbol{x}, \boldsymbol{k}) \equiv \int \frac{d\boldsymbol{p}}{(2\pi)^3} e^{i\boldsymbol{p}\boldsymbol{x}} R_i(\boldsymbol{p}, \boldsymbol{k}) \quad .$$

In both cases of elastic and inelastic scattering, r remains limited whereas $x \to \infty$. In this limiting case we have

$$|\boldsymbol{x} - \boldsymbol{x}'| \to x - \boldsymbol{n}\boldsymbol{x}' + \dots , \tag{13.95}$$

where $\boldsymbol{n}$ is the unit vector along the direction of $\boldsymbol{x}$. We introduce

$$\boldsymbol{p}_i \equiv \sqrt{2\mu_i\left(E + \kappa_{jk}^2/2\mu_{jk}\right)} \, \boldsymbol{n} \quad . \tag{13.96}$$

It is clear that as $x \to \infty$, the quantity $\boldsymbol{p}_i$ describes the momentum of relative motion in the system. After expansion (13.95) is substituted and the resulting expression is integrated, (13.94) can be written as

$$\psi^{(i)}(\boldsymbol{x}, \boldsymbol{r}) \underset{x \to \infty}{\longrightarrow} -\frac{\mu_i}{2\pi} Q_{i1}\left(\boldsymbol{p}_i; \boldsymbol{p}_1^0, E\right) \frac{\exp\left(i p_i x\right)}{x} \varphi_{\kappa_{jk}}(\boldsymbol{r}) \quad , \tag{13.97}$$

where $\varphi_{\kappa_{jk}}$ is the normalized wave function of the two-particle bound state.

Now, let us consider the contribution in (13.91) due to the pole term proportional to

$$\left(Z - \frac{p_i^2}{2\mu_i} - \frac{k_{jk}^2}{2\mu_{jk}}\right)^{-1} \quad .$$

In this case,

$$\psi(x,r) \equiv \int \frac{dp}{(2\pi)^3} \int \frac{dk}{(2\pi)^3} \exp(ipx + ikr) \frac{B_1(p,k;p_1^0,Z)}{Z - (p^2/2\mu_1) - (k^2/2\mu_{23})}$$

$$= \int dx' \int dr' \, B_1(x',r';p_1^0,E) \, \langle x',r'|G_0(Z)|r,x\rangle \quad , \tag{13.98}$$

where $\langle x',r'|G_0(Z)|r,x\rangle$ is the Green function of the three-particle system. According to (3.136),

$$\langle x',r'|G_0(Z)|r,x\rangle = -\frac{i}{2\pi^2} (\mu_1\mu_{23})^{3/2} \frac{Z}{\varrho^2} H_2^{(1)}\left(\sqrt{Z}\,\varrho\right) \quad ,$$

$$\mathrm{Im}\,\{Z\} \geq 0 \quad , \tag{13.99}$$

where $\varrho = \sqrt{2\mu_1(x - x')^2 + 2\mu_{23}(r - r')^2}$ is the distance between the points in the six-dimensional space. If $\varrho \to \infty$, then

$$\langle x',r'|G_0(Z)|r,x\rangle \to -\frac{\exp(i\pi/4)(\mu_1\mu_{23})^{3/2} Z^{3/4}}{\sqrt{2\pi}\,\pi^2\,\varrho^{5/2}} \exp(i\sqrt{Z}\,\varrho) \quad ,$$

$$\mathrm{Im}\,\{Z\} \geq 0 \quad , \tag{13.100}$$

and

$$\psi(x,r) \to -\frac{\exp(i\pi/4)(\mu_1\mu_{23})^{3/2} E^{3/4}}{\sqrt{2\pi}\,\pi^2}$$

$$\times \int dx' \int dr' \, B_1(x',r';p_1^0,Z) \frac{\exp(i\sqrt{E}\,\varrho)}{\varrho^{5/2}} \quad . \tag{13.101}$$

If scattering of particle 1 produces dissociation of the system $2 - 3$, then off the interaction region all the particles move with fixed velocities. This implies that the triangle formed by the ends of the vectors r_1, r_2, and r_3 grows with time but preserves its shape. Therefore, in the case of dissociation we have $x \to \infty$, $r \to \infty$, $x/r = \mathrm{const}$. The quantity ϱ can be expanded about the value $\left(2\mu_1 x^2 + 2\mu_{23} r^2\right)^{1/2}$, so that

$$\varrho = \sqrt{2\mu_1 x^2 + 2\mu_{23} r^2} - \frac{2\mu_1 xx' + 2\mu_{23} rr'}{\left(2\mu_1 x^2 + 2\mu_{23} r^2\right)^{1/2}} + \cdots \quad .$$

Since $x/r = (\mu_{23}/\mu_1)(p/k)$, we have

$$\sqrt{E}\,\varrho = \sqrt{2E\left(\mu_1 x^2 + \mu_{23} r^2\right)} - pn_x x' - kn_r r' + \cdots \tag{13.102}$$

and thus the asymptotic (13.101) reduces to

$$\psi(x,r) \to -\frac{\exp(i\pi/4)(\mu_1\mu_{23})^{3/2} E^{3/4}}{\sqrt{2\pi}\,\pi^2} B_1(pn_x,kn_r;p_1^0,Z)$$

$$\times \frac{\exp\left[i\sqrt{2E(\mu_1 x^2 + \mu_{23} r^2)}\right]}{\left(2\mu_1 x^2 + 2\mu_{23} r^2\right)^{5/4}} \quad , \tag{13.103}$$

where $E = p^2/2\mu_1 + k^2/2\mu_{23}$. By using (13.97, 103), one can obtain the asymptotic of the wave function (13.91) for various cases.

For elastic scattering, the dominant contribution in the wave function asymptotic is due to the term with Q_{11} in (13.82). The other contributions are appreciably smaller. Hence, the wave function can be written as

$$\Psi_1\left(x_1, r_{23}; p_1^0\right) \xrightarrow[\substack{x_1 \to \infty \\ r_{23} = \text{const}}]{} \left\{ \exp(ip_1^0 x_1) - \frac{\mu_1}{2\pi} Q_{11}\left(p_1^0 n_x; p_1^0, E\right) \right.$$

$$\left. \times \frac{\exp\left(ip_1^0 x_1\right)}{x_1} \right\} \varphi_{23}\left(r_{23}\right) \quad , \tag{13.104}$$

and in the case of particle exchange scattering we have

$$\Psi_1\left(x_i, r_{jk}; p_1^0\right) \xrightarrow[\substack{x_i \to \infty \\ r_{jk} = \text{const}}]{} -\frac{\mu_i}{2\pi} Q_{i1}\left(p_i n_x; p_1^0, E\right) \frac{\exp(ip_i x_i)}{x_i}$$

$$\times \varphi_{\kappa_{jk}}\left(r_{jk}\right) \quad , \qquad i = 2, 3 \tag{13.105}$$

where

$$p_i = \sqrt{2\mu_i \left(E + \kappa_{jk}^2/2\mu_{jk}\right)} \quad .$$

If the scattering process results in the dissociation of the scatterer, then the wave function asymptotic is determined by (13.103). We have

$$\Psi_1\left(x_1, r_{23}; p_1^2\right) \xrightarrow[\substack{x_1 \to \infty \\ r_{23} \to \infty \\ (x_1/r_{23}) = \text{const}}]{} -\frac{\exp(i\pi/4)\left(\mu_1\mu_{23}\right)^{3/2} E^{3/4}}{\sqrt{2\pi}\,\pi^2}$$

$$\times B_1\left(p_1 n_x, k_{23} n_r; p_1^0, E\right)$$

$$\times \exp\left[i\sqrt{2E\left(\mu_1 x_1^2 + \mu_{23} r_{23}^2\right)}\right]$$

$$\times \left(2\mu_1 x_1^2 + 2\mu_{23} r_{23}^2\right)^{-5/4} \quad . \tag{13.106}$$

In this case, the momenta p_1 and k_{23} satisfy the relation

$$E = \frac{p_1^2}{2\mu_1} + \frac{k_{23}^2}{2\mu_{23}} \quad .$$

14. Scattering of Spin-Possessing Particles

14.1 The Spin Wave Function and the Density Matrix

Let us consider scattering of particles which possess proper moments (spin). In the general case, the interaction potential of spin-possessing particles is noncentral, i.e., the potential of particle interaction depends not only on the relative distance between the particles, but also on the mutual orientation of particle spins and the relative distance vector. In the case of noncentral interaction, particle scattering can be accompanied by changes of particle spin orientations. [Polarization phenomena that accompany scattering of spin-possessing particles are considered in Refs. 14.1–6.]

The wave function of a spin-possessing particle depends on space coordinates and on the spin variable which conventionally is denoted as σ. It is convenient to take the spin variable to be the projection of the particle spin on some chosen direction. In contrast to continuous spatial coordinates, the spin variable takes a restricted number of discrete values. Therefore, the wave function $\psi(r, \sigma)$ of the particle that has definite spin s can be written as a $(2s + 1)$-component vector column, the components being the values of $\psi(r, \sigma)$ corresponding to different σ (σ varies from $-s$ to s). The spin operators are then described by $(2s + 1)$-row matrices.

It is clear that the spatial coordinates and the spin variable of a free particle are independent, hence the free particle wave function can be written as a product of spatial and spin wave functions. Let $\chi_{s\mu}$ be the eigenfunction of the operators of spin squared s^2 and spin projection s_z on the given direction z:

$$s^2 \chi_{s\mu} = s(s + 1) \chi_{s\mu} \quad ,$$

$$s_z \chi_{s\mu} = \mu \chi_{s\mu} \quad . \tag{14.1}$$

We assume that the functions $\chi_{s\mu}$ are normalized by the condition

$$\sum_{\sigma} \chi_{s\mu}^*(\sigma) \chi_{s\mu'}(\sigma) = \delta_{\mu\mu'} \quad . \tag{14.2}$$

Since the spin variable implies projection of the spin on the given direction, the representation in which the spin projection is well defined is proper and therefore

$$\chi_{s\mu}(\sigma) = \delta_{\mu\sigma} \quad . \tag{14.3}$$

The function $\chi_{s\mu}(\sigma)$ can be written as a column with the only nonzero (equal to one) component $\sigma = \mu$, i.e.,

$$\chi_{s\mu}(\sigma) = \begin{pmatrix} 0 \\ 0 \\ \vdots \\ 1 \\ \vdots \\ 0 \end{pmatrix} \begin{matrix} s \\ s-1 \\ \vdots \\ \mu \\ \vdots \\ -s \end{matrix} . \tag{14.4}$$

Making use of the representation (14.3), one can easily show that the functions $\chi_{s\mu}$ form a complete set, so that

$$\sum_{\mu} \chi_{s\mu}^*(\sigma) \chi_{s\mu}(\sigma') = \delta_{\sigma\sigma'} . \tag{14.5}$$

Therefore, any spin state χ can be written as a linear superposition of the states $\chi_{s\mu}$,

$$\chi(\sigma) = \sum_{\mu=-s}^{s} a_\mu \chi_{s\mu}(\sigma) . \tag{14.6}$$

A state of a system that is described by a wave function is usually referred to as a pure state. If the spin function χ is normalized to one, then

$$\sum_{\mu} |a_\mu|^2 = 1 . \tag{14.7}$$

The functions $\chi_{s\mu}$ may be regarded as basis vectors in the $(2s + 1)$-dimensional spin space. Any other spin states can be projected onto these vectors. The expansion coefficients a_μ in (14.6) can be treated as components of the spin state vector χ.

We define the polarization vector P as the ratio of the average spin vector s in the given state χ to the spin value s, i.e.,

$$P = \frac{\langle\chi|s|\chi\rangle}{s} . \tag{14.8}$$

The direction of the polarization vector determines particle spin orientation; the value of the polarization vector determines the degree of polarization, that is, the relative probability of spin orientation along a given direction. It is not difficult to verify that in the states $\chi_{s\mu=\pm s}$ the polarization vector is directed either parallel or antiparallel to the quantization axis and that its absolute value is equal to one, i.e.,

$$P \equiv \frac{1}{s} \langle\chi_{s\pm s}|s|\chi_{s\pm s}\rangle = \pm z_0 , \tag{14.9}$$

where z_0 is the unit vector along the quantization axis. In other words, in the state $\chi_{s\mu=\pm s}$ the particle spin is oriented either parallel or antiparallel to the quantization axis. For an arbitrary spin state (14.6), the polarization vector is given by

$$P = \frac{1}{2s} \sum_{\mu=-s+1}^{s} \sqrt{(s+\mu)(s-\mu+1)} \left(a_\mu^* a_{\mu-1} + a_\mu a_{\mu-1}^* \right) x_0$$

$$- \frac{i}{2s} \sum_{\mu=-s+1}^{s} \sqrt{(s+\mu)(s-\mu+1)} \left(a_\mu^* a_{\mu-1} - a_\mu a_{\mu-1}^* \right) y_0$$

$$+ \frac{1}{s} \sum_{\mu=-s}^{s} \mu |a_\mu|^2 z_0 \quad , \tag{14.10}$$

and then

$$P^2 = \frac{1}{s^2} \left\{ \left| \sum_{\mu=-s+1}^{s} \sqrt{(s+\mu)(s-\mu+1)} \, a_\mu^* a_{\mu-1} \right|^2 \right.$$

$$\left. + \left| \sum_{\mu=-s}^{s} \mu |a_\mu|^2 \right|^2 \right\} \leq 1 \quad . \tag{14.11}$$

If particle spin is equal to $1/2$, then the expression (14.10) for the polarization vector is simplified to

$$P = \left(a_{1/2}^* a_{-1/2} + a_{1/2} a_{-1/2}^* \right) x_0 - i \left(a_{1/2}^* a_{-1/2} - a_{1/2} a_{-1/2}^* \right) y_0$$

$$+ \left(|a_{1/2}|^2 - |a_{-1/2}|^2 \right) z_0 \quad . \tag{14.12}$$

One may directly verify that in this case

$$P^2 = 1 \quad . \tag{14.13}$$

Therefore, if a spin $1/2$ particle is in any pure state, then its spin is completely polarized, i.e., its spatial orientation is well defined.

Since the quadratic combination of the spin operator for a spin $1/2$ particle reduces to a linear combination, the polarization vector P completely determines the spin state of the system. If the particle spin is equal to or greater than one, then one introduces the quadratic polarization tensor. We define it as the mean value of the operator product $s_i s_j$.

The spin state of a particle with spin 1 is completely determined by the given quadratic polarization tensor $\langle s_i s_j \rangle$. Indeed, let us separate the symmetric and antisymmetric parts of the tensor $s_i s_j$:

$$s_i s_j = \tfrac{1}{2} \left(s_i s_j + s_j s_i \right) + \tfrac{1}{2} \left(s_i s_j - s_j s_i \right) \quad .$$

According to the permutation relations, the antisymmetric part is a linear function of the spin operator, i.e.,

$$s_i s_j - s_j s_i = i \varepsilon_{ijk} s_k \quad ,$$

where ε_{ijk} is a completely antisymmetric tensor of the third rank. In the symmetric part, we separate the unit tensor in such a way that the sum of diagonal elements of the residual symmetric tensor is zero. Thus, we can write

$$s_i s_j = \frac{2}{3} \delta_{ij} + s_{ij} + \frac{i}{2} \varepsilon_{ijk} s_k \quad , \tag{14.14}$$

where

$$s_{ij} \equiv \frac{1}{2} \left(s_i s_j + s_j s_i \right) - \frac{2}{3} \delta_{ij} \quad . \tag{14.15}$$

The average of the symmetric tensor s_{ij} with zero sum of diagonal elements is conventionally referred to as the polarization tensor and is denoted as P_{ij}, so that

$$2P_{ij} \equiv \langle \chi | s_{ij} | \chi \rangle \quad . \tag{14.16}$$

Thus we have

$$\langle s_i s_j \rangle = \frac{2}{3} \delta_{ij} + 2P_{ij} + \frac{i}{2} \varepsilon_{ijk} P_k \quad , \tag{14.17}$$

i.e., the spin state of a spin 1 particle is described by the polarization vector P and polarization tensor P_{ij}. For a pure state described by the wave function (14.6), the polarization vector and tensor are given by the expressions

$$P = \sqrt{2} \, \mathrm{Re} \left\{ (a_1 + a_{-1}) \, a_0^* \right\} x_0 - \sqrt{2} \, \mathrm{Im} \left\{ (a_1 - a_{-1}) \, a_0^* \right\} y_0$$
$$+ \left(|a_1|^2 - |a_{-1}|^2 \right) z_0 \quad , \tag{14.18}$$

$$2P_{ij} = \begin{pmatrix} \frac{1}{2} |a_0|^2 + \mathrm{Re} \left\{ a_1 a_{-1}^* \right\} - \frac{1}{6} & -\mathrm{Im} \left\{ a_1 a_{-1}^* \right\} \\ -\mathrm{Im} \left\{ a_1 a_{-1}^* \right\} & \frac{1}{2} |a_0|^2 - \mathrm{Re} \left\{ a_1 a_{-1}^* \right\} - \frac{1}{6} \\ \frac{1}{\sqrt{2}} \mathrm{Re} \left\{ (a_1 - a_{-1}) \, a_0^* \right\} & -\frac{1}{\sqrt{2}} \mathrm{Im} \left\{ (a_1 + a_{-1}) \, a_0^* \right\} \end{pmatrix}$$

$$\begin{matrix} \frac{1}{\sqrt{2}} \mathrm{Re} \left\{ (a_1 - a_{-1}) \, a_0^* \right\} \\ -\frac{1}{\sqrt{2}} \mathrm{Re} \left\{ (a_1 + a_{-1}) \, a_0^* \right\} \\ |a_1|^2 + |a_{-1}|^2 - \frac{1}{3} \end{matrix} \Biggr) \tag{14.19}$$

We see that the state of the system with spin 1 is described by eight real parameters – three components of the polarization vector and five components of the symmetric polarization tensor.

In the general case of a particle with arbitrary spin s, the spin state can be shown to be determined by the polarization tensor of the rank $2s$. In the laboratory measurements, one usually deals with a large particle beam that is scattered from a many-particle target rather than with individual particle-particle scattering events. It is actually impossible to describe a many-particle system in terms of wave functions. However, since experimental studies provide data averaged over a great number of particles, we can apply the density matrix approach.

It is well known that in quantum mechanics the description of physical systems in terms of wave functions is the most complete. If the wave function $\psi^{(n)}$ is given, then the mean value of any quantity associated with the system and described by the operator $\hat{Q}$ is

$$\overline{Q} = \langle \psi^{(n)}, \hat{Q}\,\psi^{(n)} \rangle \quad .$$

However, quantum mechanics deals not only with pure states described by wave functions, but also with states described by incoherent mixtures which include wave functions $\psi^{(n)}$ ($n = 1, 2, \ldots, N$) with statistical weights $w(n)$. The average value of a physical quantity described by the operator $\hat{Q}$ in such a state is determined by

$$\overline{Q} = \frac{\sum_{n=1}^{N} w(n)\left(\psi^{(n)}, \hat{Q}\,\psi^{(n)}\right)}{\sum_{n=1}^{N} w(n)\left(\psi^{(n)}, \psi^{(n)}\right)} \quad . \tag{14.20}$$

The state of the system in which the average values of the physical quantities are determined by (14.20) is referred to as a mixed state. The state of a subsystem contained in some greater system is always mixed.

We introduce a complete orthonormalized set of eigenfunctions ψ_α of some operator or operator set associated with the system. Expanding the function $\psi^{(n)}$ in terms of this set,

$$\psi^{(n)} = \sum_{\alpha} a_\alpha^{(n)}\,\psi_\alpha \quad , \tag{14.21}$$

we reduce (14.20) to

$$\overline{Q} = \frac{\sum_{\alpha,\beta} Q_{\alpha\beta}\,\varrho_{\beta\alpha}}{\sum_{\alpha} \varrho_{\alpha\alpha}} \quad , \tag{14.22}$$

where ϱ is the density matrix in the α-representation, given by

$$\varrho_{\alpha\beta} = \sum_{n=1}^{N} w(n)\,a_\alpha^{(n)}\,a_\beta^{(n)*} \quad . \tag{14.23}$$

Equation (14.22) may be rewritten in a form that is more suitable for future analysis, namely,

$$\overline{Q} = \frac{\mathrm{Tr}\{Q\varrho\}}{\mathrm{Tr}\{\varrho\}} \quad , \tag{14.24}$$

where the symbol "Tr" means the sum of the diagonal elements of a matrix. By using (14.23), we rewrite the density matrix in the representation for which the wave functions $\psi^{(n)}$ are known, so that

$$\varrho(x, x') = \sum_{n} w(n)\, \psi^{(n)}(x)\, \psi^{(n)^*}(x') \quad . \tag{14.25}$$

Once the density matrix is known, we can calculate the average value of any quantity associated with the system. Though the density matrix does not contain variables which do not correspond to the system under consideration, it nevertheless depends on the state of the entire closed system (its influence is determined by the set of statistical weights $w(n)$). A pure state is a special case of mixed one: the probability of some state is equal to one and all other probabilities vanish. Thus, the description in terms of a density matrix is the most general in quantum mechanics [14.7].

If the system consists of two noninteracting subsystems, then the density matrix ϱ of the whole system is given by the direct product of the density matrices associated with the subsystems:

$$\varrho = \varrho_1\, \varrho_2 \quad . \tag{14.26}$$

According to (14.23), the density matrix is Hermitian and positive definite. Therefore, it can always be diagonalized by means of the unitary transformation.

If the wave functions $\psi^{(n)}$ are normalized to one, and the statistical weights are taken in such a way that

$$\sum_{n=1}^{N} w(n) = 1 \quad ,$$

then the density matrix satisfies the normalization condition

$$\mathrm{Tr}\,\{\varrho\} = 1 \quad . \tag{14.27}$$

It is not difficult to show within the context of (14.23) that

$$\mathrm{Tr}\,\{\varrho^2\} = \sum_{n,n'} w(n)\, w(n') \left| \sum_{\alpha} a_{\alpha}^{(n)}\, a_{\alpha}^{(n')^*} \right|^2 \quad .$$

Then, by applying the Cauchy–Bouniakowsky inequality

$$\left| \sum_{\alpha} a_{\alpha}^{(n)}\, a_{\alpha}^{(n')^*} \right|^2 \leq \sum_{\alpha} \left| a_{\alpha}^{(n)} \right|^2 \sum_{\beta} \left| a_{\beta}^{(n')} \right|^2 \quad ,$$

we obtain

$$\mathrm{Tr}\,\{\varrho^2\} \leq \left(\mathrm{Tr}\,\{\varrho\} \right)^2 = 1 \quad . \tag{14.28}$$

The maximum value

$$\mathrm{Tr}\,\{\varrho^2\} = 1$$

is attained only in the case of a pure state when all $w(n)$ except for one vanish.

In order to describe the spin properties of a particle beam, we introduce the spin density matrix

$$\varrho = \overline{\chi \chi^+} \; . \tag{14.29}$$

(The bar implies statistical averaging.) Then polarization of the beam particles can be defined as

$$P = \frac{1}{s} \frac{\mathrm{Tr}\{\boldsymbol{s}\,\varrho\}}{\mathrm{Tr}\{\varrho\}} \; . \tag{14.30}$$

For particles with spin $\frac{1}{2}$, the polarization vector completely determines the spin state of the beam. In contrast to pure states for which $P = 1$, polarization of mixed states described by the density matrix can take any value from 0 to 1. The beam is called completely polarized if its polarization is equal to one ($P = 1$). If polarization is equal to zero ($P = 0$), the beam is called unpolarized.

The quadratic polarization tensor P_{ij} for particles with spin equal to or greater than 1 is given by

$$P_{ij} = \frac{1}{s^2} \frac{\mathrm{Tr}\{s_{ij}\,\varrho\}}{\mathrm{Tr}\{\varrho\}} \; , \tag{14.31}$$

where s_{ij} is a symmetric tensor. For an arbitrary spin s,

$$s_{ij} = \frac{1}{2} \left(s_i s_j + s_j s_i \right) - \frac{s(s+1)}{3} \delta_{ij} \; . \tag{14.32}$$

If the polarization vector P and all components of the polarization tensor P_{ij} of a beam of spin 1 particles are equal to zero, then polarization does not occur and the density matrix ϱ reduces to the unit matrix. The spin state of a beam of particles with arbitrary spin s is determined by the polarization tensor of rank $2s$.

14.2 Spin-Tensor Expansion of the Density Matrix

Operators acting upon spin variables are transformed by definite laws under the coordinate system rotations. The so-called irreducible operators have the simplest transformation proporties.

An irreducible operator of rank I is the set of quantities T_{IM} ($M = -I, -I+1, \ldots, I$) which are transformed under the coordinate system rotations as a ($2I+1$)-dimensional irreducible representation of the three-dimensional rotation group, i.e.,

$$T_{IM} = \sum_{M'} D^I_{M'M}(\theta)\, T'_{IM'} \; , \tag{14.33}$$

where $D^I_{M'M}(\theta)$ is the finite rotation matrix depending on the Euler angles θ. Any operator can be written as a linear superposition of irreducible tensor operators [14.8].

Because the spin wave functions $\chi_{s\mu}$ under the coordinate system rotations are transformed by

$$\chi_{s\mu} = \sum_{\mu'} D^s_{\mu'\mu}(\theta)\, \chi'_{s\mu'} \quad , \tag{14.34}$$

and since the complex conjugate functions $\chi^+_{s\mu}$ are transformed as functions of opposite covariance $(-1)^{s-\mu}\chi_{s-\mu}$,

$$\chi^+_{s\mu} \sim (-1)^{s-\mu}\chi_{s-\mu} \quad ,$$

the irreducible tensor operator T_{IM} can be constructed by making use of the rules for addition of angular momenta to obtain

$$T_{IM} = \alpha \sum_{\mu\mu'}(-1)^{s-\mu}\left(s'\mu's - \mu|IM\right)\chi_{s'\mu'}\,\chi^+_{s\mu} \quad , \tag{14.35}$$

where α is an arbitrary constant. Usually α is found from the normalization condition imposed on T_{IM}.

According to (14.35), the irreducible tensor operator T_{IM} in the representation determined by the spin squared and the spin projection (i.e., the quantum numbers s and μ) is described by a matrix whose elements reproduce the Clebsch–Gordan coefficients with the exception of the phase factors:

$$\langle s'\mu'|T_{IM}|s\mu\rangle = \alpha\,(-1)^{s-\mu}\,(s'\mu's - \mu|IM) \quad , \tag{14.36}$$

$$|s - s'| \le I \le s + s' \quad .$$

The matrices (14.36) with different I and M form an orthogonal set and are referred to as spin tensors. If $s' = s$, then $\langle s'\mu'|T_{IM}|s\mu\rangle$ is a square matrix.

With s fixed, we introduce an orthogonal set of matrices (spin tensors)

$$\langle s\mu'|T_{IM}|s\mu\rangle = (-1)^{s-\mu}\sqrt{2s+1}\,(s\mu's - \mu|IM) \tag{14.37}$$

$$(I = 0, 1, \ldots, 2s \quad ; \quad -I \le M \le I) \quad ,$$

which satisfy the normalization condition

$$\mathrm{Tr}\left\{T_{IM}\, T^+_{I'M'}\right\} = (2s+1)\,\delta_{II'}\,\delta_{MM'} \quad . \tag{14.38}$$

The spin tensors (14.37) form a complete set, so they may serve as an expansion basis for any operators which correspond to the system with fixed s.

Note that the spin tensors (14.37) are not Hermitian; they satisfy the condition

$$T^+_{IM} = (-1)^{-M}\, T_{I-M} \quad . \tag{14.39}$$

If we put $\alpha = (2s+1)^{1/2}\,i^I$ in the spin tensor definition (14.37), then instead of (14.39) we obtain

$$T_{IM}^{+} = (-1)^{I-M} \, T_{I-M} \quad .$$

(14.40)

With the spin tensors being defined in the latter way, (14.40), in contrast to (14.39), holds also for the vector relation.

For the normalization (14.38), the zero-rank spin tensor T_{00} coincides with the unit matrix,

$$T_{00} = 1 \quad .$$

(14.41)

The first-rank spin tensors T_{1M} are expressed in terms of the spin matrices by the linear relations

$$T_{10} = \sqrt{\frac{3}{s(s+1)}} \, s_z \quad ,$$

$$T_{1\pm 1} = \mp \sqrt{\frac{3}{2s(s+1)}} \, (s_x \pm i \, s_y) \quad .$$

(14.42)

The second-rank spin tensors T_{2M} depend on the quadratic combinations of spin matrices, so that

$$T_{20} = \sqrt{\frac{3}{2}} \, N \left(s_z^2 - \frac{s(s+1)}{3} \right) \quad ,$$

$$T_{2\pm 1} = \mp \frac{N}{2} \left[(s_x \pm i \, s_y) \, s_z + s_z \, (s_x \pm i \, s_y) \right] \quad ,$$

(14.43)

$$T_{2\pm 2} = \frac{N}{2} \, (s_x \pm i \, s_y)^2 \quad .$$

Substituting the definition (14.32) for the symmetric tensor s_{ij}, we find from (14.43) that

$$T_{20} = \sqrt{\frac{3}{2}} \, N \, s_{zz} \quad ,$$

$$T_{2\pm 1} = \mp N \, (s_{xz} \pm i \, s_{yz}) \quad ,$$

(14.44)

$$T_{2\pm 2} = \frac{N}{2} \, (s_{xx} - s_{yy} \pm 2i \, s_{xy}) \quad ,$$

where

$$N = \sqrt{\frac{30}{s(s+1)(2s-1)(2s+3)}} \quad .$$

Spin states of a system of particles with spin s are generally described by a $(2s+1)$-row density matrix. Since the density matrix is Hermitian, it contains only $(2s+1)^2$ independent parameters. If the density matrix is normalized as given by (14.26), $\mathrm{Tr}\,\{\varrho\} = 1$, then $4s(s+1)$ independent parameters exist. Instead of setting the density matrix parameters, one can expand it in terms of spin tensors (14.38) and describe the spin state of the system by the set of expansion coefficients:

$$\varrho = \frac{1}{2s+1} \sum_{I=0}^{2s} \sum_{M} \langle T_{IM}^+ \rangle T_{IM} \quad , \tag{14.45}$$

where

$$\langle T_{IM} \rangle = \text{Tr} \left\{ T_{IM} \, \varrho \right\} \quad . \tag{14.46}$$

(We assume that $\text{Tr} \{\varrho\} = 1$.) Thus, the spin state of the system of particles with spin s is determined by the set of coefficients of the density matrix expansion in terms of spin tensors T_{IM} of the rank not higher than $2s$. The parameters $\langle T_{IM}^+ \rangle$ completely determine the spin state of the system.

Multiplying (14.45) by ϱ and taking the trace of both sides of the equation, we find that

$$\text{Tr} \left\{ \varrho^2 \right\} = \frac{1}{2s+1} \sum_{I=0}^{2s} \sum_{M} \left| \langle T_{IM} \rangle \right|^2 \quad . \tag{14.47}$$

According to (14.28), $\text{Tr} \{\varrho^2\} = 1$ if the state is pure, that is, described by a wave function, and $\text{Tr} \{\varrho^2\} < 1$ if the state of the system is mixed. Thus, the allowed average values of the spin tensors describing the spin state of the system must satisfy the condition

$$\frac{1}{2s+1} \sum_{I=0}^{2s} \sum_{M} \left| \langle T_{IM} \rangle \right|^2 \leq 1 \quad . \tag{14.48}$$

Maximum polarization is attained only in pure states; it corresponds to the equality sign in (14.48). If the mean values of all the spin tensor components with $I > 0$ are equal to zero, then the density matrix reduces to the unit matrix and polarization does not occur in the system.

In the special case of spin $\frac{1}{2}$ particles, the density matrix must be expanded in terms of the spin tensors

$$T_{00} = 1 \quad , \qquad T_{10} = \sigma_z \quad , \qquad T_{1\pm1} = \mp \frac{1}{\sqrt{2}} \left(\sigma_x \pm i \sigma_y \right) \quad . \tag{14.49}$$

Here σ_x, σ_y, and σ_z are the Pauli matrices. The expansion is given by

$$\varrho = \frac{1}{2} \left(1 + \sum_{M=-1}^{1} \langle T_{1M}^+ \rangle T_{1M} \right) \quad . \tag{14.50}$$

The expansion coefficients are functions of the beam particle density (for the accepted normalization (14.26), it is equal to one) and the polarization vector, i.e.,

$$\langle T_{10}\rangle = P_z \quad , \qquad \langle T_{1\pm1}\rangle = \mp \frac{1}{\sqrt{2}}\left(P_x \mp iP_y\right) \quad . \tag{14.51}$$

Therefore, (14.50) can be rewritten as

$$\varrho = \tfrac{1}{2}(1 + \boldsymbol{P\sigma}) \quad . \tag{14.52}$$

Multiplying (14.52) by ϱ and taking the trace of both sides of the resulting equation, we find that for a pure state we have

$$\tfrac{1}{2}\left(1 + P^2\right) = 1 \quad , \tag{14.53}$$

which yields $P^2 = 1$, which was obtained earlier. For a mixed state, $P^2 < 1$.

In the case of particles with spin 1, the density matrix is expanded in terms of irreducible zero-, first-, and second-rank tensors T_{00}, T_{1M}, and T_{2M}, i.e.,

$$T_{00} = 1 \quad , \qquad T_{10} = \sqrt{\frac{3}{2}}\, s_z \quad , \qquad T_{1\pm1} = \mp\frac{\sqrt{3}}{2}\left(s_x \pm is_y\right) \quad ,$$

$$T_{20} = \frac{3}{\sqrt{2}}\, s_{zz} \quad , \qquad T_{2\pm1} = \mp\sqrt{3}\left(s_{zz} \pm is_{yz}\right) \quad , \tag{14.54}$$

$$T_{2\pm2} = \frac{\sqrt{3}}{2}\left(s_{xx} - s_{yy} \pm 2is_{xy}\right) \quad .$$

The expansion coefficients are functions of the beam particle density, polarization vector, and quadratic polarization tensor. Note that the $\langle T_{2M}\rangle$ are more suitable for the description of spin states than the polarization tensor $\langle s_i s_j\rangle$ because under coordinate system rotations $\langle T_{2M}\rangle$ are transformed as the irreducible representation of the rotation group.

The spin tensor expansion of the density matrix in the case of spin 1 is given by

$$\varrho = \frac{1}{3}\sum_{I=0}^{2}\sum_{M=-I}^{I}\langle T_{IM}^{+}\rangle\, T_{IM} \quad . \tag{14.55}$$

An equivalent expansion can be written as

$$\varrho = \tfrac{1}{3}\left(1 + \tfrac{3}{2}\,\boldsymbol{Ps} + 3P_{ij}\,s_{ij}\right) \quad . \tag{14.56}$$

If the particle state is described by the wave function, then the mean values of spin tensor components are given by

$$\langle T_{10}\rangle = \sqrt{\frac{3}{2}}\left(|a_1|^2 - |a_{-1}|^2\right) \quad ,$$

$$\langle T_{11}\rangle = -\sqrt{\frac{3}{2}}\left(a_1^* a_0 - a_0^* a_{-1}\right) \quad , \qquad \langle T_{1-1}\rangle = -\langle T_{11}\rangle^* \quad ,$$

$$\langle T_{20} \rangle = \frac{3}{\sqrt{2}} \left(|a_1|^2 + |a_{-1}|^2 - \tfrac{2}{3} \right) \quad , \tag{14.57}$$

$$\langle T_{21} \rangle = -\sqrt{\frac{3}{2}} \left(a_1^* a_0 - a_0^* a_{-1} \right) \quad , \qquad \langle T_{2-1} \rangle = -\langle T_{21} \rangle^* \quad ,$$

$$\langle T_{22} \rangle = \sqrt{3}\, a_1^* a_{-1} \quad , \qquad \langle T_{2-2} \rangle = \langle T_{22} \rangle^* \quad .$$

Because polarization is at maximum for pure states, the above formulas determine maximum attainable values of the relevant spin tensor components. According to (14.28), for pure states we have

$$\frac{1}{3} + \frac{1}{3} \sum_{M=-1}^{1} |\langle T_{1M} \rangle|^2 + \frac{1}{3} \sum_{M=-2}^{2} |T_{2M}|^2 = 1 \quad . \tag{14.58}$$

Individual terms entering the left-hand part of (14.58) can easily be found by using (14.57). We obtain

$$\frac{1}{3} \sum_{M=-1}^{1} |\langle T_{1M} \rangle|^2 = \frac{1}{2} \left(|a_1|^2 - |a_{-1}|^2 \right)^2 + |a_1^* a_0 + a_0^* a_{-1}|^2 \quad , \tag{14.59}$$

$$\frac{1}{3} \sum_{M=-2}^{2} |\langle T_{2M} \rangle|^2 = \frac{3}{2} \left(|a_1|^2 + |a_{-1}|^2 - \frac{2}{3} \right)^2$$

$$+ |a_1^* a_0 - a_0^* a_{-1}|^2 + 2 |a_1|^2 |a_{-1}|^2 \quad . \tag{14.60}$$

Substituting (14.59, 60) into (14.58) proves the validity of the latter equation for pure states. It is clear that the quantities given by (14.59, 60) may be regarded as vector and tensor polarization weights, respectively. If (14.59) vanishes, then vector polarization of the beam does not occur. If (14.60) vanishes, then tensor polarization is absent. Expressions (14.59, 60) determine maximum probable weights of vector and tensor polarization which are attained only in pure states of the system.

14.3 The Scattering Amplitude in the Case of Spin-Possessing Particles

First of all we consider the simplest case when a particle with spin s is scattered by a spinless one. The interaction potential $\hat{V}$ entering the Hamiltonian

$$H = H_0 + \hat{V} \tag{14.61}$$

depends both on the relative distance between the particles and the spin; therefore, $\hat{V}$ is a matrix in the spin space. Let us find the scattering amplitude which is also a matrix in the spin space.

We assume that the initial wave function φ of the system is a product of the plane wave describing relative motion of particles with momentum k, and the spin function associated with the projection μ of particle spin onto the z axis, i.e.,

$$\varphi_{k\mu} = e^{ikr} \chi_\mu \quad . \tag{14.62}$$

Then, as follows from the general formula (3.3), the wave function $\psi_k^{(+)}$ which satisfies the Schrödinger equation and describes particle scattering can be written as

$$\psi_k^{(+)}(r) = \varphi_{k\mu}(r) + \int dr \, \hat{G}_0(r - r') \, \hat{V}(r') \, \psi_{k\mu}^{(+)}(r') \quad , \tag{14.63}$$

or, in more detail, as

$$\psi_{k\mu}^{(+)}(r, \sigma) = \varphi_{k\mu}(r, \sigma)$$
$$+ \sum_{\sigma'\sigma''} \int dr' \, G_{\sigma\sigma'}^0(r - r') \, V_{\sigma'\sigma''}(r') \, \psi_{k\mu}^{(+)}(r, \sigma) \quad , \tag{14.64}$$

where $G_0(r - r')$ is the Green function defined by (3.28). In the coordinate representation it is given by

$$G_{\sigma\sigma'}^0(r - r') = -\frac{\mu}{2\pi\hbar^2} \frac{\exp\left(ik\,|r - r'|\right)}{|r - r'|} \sum_{\mu'} \chi_{\mu'}(\sigma) \, \chi_{\mu'}^+(\sigma') \quad . \tag{14.65}$$

By virtue of the completeness condition for the spin functions (14.5), the Green function is diagonal in the spin space, i.e.,

$$G_{\sigma\sigma'}^0(r - r') = -\frac{\mu}{2\pi\hbar^2} \frac{\exp\left(ik\,|r - r'|\right)}{|r - r'|} \delta_{\sigma\sigma'} \quad . \tag{14.66}$$

Making use of the definition of the transition operator $\hat{t}$, we find from (3.70) that

$$\hat{V} \, \psi_{k\mu}^{(+)} = \hat{t} \, \varphi_{k\mu} \quad , \tag{14.67}$$

then we can separate in the right-hand part of (14.63) the initial spin function

$$\psi_{k\mu}^{(+)}(r) = \left\{ e^{ikr} + \int dr' \, \hat{G}_0(r - r')\hat{t} \, e^{ikr} \right\} \chi_\mu \quad . \tag{14.68}$$

The expression within the curly braces is an operator in the spin space. By using (14.66), the asymptotic of the wave function (14.68) can be written as

$$\psi_k^{(+)}(r) \rightarrow \left\{ e^{ikr} + \frac{e^{ikr}}{r} \, \hat{f}(k, k') \right\} \chi_\mu \qquad (r \rightarrow \infty) \quad , \tag{14.69}$$

where the scattering amplitude $\hat{f}(k, k')$ is given by

$$\hat{f}(\boldsymbol{k}, \boldsymbol{k}') = -\frac{\mu}{2\pi\hbar^2} \langle \boldsymbol{k}' | \hat{t} | \boldsymbol{k} \rangle \quad . \tag{14.70}$$

Similarly to the transition operator $\hat{t}$, the scattering amplitude $\hat{f}(\boldsymbol{k}, \boldsymbol{k}')$ is an operator (matrix) in the spin space.

The function produced by the action of the operator $\hat{f}$ on the initial spin function χ_μ,

$$\chi' = \hat{f} \chi_\mu \quad , \tag{14.71}$$

may be regarded as the final spin function of the particle, i.e., the scattering operator $\hat{f}$ transforms the initial spin state χ_μ into the final spin state χ'. Expanding the spin function χ' in terms of the complete set of functions $\chi_{\mu'}$, we obtain

$$\chi' = \sum_{\mu'} f_{\mu'\mu} \chi_{\mu'} \quad , \tag{14.72}$$

where $f_{\mu'\mu}$ is the amplitude of elastic scattering accompanied by the change of particle spin projection, i.e.,

$$f_{\mu'\mu} = -\frac{\mu}{2\pi\hbar^2} \langle \boldsymbol{k}'\mu' | t | \boldsymbol{k}\mu \rangle \quad . \tag{14.73}$$

Here μ and μ' are the initial and final spin projections, respectively.

If the particle interaction depends on the spin orientation, then the scattering pattern shows no azimuthal symmetry. In this case, the scattering amplitude $f_{\mu'\mu}(\vartheta, \varphi)$ depends both on the scattering angle ϑ and the azimuthal angle φ. We take the latter to be the angle between the plane of the quantization axis and $\boldsymbol{k}$, and the scattering plane (Fig. 14.1).[1]

Making use of (14.67), we express the scattering amplitude (14.73) in terms of the potential to obtain

$$f_{\mu'\mu}(\vartheta, \varphi) = -\frac{\mu}{2\pi\hbar^2} \int d\boldsymbol{r} \, \chi_{\mu'}^+ \, e^{-i\boldsymbol{k}'\boldsymbol{r}} \, \hat{V}(\boldsymbol{r}) \, \psi_{\boldsymbol{k}\mu}^{(+)}(\boldsymbol{r}) \quad . \tag{14.74}$$

This form is especially suitable for the study of sufficiently weak particle interactions when perturbation theory can be applied.

The differential scattering cross section for particle transitions from an initial state with spin projection μ into a final state with spin projection μ' is determined by the squared modulus of the amplitude (14.73):

$$\sigma_{\mu \to \mu'}(\vartheta, \varphi) = \left| f_{\mu'\mu}(\vartheta, \varphi) \right|^2 \quad . \tag{14.75}$$

If incident particles are unpolarized and the final spin projection is not fixed, then we have to average the cross section (14.75) over probable initial spin projections and sum over all probable final spin projections. Thus,

---

[1] Usually one takes the azimuthal angle $\bar{\varphi}$ to be the angle between the quantization axis (the chosen direction of incident particle polarization) and the normal to the scattering plane, i.e., $\cos \bar{\varphi} = \sin \vartheta_0 \cos [(\pi/2) + \varphi]$, where ϑ_0 is the angle between the vector $\boldsymbol{k}$ and the quantization axis. If $\vartheta_0 = \pi/2$, then $\bar{\varphi} = \pi/2 + \varphi$.

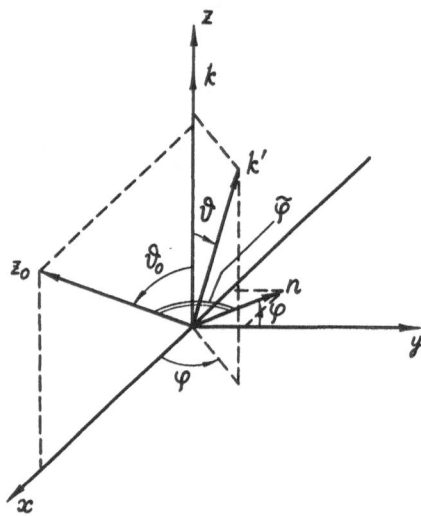

Fig. 14.1. Scattering of polarized particles. k, k' are the initial and final wave vectors; θ, θ_0 are the deflection angle in the scattering plane and the angle between k and the quantization axis; z_0 is the direction of polarization of the incident particle; n is the normal to the scattering plane; φ is the azimuthal angle between the scattering plane and the plane determined by k and the polarization vector z_0; $\bar{\varphi}$ is the azimuthal angle between the vector z_0 and n

$$\sigma(\vartheta) = \frac{1}{2s+1} \sum_{\mu\mu'} |f_{\mu'\mu}(\vartheta,\varphi)|^2 \quad . \tag{14.76}$$

Since no direction is distinguished in the case of unpolarized particles, the averaged cross section (14.76) depends only on the scattering angle.

The above analysis can easily be extended to the case when the target also possesses spin. Let the incident particle spin be s_1 and that of the scatterer be s_2. The initial wave function may be written as

$$\varphi_{k\mu_1\mu_2} = \mathrm{e}^{\mathrm{i}kr} \chi_{s_1\mu_1} \chi_{s_2\mu_2} \quad , \tag{14.77}$$

where μ_1 and μ_2 are particle spin projections. We introduce the channel spin wave functions $\chi_{s\mu}$ by means of the usual addition rules

$$\chi_{s\mu} = \sum_{\mu_1\mu_2} \left(s_1\mu_1 \, s_2\mu_2 | s\mu \right) \chi_{s_1\mu_1} \chi_{s_2\mu_2} \quad .$$

Having inverted the latter equation, we obtain

$$\chi_{s_1\mu_1} \chi_{s_2\mu_2} = \sum_{s\mu} \left(s_1\mu_1 \, s_2\mu_2 | s\mu \right) \chi_{s\mu} \quad . \tag{14.78}$$

We substitute (14.78) into (14.77) bearing in mind that different channel spin functions are orthogonal. Then scattering from different entrance channels occurs independently and can be described in the same manner as scattering of a particle with spin s by a spinless one. Hence the scattering wave function associated with a given entrance channel can be written as

$$\psi_{ks\mu}^{(+)}(\boldsymbol{r}) = e^{i\boldsymbol{k}\boldsymbol{r}}\,\chi_{s\mu}$$

$$-\frac{\mu}{2\pi\hbar^2}\sum_{s'\mu'}\chi_{s'\mu'}\int d\boldsymbol{r}'\,\frac{\exp\left(ik\,|\boldsymbol{r}-\boldsymbol{r}'|\right)}{|\boldsymbol{r}-\boldsymbol{r}'|}$$

$$\times\left(\chi_{s'\mu'},\hat{V}\,\psi_{ks\mu}^{(+)}(\boldsymbol{r}')\right)\quad,\tag{14.79}$$

and the elastic scattering amplitude is given by

$$f_{s'\mu's\mu}(\vartheta,\varphi) = -\frac{\mu}{2\pi\hbar^2}\int d\boldsymbol{r}\,\chi_{s'\mu'}^{+}\,e^{-i\boldsymbol{k}'\boldsymbol{r}}\,\hat{V}(\boldsymbol{r})\,\psi_{ks\mu}^{(+)}(\boldsymbol{r})\quad.\tag{14.80}$$

Note that channel spin is not conserved under scattering and in the general case $s' \neq s$. Since spin projections of interacting particles, rather than channel spins, are observable quantities, the amplitude $f_{\mu_1'\mu_2'\mu_1\mu_2}$, rather than $f_{s'\mu's\mu}$, is physically meaningful. The former can be expressed in terms of the latter by using (14.78):

$$f_{\mu_1'\mu_2'\mu_1\mu_2}(\vartheta,\varphi)$$

$$= \sum_{s\mu s'\mu'}\left(s_1\mu_1\,s_2\mu_2\,|s\mu\right)\left(s_1\mu_1'\,s_2\mu_2'\,|s'\mu'\right)f_{s'\mu's\mu}(\vartheta,\varphi)\quad.\tag{14.81}$$

Summation in (14.81) extends over all allowed values of the channel spins s and s'.

If both initial and final particle spin projections are fixed, then the differential scattering cross section is determined by the square modulus of the amplitude (14.81)

$$\sigma_{\mu_1\mu_2\rightarrow\mu_1'\mu_2'}(\vartheta,\varphi) = \left|f_{\mu_1'\mu_2'\mu_1\mu_2}(\vartheta,\varphi)\right|^2\quad.\tag{14.82}$$

If spin projections of interacting particles are not fixed, then the cross section (14.82) must be averaged over initial and summed over final spin projections. By virtue of the orthogonality condition for the Clebsch–Gordan coefficients, we thus obtain

$$\sigma(\vartheta) = \frac{1}{(2s_1+1)(2s_2+1)}\sum_{s\mu s'\mu'}\left|f_{s'\mu's\mu}(\vartheta,\varphi)\right|^2\quad.\tag{14.83}$$

In the previous analysis we assumed the system under consideration to be described by a wave funtion. In the general case, the spin states of the incident and scattered beams must be described by a density matrix. The relation between the initial and final density matrices can be derived from (14.71). If the incident beam is described by the density matrix

$$\varrho = \overline{\chi\chi^{+}}$$

(the bar denotes statistical averaging), then the final density matrix of the system is given by

$$\varrho' = \overline{\chi' \chi'^+} = f \varrho f^+ \tag{14.84}$$

It is not difficult to show that for the accepted normalization of the wave function (14.71), the trace of the density matrix ϱ' determines the differential scattering cross section, i.e.,

$$\sigma(\vartheta, \varphi) = \frac{\mathrm{Tr}\{\varrho'\}}{\mathrm{Tr}\{\varrho\}} = \frac{\mathrm{Tr}\{f \varrho f^+\}}{\mathrm{Tr}\{\varrho\}} \quad . \tag{14.85}$$

If the incident beam is unpolarized, then the initial density matrix reduces to the unit matrix and the differential scattering cross section is given by

$$\sigma(\vartheta) = \frac{\mathrm{Tr}\{f f^+\}}{\mathrm{Tr}\{\varrho\}} \quad . \tag{14.86}$$

To verify that this formula reproduces (14.83) is a simple matter. If incident particles are polarized, then the cross section (14.85) depends on the direction of the polarization vector.

In general, scattering produces particle spin polarization even if the incident particles are unpolarized. The final particle polarization is described by the general formula

$$P = \frac{1}{s} \frac{\mathrm{Tr}\{s\varrho'\}}{\mathrm{Tr}\{\varrho'\}} \quad . \tag{14.87}$$

If incident particles are unpolarized and the initial density matrix reduces to the unit matrix, then (14.87) describes particle polarization produced by the scattering process.

Equations (14.85, 87) hold both for the case when spin-possessing particles are scattered by spinless ones and when both the probe and the target possess spins. In the latter case, the density matrix is given by the direct product of probe and target density matrices,

$$\varrho = \varrho_1 \varrho_2 \tag{14.88}$$

Polarization of recoil particles is given by an expression analogous to (14.87).

14.4 Addition of Spin and Angular Momentum and Diagonalization of the S Matrix

If the interaction is noncentral, that is the potential depends both on the relative distance between particles and the spin vector, then neither spin nor angular momentum are conserved after scattering. The integral of motion (the potential is assumed to be invariant with respect to rotations) is the total moment j which by definition is the vector sum of the angular momentum l and spin s, i.e.,

$$j = l + s \quad . \tag{14.89}$$

Suppose $\chi_{s\mu}$ is the spin operator eigenfunction, and $i^l Y_{lm}(n)$ is the eigenfunction of the angular momentum operator (n is the unit vector along the coordinate vector r; the additional phase factor is introduced in order to ensure that under time inversion the functions must be transformed according to (6.43)). We put

$$\mathcal{Y}_{lsjM}(\boldsymbol{n}_r, \sigma) = \sum_{m\mu} (lms\mu \,|\, jM)\, i^l Y_{lm}(\boldsymbol{n}_r)\, \chi_{s\mu}(\sigma) \quad , \tag{14.90}$$

where $(lms\mu \,|\, jM)$ is the Clebsch–Gordan coefficient. The spin-angle function $\mathcal{Y}_{lsjM}$ is the eigenfunction of the operators of total moment squared and its projection on a chosen direction z, so that

$$\boldsymbol{j}^2 \,\mathcal{Y}_{lsjM} = j(j+1)\, \mathcal{Y}_{lsjM} \quad , \tag{14.91}$$

$$j_z \,\mathcal{Y}_{lsjM} = M\, \mathcal{Y}_{lsjM} \quad . \tag{14.92}$$

At the same time, $\mathcal{Y}_{lsjM}$ is the eigenfunction of the operators of angular momentum squared,

$$\boldsymbol{l}^2 \,\mathcal{Y}_{lsjM} = l(l+1)\, \mathcal{Y}_{lsjM} \quad , \tag{14.93}$$

and spin squared,

$$\boldsymbol{s}^2 \,\mathcal{Y}_{lsjM} = s(s+1)\, \mathcal{Y}_{lsjM} \quad . \tag{14.94}$$

By virtue of the orthonormalization conditions for the spherical functions Y_{lm} and spin functions $\chi_{s\mu}$, the functions $\mathcal{Y}_{lsjM}$ satisfy the orthonormalization condition

$$\sum_{\sigma} \int do\, \mathcal{Y}^*_{l's'j'M'}\, \mathcal{Y}_{lsjM} = \delta_{ll'}\, \delta_{ss'}\, \delta_{jj'}\, \delta_{MM'} \quad . \tag{14.95}$$

Let us consider the scattering problem. Suppose the incident wave in the entrance channel is given by the product of the plane wave describing relative motion of particles with momentum k, and the spin function s and its projection μ on a chosen axis:

$$\varphi_{ks\mu}(r, \sigma) = e^{ikr}\, \chi_{s\mu}(\sigma) \quad . \tag{14.96}$$

We expand the wave function of the entrance channel $\varphi_{ks\mu}$ in terms of the complete set of spin-angular functions $\mathcal{Y}_{lsjM}$ to obtain

$$\varphi_{ks\mu}(r, \sigma) = \sum_{l's'jM} a^0_{l's'jM}(k, s, \mu; r)\, \mathcal{Y}_{l's'jM}(\boldsymbol{n}_r, \sigma) \quad . \tag{14.97}$$

To find the expansion coefficients $a^0_{l's'jM}$, we employ the expansion (4.20) of the plane wave in terms of spherical functions and the orthonormalization conditions for spherical and spin functions. The result is

$$a^0_{l's'jM}(k, s, \mu; r) = 4\pi \sum_{lm} \delta_{ll'} \delta_{ss'} j_l(kr)(lms\mu | jM) Y^*_{lm}(n_k) \quad , \quad (14.98)$$

where n_k is the unit vector directed along k. Thus we have

$$\varphi_{ks\mu}(r, \sigma) = 4\pi \sum_{jM} \sum_{lm} j_l(kr)(lms\mu | jM) Y^*_{lm}(n_k)$$
$$\times \mathcal{Y}_{lsjM}(n_r, \sigma) \quad . \quad (14.99)$$

An analogous procedure yields the scattering wave function $\psi_{ks\mu}$

$$\psi_{ks\mu}(r, \sigma) = \sum_{l's'jM} a_{l's'jM}(k, s, \mu; r) \mathcal{Y}_{l's'jM}(n_r, \sigma) \quad , \quad (14.100)$$

where

$$a_{l's'jM}(k, s, \mu; r) = \sum_{lm} \psi^j_{l's',ls}(r)(lms\mu | jM) Y^*_{lm}(n_k) \quad ,$$

and hence

$$\psi_{ks\mu}(r, \sigma) = \sum_{lm} \sum_{l's'jM} \psi^j_{l's',ls}(r)(lms\mu | jM) Y^*_{lm}(n_r, \sigma)\mathcal{Y}_{l's'jM}(n_r, \sigma) \quad . \quad (14.101)$$

Since neither spin nor angular momentum are conserved after scattering, the expansion coefficients $\psi^j_{l's',ls}$ entering (14.101) are not diagonal, in contrast to (14.99). The subscripts l' and s' of the expansion coefficient $\psi^j_{l's',ls}$ indicate that the corresponding component in (14.101) is associated with definite spin and angular momentum, i.e., it determines the amplitude of probability of finding the system with given angular momentum l' and spin s'. The subscripts l and s correspond to the entrance channel spin and angular momentum and are determined by the initial conditions of the problem. In the case of a centrosymmetric interaction potential, we have

$$\psi^j_{l's',ls} = \delta_{ll'} \delta_{ss'} \psi^j_{ls} \quad , \quad (14.102)$$

and the expansion (14.101) reduces to (4.21).

Substituting (14.101) for the wave function in the Schrödinger equation, we obtain a set of coupled equations for the radial functions $\psi^j_{l's',ls}$, i.e.,

$$\left\{ \frac{1}{r^2} \frac{d}{dr} \left(r^2 \frac{d}{dr} \right) - \frac{l'(l'+1)}{r^2} + k^2 \right\} \psi^j_{l's',ls}(r)$$
$$= \sum_{l''s''} v^j_{l's',l''s''}(r) \psi^j_{l''s'',ls}(r) \quad , \quad (14.103)$$

where

$$v^j_{l's',ls}(r) = \frac{2\mu}{\hbar^2} \sum_\sigma \int do \, \mathcal{Y}^*_{l's'jM} \hat{V}_r(r) \mathcal{Y}_{lsjM} \quad . \quad (14.104)$$

The matrix element of the potential V_r does not depend on M because interaction is invariant with respect to rotations. The allowed values of the channel spin s and s' are determined by the spins s_1 and s_2 of the particles involved in the collision. With fixed j, the allowed l and l' are determined by the summation rules which follow from the properties of the Clebsch–Gordan coefficients.

Substituting the expansions (14.99, 101) into (14.79), we obtain a set of coupled integral equations for the radial functions $\psi^j_{l's',ls}$:

$$\psi^j_{l's',ls}(r) = 4\pi\, j_l(kr)\, \delta_{ll'}\, \delta_{ss'} \tag{14.105}$$

$$+ \sum_{l''s''} \int_0^\infty dr'\, r'^2\, g^0_{l,k}(r,r')\, v^j_{l's',l''s''}(r')\, \psi^j_{l''s'',ls}(r') \quad,$$

where $g^0_{l,k}(r,r')$ is the Green function,

$$g^0_{l,k}(r,r') = -ik \begin{cases} j_l(kr)\, h^{(+)}_l(kr') & r < r' \\ h^{(+)}_l(kr)\, j_l(kr') & r > r' \end{cases} . \tag{14.106}$$

In view of the integral representation (14.105), the asymptotics of the radial function $\psi^j_{l's',ls}$ can be written as

$$\psi^j_{l's',ls}(r) \xrightarrow[r\to\infty]{} \frac{2\pi i}{kr} \left\{ \delta_{ll'}\, \delta_{ss'}\, \exp\left[-i\left(kr - \frac{l\pi}{2}\right)\right] \right.$$

$$\left. - S^j_{l's',ls}\, \exp\left[i\left(kr - \frac{l\pi}{2}\right)\right] \right\} \quad, \tag{14.107}$$

where

$$S^j_{l's',ls} = \delta_{ll'}\, \delta_{ss'} - 2\pi i\, t^j_{l's',ls} \quad, \tag{14.108}$$

$$t^j_{l's',ls} = \frac{k}{4\pi^2} \sum_{l''s''} \int_0^\infty dr\, r^2\, j_{l'}(kr)\, v^j_{l's',l''s''}(r)\, \psi^j_{l''s'',ls}(r) \quad. \tag{14.109}$$

The quantities $S^j_{l's',ls}$ immediately determine the elastic scattering amplitude. Indeed, making use of (14.73) and transforming to the representation determined by l, s, j, and M, we obtain

$$f_{s'\mu'su}(k,k') = -\frac{4\pi^2}{k} \langle n_{k'}s'\mu' |t| n_k s\mu \rangle$$

$$= -\frac{4\pi^2}{k} \sum_{\substack{l'm'j'M' \\ lmjM}} \langle n_{k'}|l'm'\rangle\, \langle l'm's'\mu'|l's'j'M'\rangle$$

$$\times \langle l's'j'M' |t| lsjM \rangle\, \langle lsjM|lms\mu\rangle\, \langle lm|n_k\rangle \quad. \tag{14.110}$$

Because the potential is invariant with respect to rotations, the t matrix, similarly to the S matrix, is diagonal in the j, M-representation; the matrix elements are independent of the magnetic quantum number M, so that

$$\langle l's'j'M'\,|t|\,lsjM\rangle = t^j_{l's',ls}\,\delta_{jj'}\,\delta_{MM'} \quad . \tag{14.111}$$

By virtue of $\langle n'|l'm'\rangle = Y_{l'm'}(n')$, $\langle lm|n\rangle = Y^*_{lm}(n)$, and (14.108), we thus find that

$$f_{s'\mu'\,s\mu}(k, k') = \frac{2\pi i}{k} \sum_{\substack{lml'm'\\jM}} (lms\mu|jM)\,(l'm's'\mu'|jM)$$

$$\times \left\{\delta_{ll'}\,\delta_{ss'} - S^j_{l's',ls}\right\} Y^*_{lm}(n_k)\,Y_{lm}(n_{k'}) \quad . \tag{14.112}$$

If the interaction potential is invariant with respect to spatial inversion and hence, conserves parity, l and l' in (14.112) must be of equal parity. If the interaction is also invariant with respect to time inversion, then we find from the reciprocity condition that

$$S^j_{l's',ls} = S^j_{ls,l's'} \quad , \tag{14.113}$$

i.e., the matrix $S^j_{l's',ls}$ is symmetric.

In order to diagonalize the symmetric unitary matrix $S^j_{l's',ls}$, we introduce an orthogonal real matrix U $(U^T = U^{-1})$ such that

$$S^j_{l's',ls} = \sum_\alpha U^j_{\alpha,l's'}\,\exp\left(2i\,\delta^j_\alpha\right) U^j_{\alpha,ls} \quad . \tag{14.114}$$

Here $\exp\left(2i\,\delta^j_\alpha\right)$ are the diagonal elements of the S matrix expressed in terms of the proper phase shifts δ^j_α which are real due to unitarity of the S matrix.

The representation (14.114) enables us to find how many parameters are required in order to determine the matrix $S^j_{l's',ls}$. Suppose the pairs l, s form n combinations, i.e., the matrix S^j is $(n \times n)$-dimensional. In the general case, any n-dimensional complex matrix is determined by $2n^2$ parameters. However, $n(n+1)/2$ real parameters are sufficient for (14.114). Indeed, a real orthogonal matrix depends on $n(n-1)/2$ real numbers. Adding to these n proper phase shifts δ^j_α, we obtain just $n(n+1)/2$ real parameters.

For example, we consider scattering in the system where spin is conserved and equal to one, $s = s' = 1$ (e.g., nucleon-nucleon scattering in the triplet state). For j fixed, each angular momentum l and l' can take three values $j-1$, j, and $j+1$. If l is equal to $j+1$ or $j-1$, then, due to parity conservation, the probable values of l' are also $j+1$ or $j-1$. Therefore, the dimensionality of the matrix S^j, and hence of the orthogonal matrix U^j, is 2×2. A (2×2)-dimensional real orthogonal matrix in the general case is given by

$$U^j = \begin{pmatrix} \cos \varepsilon^j & \sin \varepsilon^j \\ -\sin \varepsilon^j & \cos \varepsilon^j \end{pmatrix} , \tag{14.115}$$

where ε^j is a real parameter. Thus we have

$$S^j_{l'1,l1} = \begin{pmatrix} \cos^2 \varepsilon\,e^{2i\delta_1} + \sin^2 \varepsilon\,e^{2i\delta_2} & \sin \varepsilon \cos \varepsilon \left(e^{2i\delta_1} - e^{2i\delta_2}\right) \\ \sin \varepsilon \cos \varepsilon \left(e^{2i\delta_1} - e^{2i\delta_2}\right) & \sin^2 \varepsilon\,e^{2i\delta_1} + \cos^2 \varepsilon\,e^{2i\delta_2} \end{pmatrix} \tag{14.116}$$

$$l = j+1, \ j-1 \ ; \qquad l' = j+1, \ j-1 \ .$$

The quantity ε^j is referred to as the mixing parameter. The matrix S^j in the case under consideration is determined by three real parameters: two proper scattering phase shifts, δ_1^j and δ_2^j, and the mixing parameter ε^j. If $l = j$, then $l' = j$ by virtue of parity conservation. Since the S matrix is unitary, we have

$$S_{j1,j1}^{j} = \exp\left(2i\,\delta_3^j\right) \ . \tag{14.117}$$

The wave function $\psi_{l's',\alpha}^{j}$ produced by the orthogonal transformation that results in

$$\psi_{l's',\alpha}^{j}(r) = \sum_{l''s''} U_{\alpha,l''s''}^{j} \ \psi_{l's',l''s''}^{j}(r) \ , \tag{14.118}$$

satisfies (14.103) and boundary conditions at infinity. However, whereas $\psi_{l's',ls}^{j}$ contains only converging waves with angular momentum l and spin s, the function $\psi_{l's',\alpha}^{j}$ contains converging waves with all quantum numbers allowed for a given j. Then the asymptotic of the wave function (14.118) is given by

$$\psi_{l's',\alpha}^{j}(r) \underset{r\to\infty}{\longrightarrow} 4\pi \, U_{\alpha,l's'}^{j} \ \exp\left(i\,\delta_\alpha^j\right) \frac{\sin\left[kr - (l'\pi/2) + \delta_\alpha^j\right]}{kr} \ , \tag{14.119}$$

that is, all components are associated with the same scattering phase shift.

14.5 Spin $\frac{1}{2}$ – Spin 0 Particle Scattering

Let us consider the simplest example – scattering of a particle with spin $\frac{1}{2}$ by a spinless particle (e.g., nucleon scattering by a spinless nucleus). The spin operator s of a spin $\frac{1}{2}$ particle is determined by the Pauli matrices σ, i.e.,

$$s = \frac{1}{2}\,\sigma \ , \tag{14.120}$$

where

$$\sigma_x = \begin{pmatrix} 0 & 1 \\ 1 & 0 \end{pmatrix} \ , \qquad \sigma_y = \begin{pmatrix} 0 & -i \\ i & 0 \end{pmatrix} \ , \qquad \sigma_z = \begin{pmatrix} 1 & 0 \\ 0 & -1 \end{pmatrix} \ . \tag{14.121}$$

The eigenfunctions of the spin squared and spin projection operators are

$$\chi_{1/2,1/2} = \begin{pmatrix} 1 \\ 0 \end{pmatrix} \ , \qquad \chi_{1/2,-1/2} = \begin{pmatrix} 0 \\ 1 \end{pmatrix} \ . \tag{14.122}$$

Suppose the incident particle beam is described by the density matrix ϱ. For spin $\frac{1}{2}$, it is a two-row matrix. Because any two-row matrix can be written as a linear combination of the Pauli and unit matrices, for the density matrix ϱ we have

$$\varrho = a1 + b\boldsymbol{\sigma} \quad .$$

The physical meaning of the constants a and b is very simple. The quantity a is associated with the beam particle density I, so that

$$I = \mathrm{Tr}\{\varrho\} = 2a \quad ,$$

and b can be easily shown to be proportional to the polarization vector P, i.e.,

$$P = \frac{\mathrm{Tr}\{\sigma \varrho\}}{\mathrm{Tr}\{\varrho\}} = \frac{2b}{I} \quad .$$

Thus, the density matrix of a particle beam with arbitrary polarization can be represented in the form

$$\varrho = \frac{I}{2}(1 + \boldsymbol{p}\boldsymbol{\sigma}) \quad . \tag{14.123}$$

If the beam is completely polarized along the z axis, then

$$\varrho = \begin{pmatrix} I & 0 \\ 0 & 0 \end{pmatrix} \quad .$$

An unpolarized beam with $P = 0$ is described by the unit density matrix,

$$\varrho = \frac{I}{2} \quad .$$

In the general case, a particle beam with spin $\frac{1}{2}$ is described by four real measurable variables I and P.

The density matrix ϱ' of the scattered beam is expressed in terms of the incident beam density matrix ϱ and the scattering amplitude f according to (14.84),

$$\varrho' = f \varrho f^+ \quad . \tag{14.124}$$

The scattering amplitude, similarly to the density matrix ϱ, can be expressed in terms of a linear combination of the unit and Pauli matrices,

$$f = g1 + h\boldsymbol{\sigma} \quad , \tag{14.125}$$

where g and h depend only on the collision geometry, i.e., the initial and final momenta $\boldsymbol{k}$ and $\boldsymbol{k}'$ and the potential. Since the amplitude (14.125) must not depend on the choice of the coordinate system (i.e., must be invariant under the rotations and reflections of the coordinate system), the coefficient g must be scalar, and h must be pseudoscalar (we remind the reader that the spin $\boldsymbol{\sigma}$, as well as $\boldsymbol{l} = \boldsymbol{r} \times \boldsymbol{p}$, is a pseudovector in the usual three-dimensional space). Since $\boldsymbol{k}$ and $\boldsymbol{k}'$ cannot form any pseudovectors other than $\boldsymbol{k} \times \boldsymbol{k}'$, h can be written as

$$\boldsymbol{h} = h\boldsymbol{n} \quad ,$$

where h is a scalar and n is the unit vector perpendicular to the scattering plane,

$$n = \frac{k \times k'}{|k \times k'|} \quad . \tag{14.126}$$

Thus, the spin $\frac{1}{2}$ – spin 0 particle scattering amplitude can finally be written in the form

$$f = g + h\,n\,\sigma \quad , \tag{14.127}$$

where the scalar amplitudes g and h (in the general case they are complex) depend on the scattering angle, particle energies, and the type of interaction.

Let us express the scalar amplitudes g and h in terms of scattering phase shifts. First of all we note that for a state with fixed total moment quantum number j, the angular momentum quantum number can take two values $l = j + \frac{1}{2}$ and $l = j - \frac{1}{2}$ associated with different parities. Therefore, if the interaction conserves parity, then the scattering matrix (14.113) is diagonal with respect to the quantum numbers l and l', i.e.,

$$S^j_{l'\,1/2,l\,1/2} = \delta_{ll'}\, S^j_l \quad . \tag{14.128}$$

If only elastic scattering occurs, then by virtue of the unitarity condition for the scattering matrix we have

$$S^j_l = \exp\left(2i\,\delta^j_l\right) \quad , \tag{14.129}$$

where δ^j_l are real scattering phase shifts in the states with fixed j and l. If some channels other than elastic scattering are also open, then the phase shifts δ^j_l are complex.

To find explicit expressions for the scalar amplitudes g and h, we employ the partial wave expansion of the scattering amplitude f and take into account the spin dependence of the scattering operator $\hat{S}_l$. Since any fixed l can correspond to only two values of j, we can introduce the operators of projection on the state with $j = l + \frac{1}{2}$ and $j = l - \frac{1}{2}$, $\prod_{l+(1/2)}$ and $\prod_{l-(1/2)}$, to obtain

$$\hat{S}_l = S^{j=l+(1/2)}_l \prod_{l+(1/2)} + S^{j=l-(1/2)}_l \prod_{l-(1/2)} \quad . \tag{14.130}$$

Since

$$l\sigma = \begin{cases} l & j = l + \frac{1}{2} \,, \\ -(l+1) & j = l - \frac{1}{2} \,, \end{cases}$$

where $l = -i\,n\,\partial/\partial\vartheta$ is the angular momentum operator, the projection operators $\prod_{l+(1/2)}$ and $\prod_{l-(1/2)}$ reduce to

$$\prod_{l+(1/2)} = \frac{l+1+l\sigma}{2l+1} \quad , \qquad \prod_{l-(1/2)} = \frac{l-l\sigma}{2l+1} \quad . \tag{14.131}$$

Note that

$$\prod_{l+(1/2)} + \prod_{l-(1/2)} = 1 \quad . \tag{14.132}$$

Substituting (14.130) into (3.75) and using (14.131), we reduce the scattering amplitude f to the form of (14.127), with g and h being determined by the expressions

$$g(\vartheta) = \frac{i}{2k} \sum_{l=0}^{\infty} \left\{ 2l + 1 - (l+1) \exp\left(2i\,\delta_l^{l+(1/2)}\right) \right.$$

$$\left. - l \exp\left(2i\,\delta_l^{l-(1/2)}\right) \right\} P_l(\cos\vartheta) \quad , \tag{14.133}$$

$$h(\vartheta) = -\frac{1}{2k} \sum_{l=0}^{\infty} \left\{ \exp\left(2i\,\delta_l^{l+(1/2)}\right) - \exp\left(2i\,\delta_l^{l-(1/2)}\right) \right\}$$

$$\times \frac{\partial}{\partial\vartheta} P_l(\cos\vartheta) \quad . \tag{14.134}$$

These expressions can also be derived from (4.112) by using (4.128, 129) and taking into account that the central and spin components of the amplitude, represented by g and h, describe scattering without and with particle spin reorientation, respectively.

If the incident beam is unpolarized, then its density matrix reduces to the unit matrix. In this case, the scattered beam density matrix is expressed in terms of the scattering amplitude only, i.e.,

$$\varrho = f f^{+} \quad . \tag{14.135}$$

The differential scattering cross section is given by the ratio of the traces of the scattered and incident beam density matrices,

$$\sigma(\vartheta) = \frac{\text{Tr}\,\{\varrho'\}}{\text{Tr}\,\{\varrho\}} \quad . \tag{14.136}$$

Making use of (14.127, 135) we find that

$$\sigma(\vartheta) = |g(\vartheta)|^2 + |h(\vartheta)|^2 \quad . \tag{14.137}$$

Particle polarization produced by scattering is given by

$$P = \frac{\text{Tr}\,\{\sigma\,\varrho'\}}{\text{Tr}\,\{\varrho'\}} \quad . \tag{14.138}$$

Substituting (14.135) for the density matrix ϱ' in (14.138), we find the polarization vector to be directed along n,

$$P = P\,n \quad , \tag{14.139}$$

and the value of polarization to be given by

$$P = \frac{2\,\mathrm{Re}\,\{gh^*\}}{|g|^2 + |h|^2} \quad . \tag{14.140}$$

Being an interference effect, polarization vanishes if one of the amplitudes, g or h, is equal to zero. Polarization also disappears if one amplitude is real and the other one is pure imaginary. According to (14.139), scattering can produce particle polarization only along the direction perpendicular to the scattering plane.

If the incident beam polarization is P_0, and the density matrix is given by

$$\varrho = \frac{I}{2}\left(1 + P_0\,\sigma\right) \quad , \tag{14.141}$$

then the scattering cross section and polarization of scattered particles are described by formulas more complicated than (14.137, 139, 140). Substituting (14.141) into (14.124) and taking into account (14.127), we obtain for the density matrix

$$\varrho' = \frac{I}{2}\left(g + h\,n\,\sigma\right)\left(1 + P_0\,\sigma\right)\left(g^* + h^*\,n\,\sigma\right) \quad . \tag{14.142}$$

The differential scattering cross-section is determined by the trace of (14.142). It is convenient to calculate the latter by using

$$(a\sigma)\,(b\sigma) = (ab) + \mathrm{i}(a \times b)\,\sigma \quad , \tag{14.143}$$

which follows from the permutation relations for the Pauli matrices. Since the trace of the terms of (14.142) containing odd products of Pauli matrices vanishes, and

$$\mathrm{Tr}\,\{a\sigma\}\,(b\sigma) = 2ab \quad ,$$

we find the differential cross section of scattering of arbitrarily polarized spin $\frac{1}{2}$ particles by spinless ones to be given by

$$\sigma(\vartheta, \varphi) = \left(|g|^2 + |H|^2\right)\left(1 + P_0 P\right) \quad , \tag{14.144}$$

where P is the polarization vector which would be produced by scattering, were the incident particles unpolarized. The vector P is determined by (14.139, 140).

The first factor in (14.144) coincides with the differential scattering cross section (14.137) for unpolarized incident particles. The second factor,

$$\left(1 + P_0 P\right) = \left(1 + P_0 P \cos \varphi\right) \quad ,$$

where φ is the angle between the incident particle polarization and the normal to the scattering plane, is responsible for the azimuthal asymmetry that occurs after scattering of polarized particles. The quantity

$$e = P_0 P \tag{14.145}$$

is usually referred to as the azimuthal asymmetry coefficient. If polarization P_0 of incident particles is known, then the azimuthal asymmetry coefficient determines the polarization P due to scattering of unpolarized particles.

If $\boldsymbol{P}_0$ is the initial particle polarization vector, then the scattered particle polarization is

$$P' = \frac{P + \hat{\chi} P_0}{1 + PP_0} \quad , \tag{14.146}$$

with the scalar product $\hat{\chi} P_0$ being given by

$$\hat{\chi} P_0 = P_0 - \frac{2|h|^2}{|g|^2 + |h|^2} \left\{ P_0 - (P_0 n)n \right\}$$
$$+ \frac{2 \operatorname{Im} \{gh^*\}}{|g|^2 + |h|^2} \, n \times P_0 \quad . \tag{14.147}$$

If the incident polarization vector $\boldsymbol{P}_0$ lies in the scattering plane, then $\hat{\chi} P_0$ also lies in this plane, and the component of $\boldsymbol{P}'$ normal to the scattering plane does not depend on $\boldsymbol{P}_0$, so that

$$P'n = P \quad . \tag{14.148}$$

If the incident particle polarization vector $\boldsymbol{P}_0$ is perpendicular to the scattering plane, then

$$\hat{\chi} P_0 = P_0 \quad ,$$

and the polarization vector (14.146) of scattered particles is also perpendicular to the scattering plane, so that

$$P' = \frac{P + P_0}{1 + PP_0} \, n \quad . \tag{14.149}$$

Polarization produced by scattering of unpolarized particles by a spinless target can be measured by means of a double scattering technique. The incident unpolarized beam acquires polarization after being scattered by the first target. Then scattering by the second target produces azimuthal asymmetry which can be measured. Thus the scattering-induced polarization is found (Fig. 14.2).

Suppose a beam of particles with momenta $\boldsymbol{k}$ is incident on a (first) target. If the beam is unpolarized, then the density matrix reduces to the unit matrix,

$$\varrho = \frac{I}{2} \quad ,$$

where I is the beam particle density. We consider scattering at given angle ϑ_1 and denote the particle momenta after this first scattering by $\boldsymbol{k}_1$. The amplitude of particle scattering by the first target is given by

$$f_1 = g_1 + h_1 \, \boldsymbol{n}_1 \, \boldsymbol{\sigma} \quad , \tag{14.150}$$

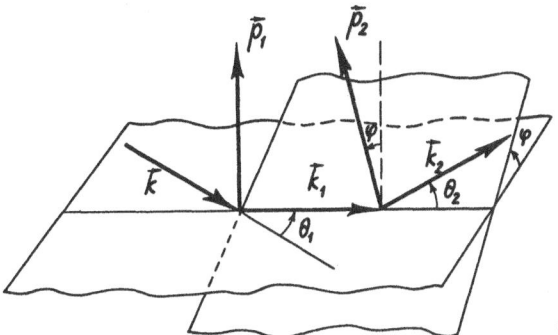

Fig. 14.2. Double scattering

where g_1 and h_1 are the scalar amplitudes depending on k, k_1, and ϑ_1; n_1 is the unit vector perpendicular to the first scattering plane. The particles escaping at the angle ϑ_1 form the scattered beam that is described by the density matrix

$$\varrho_1 = f_1 \, \varrho \, f_1^+ \quad ,$$

which can be written as

$$\varrho_1 = \frac{I}{2} \left(|g_1|^2 + |h_1|^2 \right) \left(1 + \boldsymbol{P}_1 \boldsymbol{\sigma} \right) \quad , \tag{14.151}$$

where $\boldsymbol{P}_1$ is the polarization vector given rise to by particle scattering by the first target,

$$\boldsymbol{P}_1 = \frac{2 \, \mathrm{Re} \left\{ g_1 h_1^* \right\}}{|g_1|^2 + |h_1|^2} \, \boldsymbol{n}_1 \quad . \tag{14.152}$$

The differential scattering cross section for the first target is given by

$$\sigma_1 \left(\vartheta_1 \right) = |g_1|^2 + |h_1|^2 \quad ; \tag{14.153}$$

it is azimuthally symmetric.

Let us denote the particle momentum after scattering by the second target as k_2 and the angle between k_2 and k_1 as ϑ_2. The second scattering amplitude can be written as

$$f_2 = g_2 + h_2 \, \boldsymbol{n}_2 \, \boldsymbol{\sigma} \quad , \tag{14.154}$$

where g_2 and h_2 are the scalar amplitudes depending on k_1, k_2, and the angle ϑ_2; n_2 is the unit vector perpendicular to the second scattering plane. Since the incident particles are now polarized with (14.151) being their initial density matrix, the final density matrix after the second scattering event is

$$\varrho_2 = f_2 \, \varrho_1 \, f_2^+ \quad . \tag{14.155}$$

The differential cross section of the second scattering is determined by the ratio of the traces of the final and initial density matrices. Thus we have

$$\sigma_2(\vartheta_2,\varphi) = \left(|g_2|^2 + |h_2|^2\right)\left(1 + P_1 P_2\right) \quad , \tag{14.156}$$

where P_2 is the polarization vector that would be produced under scattering by the second target, were the incident particles unpolarized, i.e.,

$$P_2 = \frac{2\,\mathrm{Re}\left\{g_2 h_2^*\right\}}{|g_2|^2 + |h_2|^2}\, n_2 \quad . \tag{14.157}$$

According to (14.156), the second scattering cross section is azimuthally asymmetric because it depends on the angle between the vectors P_2 and P_1, i.e., the angle between the first and second scattering planes. The final polarization vector P_2' is described by the expression

$$P_2' = \frac{P_2 + \chi_2\, P_1}{1 + P_1 P_2} \quad , \tag{14.158}$$

where χ depends on g_2, h_2, and n_2. In contrast to (14.152), the polarization vector after the second scattering has a longitudinal component directed along the momentum k_2.

If both targets consist of identical particles, then we can put $\vartheta_2 = \vartheta_1$ and take P_2 to be approximately equal to P_1 (they can be equal only approximately because of the recoil losses of scattered particle energy). Assuming that both collisions occur in the same plane, we can write the differential scattering cross section as

$$\sigma_2 = \left(|g|^2 + |h|^2\right)(1 \pm e) \quad , \tag{14.159}$$

where the plus and minus signs correspond to parallel and antiparallel orientations of the vectors n_1 and n_2 which, in turn, depend on whether the second collision occurs to the left or to the right with respect to the direction of the momentum k_1. This left-right asymmetry e determines the polarization, so that

$$e = \frac{\sigma^l - \sigma^r}{\sigma^l + \sigma^r} = P_1 P_2 \quad . \tag{14.160}$$

If experimental conditions are such that $P_1 = P_2$, then (14.160) immediately yields

$$P = e^{1/2} \quad . \tag{14.161}$$

Note that scattering of spin $\frac{1}{2}$ particles by unpolarized particles with arbitrary spin can be considered in a similar manner. In particular, the scattering cross section and polarization produced by scattering are described by the expressions of the type (14.137, 139, 140).

14.6 Spin 1 – Spin 0 Particle Scattering

Let us consider scattering of a particle with spin 1 by a spinless particle (e.g., deuteron scattering by a spinless nucleus) [14.9–11].

The spin wave function of a spin 1 particle has three components and the spin matrices have three rows. In the representation of the spin squared and its projection on the z axis, the spin matrices are given by

$$S_x = \frac{1}{\sqrt{2}} \begin{pmatrix} 0 & 1 & 0 \\ 1 & 0 & 1 \\ 0 & 1 & 0 \end{pmatrix} \quad , \quad S_y = \frac{1}{\sqrt{2}} \begin{pmatrix} 0 & -i & 0 \\ i & 0 & -i \\ 0 & i & 0 \end{pmatrix} \quad ,$$

$$S_z = \begin{pmatrix} 1 & 0 & 0 \\ 0 & 0 & 0 \\ 0 & 0 & -1 \end{pmatrix} \quad . \tag{14.162}$$

With regard to probable values of the spin projections, we can write the characteristic equation for the spin matrices S_i $(i = x, y, z)$ as

$$\|S_i - \lambda I\| = \lambda(\lambda + 1)(\lambda - 1) = 0 \quad . \tag{14.163}$$

Since each matrix is a root of its characteristic equation, it follows from (14.163) that a product of three spin matrices can always be reduced to a combination of quadratic products and a linear combination of spin projections, i.e.,

$$S_i S_j S_k = \frac{i}{2} \left(\varepsilon_{ijl} S_k S_l + \varepsilon_{ikl} S_l S_j + \varepsilon_{jkl} S_i S_l \right)$$
$$+ \frac{1}{2} \left(\delta_{jk} S_i + \delta_{ij} S_k \right) \quad , \tag{14.164}$$

where ε_{ijk} is a completely antisymmetric tensor of the third rank. Therefore, a complete set of three-row matrices can be formed by the unit matrix 1, three spin matrices S_i, and quadratic products of spin matrices which in the usual space form the symmetric tensor S_{ij} with a vanishing sum of diagonal elements,

$$S_{ij} = \frac{1}{2} \left(S_i S_j + S_j S_i \right) - \frac{2}{3} \delta_{ij} \quad . \tag{14.165}$$

The tensor S_{ij} has five independent components.

Another complete set of three-row matrices is formed by the spin tensor components T_{IM} with $I = 0, 1, 2$ and $-I \leq M \leq I$.

Let ϱ be the density matrix of the incident beam of spin 1 particles. The scattered beam density matrix ϱ' is given by

$$\varrho' = f \varrho f^+ \quad , \tag{14.166}$$

where f is the scattering amplitude. In the case of spin 1 – spin 0 particle scattering, f is a three-row matrix.

The differential scattering cross section is equal to

$$\sigma = \frac{\text{Tr}\{\varrho'\}}{\text{Tr}\{\varrho\}} \quad . \tag{14.167}$$

Polarization of scattered particles is determined by the mean values of the spin tensor components T_{IM}, i.e.,

$$\langle T_{IM} \rangle = \frac{\text{Tr}\{T_{IM}\,\varrho\}}{\text{Tr}\{\varrho'\}} \quad , \qquad -I \le M \le I \quad , \qquad I = 0,\,1,\,2 \quad . \tag{14.168}$$

We expand the scattering amplitude f in terms of the complete set of three-row matrices 1, S_i, and S_{ij} to obtain

$$f = A + \sum_i B_i S_i + \sum_{ij} C_{ij} S_{ij} \quad . \tag{14.169}$$

The expansion coefficients A, B_i, and C_{ij} are governed by the collision geometry, i.e., the vectors k and k' and the nature of interaction.

It is convenient to introduce three mutually orthogonal unit vectors

$$n = \frac{k \times k'}{|k \times k'|} \quad , \qquad p = \frac{k + k'}{|k + k'|} \quad , \qquad q = n \times p \quad . \tag{14.170}$$

Under time inversion, n and p change their signs, whereas q is invariant.

We assume that the interaction potential is invariant with respect to time inversion and spatial rotations and reflections. The amplitude (14.168) must possess an analogous invariance property, which imposes the following requirements: (i) A must be a scalar that can depend only on the moduli of k and k' and the scattering angle ϑ; (ii) B_i must be the only pseudoscalar that can be formed by the vectors k and k', i.e.,

$$B = Bn \quad , \tag{14.171}$$

where B is a scalar function of the moduli of k and k' and the scattering angle ϑ; (iii) C_{ij} must be a symmetric tensor invariant with respect to spatial reflections and time inversion. Making use of the definition (14.170), we can write the tensor with these required properties as

$$C_{ij} = C_1 n_i n_j + C_2 p_i p_j + C_3 q_i q_j \quad , \tag{14.172}$$

where C_1, C_2, and C_3 are scalar coefficients. However, since the vectors (14.170) form an orthogonal set, we have

$$n_i n_j + p_i p_j + q_i q_j + \delta_{ij} \quad , \tag{14.173}$$

and therefore, (14.172) contains only two independent combinations and C_{ij} can be taken to be

$$C_{ij} = c\, p_i p_j + d\, q_i q_j \quad , \tag{14.174}$$

where c and d are scalar coefficients depending on the moduli of vectors k and k' and the scattering angle ϑ. For the sake of convenience, we introduce linear combinations of the amplitudes, i.e.,

$$C = \tfrac{1}{2}(c + d) \quad , \qquad D = \tfrac{1}{2}(d - c) \quad . \tag{14.175}$$

In the coordinate system with the x axis directed along q, the y axis along n, and the z axis along p, the scattering amplitude (14.169) can be written as

$$f = A + B S_y + (C + D)\left(S_z^2 - \tfrac{2}{3}\right) + (C - D)\left(S_z^2 - \tfrac{2}{3}\right) \quad . \tag{14.176}$$

The scalar amplitudes entering (14.176) can be expressed in terms of the scattering phase shifts δ_1^j, δ_2^j, δ_3^j, and the mixing coefficients ε^j. We obtain

$$A = \frac{1}{3k} \sum_{l=0}^{\infty} \Big[(l+2)\, a_l^{l+1} + (2l+1)\, a_l^l + (l-1)\, a_l^{l-1}$$

$$+ (l+1)\, b_l^{l+1} + l b_l^{l-1} \Big]\, P_l(\cos \vartheta) \quad ,$$

$$B = \frac{i}{2k} \sum_{l=1}^{\infty} \left[\frac{l+2}{l-1}\, a_l^{l+1} - \frac{2l+1}{l(l+1)}\, a_l^l - \frac{l-1}{l}\, a_l^{l-1} \right.$$

$$\left. + b_l^{l+1} - b_l^{l-1} \right] P_l^1(\cos \vartheta) \quad ,$$

$$C = \frac{1}{4k} \left\{ \sum_{l=0}^{\infty} \Big[(l+2)\, a_l^{l+1} + (2l+1)\, a_l^l + (l-1)\, a_l^{l-1} \right.$$

$$- 2(l+1)\, b_l^{l+1} - 2l b_l^{l-1} \Big]\, P_l(\cos \vartheta) \tag{14.177}$$

$$\left. + 3 \sum_{l=2}^{\infty} \left[\frac{1}{l+1}\, a_l^{l+1} - \frac{2l+1}{l(l+1)}\, a_l^l + \frac{1}{l}\, a_l^{l-1} \right] P_l^2(\cos \vartheta) \right\} \quad ,$$

$$D = \frac{1}{4k \cos \vartheta} \left\{ \sum_{l=0}^{\infty} \Big[(l+2)\, a_l^{l+1} + (2l+1)\, a_l^l + (l-1)\, a_l^{l-1} \right.$$

$$- 2(l-1)\, b_l^{l+1} - 2l b_l^{l-1} \Big]\, P_l(\cos \vartheta)$$

$$\left. - \sum_{l=2}^{\infty} \left[\frac{1}{l+1}\, a_l^{l+1} - \frac{2l+1}{l(l+1)}\, a_l^l + a_l^{l-1} \right] P_l^2(\cos \vartheta) \right\} \quad ,$$

where

$$P_l^m(\cos \vartheta) = \sin^m \vartheta \, \frac{d^m}{(d\cos \vartheta)^m}\, P_l(\cos \vartheta)$$

and

$$a_l^{j=l+1} \equiv \alpha^j \cos^2 \varepsilon^j + \beta^j \sin^2 \varepsilon^j + \frac{1}{2} \sqrt{\frac{j}{j+1}} \, (\alpha^j - \beta^j) \sin 2\varepsilon^j \quad ;$$

$$b_l^{j=l+1} \equiv \alpha^j \cos^2 \varepsilon^j + \beta^j \sin^2 \varepsilon^j - \frac{1}{2} \sqrt{\frac{j+1}{j}} \, (\alpha^j - \beta^j) \sin 2\varepsilon^j \quad ;$$

$$a_l^{j=l-1} \equiv \alpha^j \sin^2 \varepsilon^j + \beta^j \cos^2 \varepsilon^j + \frac{1}{2} \sqrt{\frac{j+1}{j}} \, (\alpha^j - \beta^j) \sin 2\varepsilon^j \quad , \quad (14.178)$$

$$b_l^{j=l-1} \equiv \alpha^j \sin^2 \varepsilon^j + \beta^j \cos^2 \varepsilon^j - \frac{1}{2} \sqrt{\frac{j}{j+1}} \, (\alpha^j - \beta^j) \sin 2\varepsilon^j \quad ;$$

$$\alpha^j \equiv \sin \delta_1^j \exp(i\delta_1^j) \quad , \qquad \beta^j \equiv \sin \delta_2^j \exp(i\delta_2^j) \quad .$$

If $j = 0$, then $\varepsilon^0 = 0$, $\delta_1^0 = 0$, and $\delta_2^0 = \delta(P_0)$;

$$a_l^{j=l} \equiv \sin \delta_3^j \exp(i\delta_3^j) \quad .$$

If the incident particles are unpolarized, then the differential cross section is given by

$$\sigma(\vartheta) = |A|^2 + \tfrac{2}{3}|B|^2 + \tfrac{2}{9}|C|^2 + \tfrac{2}{3}|D|^2 \quad . \tag{14.179}$$

In the coordinate system under consideration, the mean values of the spin tensor components T_{IM} associated with polarization of scattered particles are

$$\langle T_{10} \rangle = 0 \quad ,$$

$$\langle T_{11} \rangle = -\frac{2i}{\sqrt{3}\,\sigma(\vartheta)} \, \mathrm{Re} \left\{ \left(A - \tfrac{1}{3} C \right) B^* \right\} \quad ,$$

$$\langle T_{20} \rangle = \frac{\sqrt{2}}{3\sigma(\vartheta)} \left\{ \mathrm{Re} \left\{ A(C - 3D)^* \right\} + \mathrm{Re} \left\{ CD^* \right\} \right.$$
$$\left. - \frac{1}{2}|B|^2 - \frac{1}{6}|C|^2 + \frac{1}{2}|D|^2 \right\} \quad ,$$

$$\langle T_{21} \rangle = \frac{2}{\sqrt{3}\,\sigma(\vartheta)} \, \mathrm{Im} \left\{ BD^* \right\} \quad , \tag{14.180}$$

$$\langle T_{22} \rangle = \frac{1}{\sqrt{3}\,\sigma(\vartheta)} \left\{ \mathrm{Re} \left\{ A(C + D)^* \right\} - \frac{1}{3} \mathrm{Re} \left\{ CD^* \right\} \right.$$
$$\left. - \frac{1}{2}|B|^2 + \frac{1}{6}|C|^2 - \frac{1}{2}|D|^2 \right\} \quad .$$

According to (14.154), the particle polarization vector P is directly determined by the component $\langle T_{11} \rangle$, so we have

$$P = \frac{4}{3} \frac{\mathrm{Re} \left\{ \left(A - \tfrac{1}{3} C \right) B^* \right\}}{\sigma(\vartheta)} \, n \quad . \tag{14.181}$$

If the incident particle beam is polarized and the initial density matrix ϱ is equal to

$$\varrho = \frac{1}{3} \sum_{I=0}^{2} \sum_{M} \langle T_{IM}^{+} \rangle_0 \, T_{IM} \quad , \tag{14.182}$$

where $\langle T_{IM} \rangle_0$ are the mean values of the spin tensor describing the incident beam polarization state, then the differential scattering cross section is determined by the expression

$$\sigma(\vartheta, \varphi) = \sigma(\vartheta) \sum_{I=0}^{2} \sum_{M} \langle T_{IM}^{+} \rangle_0 \, \langle T_{IM} \rangle \quad , \tag{14.183}$$

where the cross section $\sigma(\vartheta)$ and the polarization $\langle T_{IM} \rangle$ are given by (14.179, 180) corresponding to scattering of unpolarized particles.

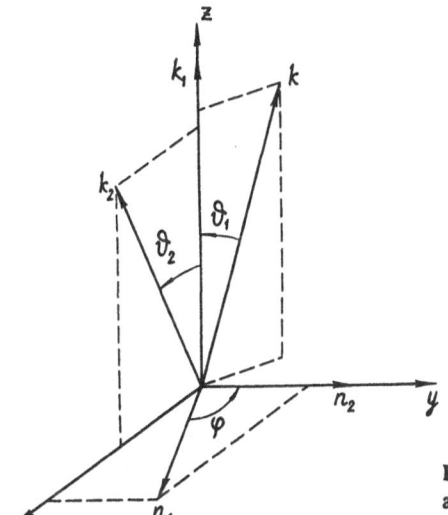

Fig. 14.3. Coordinate system showing the azimuthal angle φ between the first and the second scattering planes with normals n_1 and n_2

In conclusion, we derive the formula for the differential cross section of double scattering of spin 1 particles. It is not difficult to show that the double scattering cross section is described by (14.183), where $\langle T_{IM} \rangle_0$ must be regarded as the polarization $\langle T_{IM} \rangle_1'$ produced by the first collision and $\langle T_{IM} \rangle$ – as polarization $\langle T_{IM} \rangle_2'$ that occurs in the second collision, were the particles incident on the second target unpolarized.

If the expansion of the double scattering cross section (14.183) is written for the coordinate system (Fig. 14.3) with the z axis directed along the vector k_1 and the y axis along n_2, then we have to express $\langle T_{IM} \rangle_1'$ in terms of the spin tensor $\langle T_{IM} \rangle_1$ that is given by (14.180) in the coordinate system with the z axis directed along $(k + k_1)/2$ and y axis along n_1. The spin tensor $\langle T_{IM} \rangle'_2$ must be expressed in terms of $\langle T_{IM} \rangle_2$ that is given in the coordinate system with the z axis along $(k_1 + k_2)/2$ and y axis along n_2. The conversion from $\langle T_{IM} \rangle_1$ to $\langle T_{IM} \rangle'_1$ is associated with the coordinate system rotation by the Euler angles $\vartheta_1/2$ and φ (φ is the angle between the first and second scattering planes), i.e.,

$$\langle T_{IM} \rangle'_1 = \sum_{M'} D^{I^*}_{MM'}\left(0, \tfrac{1}{2}\vartheta_1, \varphi\right) \langle T_{IM'} \rangle_1 \quad , \tag{14.184}$$

and passage from $\langle T_{IM} \rangle_2$ to $\langle T_{IM} \rangle'_2$ corresponds to the rotation by an angle $\vartheta_2/2$,

$$\langle T_{IM} \rangle'_2 = \sum_{M''} D^{I^*}_{MM''}\left(0, \tfrac{1}{2}\vartheta_2, 0\right) \langle T_{IM''} \rangle_2 \quad . \tag{14.185}$$

Thus, the differential cross section after the second collision can be written as

$$\sigma(\vartheta_2, \varphi) = \sigma(\vartheta_2) \sum_{I=0}^{2} \sum_{MM'M''} D^{I}_{MM'}\left(0, \frac{\vartheta_1}{2}, \varphi\right) D^{I^*}_{MM''}\left(0, \frac{\vartheta_2}{2}, 0\right)$$
$$\times \langle T^+_{IM'} \rangle_1 \langle T_{IM''} \rangle_2 \quad . \tag{14.186}$$

Making use of the relation $D^{I}_{MM'}(0, \vartheta_1/2, \varphi) = \exp(iM\varphi)\, d^{I}_{MM'}\left(\vartheta_1/2\right)$ and substituting the explicit expressions for $d^{1}_{MM'}(\vartheta)$ and $d^{2}_{MM'}(\vartheta)$ (given in the Appendix), we obtain

$$\sigma(\vartheta_2, \varphi) = \sigma(\vartheta_2)\, (a + b \cos\varphi + c \cos 2\varphi) \quad , \tag{14.187}$$

where

$$a = 1 + \left(\sqrt{\frac{3}{2}} \sin^2 \frac{\vartheta_1}{2} \langle T_{22} \rangle_1 - \sqrt{\frac{3}{2}} \sin\vartheta_1 \langle T_{21} \rangle_1 \right.$$
$$\left. + \frac{3\cos^2(\vartheta_1/2) - 1}{2} \langle T_{20} \rangle_1 \right) \left(\sqrt{\frac{3}{2}} \sin^2 \frac{\vartheta_2}{2} \langle T_{22} \rangle_2 \right.$$
$$\left. - \sqrt{\frac{3}{2}} \sin\vartheta_2 \langle T_{21} \rangle_2 + \frac{3\cos^2(\vartheta_2/2) - 1}{2} \langle T_{20} \rangle_2 \right) \quad ,$$

$$b = 2 \langle T^+_{11} \rangle_1 \langle T_{11} \rangle_2 + \frac{1}{2} \left(\sin\vartheta_1 \langle T_{22} \rangle_1 - 2\cos\vartheta_1 \langle T_{21} \rangle_1 \right.$$
$$\left. - \sqrt{\frac{3}{2}} \sin\vartheta_1 \langle T_{20} \rangle_1 \right) \tag{14.188}$$
$$\times \left(\sin\vartheta_2 \langle T_{22} \rangle_2 - 2\cos\vartheta_2 \langle T_{21} \rangle_2 - \sqrt{\frac{3}{2}} \sin\vartheta_2 \langle T_{20} \rangle_2 \right) \quad ,$$

$$c = \frac{1}{2} \left(\frac{3 + \cos \vartheta_1}{2} \langle T_{22} \rangle_1 - \sin \vartheta_1 \langle T_{21} \rangle_1 + \sqrt{\frac{3}{2}} \frac{1 - \cos \vartheta_1}{2} \langle T_{20} \rangle_1 \right)$$

$$\times \left(\frac{3 + \cos \vartheta_2}{2} \langle T_{22} \rangle_2 - \sin \vartheta_2 \langle T_{21} \rangle_2 + \sqrt{\frac{3}{2}} \frac{1 + \cos \vartheta_2}{2} \langle T_{20} \rangle_2 \right) \quad .$$

Note that, in contrast to the case of spin $\frac{1}{2}$ particles, the spin 1 particle scattering cross section contains terms proportional to both $\cos \varphi$ and $\cos 2\varphi$. The proportionality coefficients can be found from azimuthal asymmetry in experimental data.

Problems

14.1   Derive the general form of the elastic scattering amplitude for the interaction of two spin $\frac{1}{2}$ particles [14.5].

The scattering amplitude obviously must depend on the initial and final relative momenta k and k' and the spins of interacting particles σ_1 and σ_2. The functional form of the amplitude $f(k, k'; \sigma_1, \sigma_2)$ is governed by the interaction symmetry.

Invariance of the interaction potential with respect to rotation leads to the requirement that the scattering amplitude must be a scalar function of its arguments. Invariance of the potential with respect to the spatial inversion results in the condition

$$f(k, k'; \sigma_1, \sigma_2) = f(-k, -k'; \sigma_1, \sigma_2) \quad . \tag{14.189}$$

Lastly, invariance with respect to time inversion yields the symmetry property

$$f(k, k'; \sigma_1, \sigma_2) = f(-k, -k'; -\sigma_1, -\sigma_2) \quad . \tag{14.190}$$

These relations make it possible to derive the explicit σ_1 and σ_2 dependence of the scattering amplitude. By virtue of the permutation properties of the Pauli matrices, $f(k, k'; \sigma_1, \sigma_2)$ can only be a linear function of σ_1 and σ_2, therefore, it can always be written as

$$f(k, k'; \sigma_1, \sigma_2) = A + B \sigma_1 \quad , \tag{14.191}$$

where A is a scalar and B is a pseudovector, both linear functions of σ_2. Moreover, A and B can depend on the vectors k and k' as well as on the pseudovector $k \times k'$.

For the sake of convenience, the amplitude is usually written in terms of three mutually orthogonal unit vectors l, m, and n defined by

$$l = \frac{k + k'}{|k + k'|} \quad , \qquad m = \frac{k - k'}{|k - k'|} \quad , \qquad n = \frac{k \times k'}{|k \times k'|} \quad . \tag{14.192}$$

Note that the basis vectors l and n change their signs under time inversion, whereas m is invariant. An advantage of the coordinate system (14.192) is that if the particles involved in the interaction have equal masses, then its basis vectors in the laboratory system are parallel to both the momenta of the scattered particles and the normal to the scattering plane.

Of course, the most general form of the linear σ_2 dependence of the scalar function A is

$$A = \alpha + a n\,\sigma_2 \quad , \tag{14.193}$$

and the coefficients α and a can depend only on the scattering angle and energy.

It is convenient to expand the pseudovector B in terms of the basis vectors l, m, and n, so that

$$B = bl + cm + dn \quad , \tag{14.194}$$

where, because l and m are vectors and n is a pseudovector, the coefficients b and c are pseudoscalars and d is a scalar. Amplitude invariance with respect to time inversion imposes the requirement that the pseudoscalars b and c (which are linear functions of σ_2) must have the form

$$b = \varepsilon\,l\sigma_2 \quad , \qquad c = \delta\,m\sigma_2 \quad , \tag{14.195}$$

where ε and δ can depend only on energy and the scattering angle. Analogous reasoning suggests the following formula for the scalar d:

$$d = d' + \beta\,n\sigma_2 \quad , \tag{14.196}$$

with d' and β also depending only on energy and the scattering angle.

Substituting (14.193, 194) into (14.191) after noting (14.195, 196) and introducing the notation

$$d' + a = 2\mathrm{i}\gamma \quad , \qquad d' - a = 2\mathrm{i}\gamma' \quad ,$$

we find the spin $\frac{1}{2}$ – spin $\frac{1}{2}$ particle scattering amplitude to be

$$\begin{aligned}
f(k, k'; \sigma_1, \sigma_2) &= \alpha + \beta\,n\sigma_1 \cdot n\sigma_2 + \mathrm{i}\gamma\,n(\sigma_1 + \sigma_2) \\
&\quad + \mathrm{i}\gamma'\,n(\sigma_1 - \sigma_2) + \delta\,m\sigma_1 \cdot m\sigma_2 + \varepsilon\,l\sigma_1 \cdot l\sigma_2 \quad .
\end{aligned} \tag{14.197}$$

This expression gives the most general representation of the scattering amplitude compatible with the requirements of parity conservation and invariance with respect to time inversion. Note that the scattering amplitude (14.197) is a four-row matrix which depends only on ten independent four-row matrices

$$1, \ \sigma_1, \ \sigma_2, \ n\sigma_1 \cdot n\sigma_2, \ m\sigma_1 \cdot m\sigma_2, \ l\sigma_1 \cdot l\sigma_2 \quad .$$

The expansion coefficients before the other six independent four-row matrices vanish for the symmetry reasons mentioned above.

Additional symmetry conditions only simplify the form of the amplitude (14.197). Thus, if the particles involved in the scattering process are identical, then the amplitude must be symmetric with respect to permutations of particle spins σ_1 and σ_2 and therefore, we have to set $\gamma' = 0$ in (14.197).

14.2   Express the coefficients in the scattering amplitude for two identical spin $\frac{1}{2}$ particles in terms of the scattering phase shifts δ_α^j and the mixing parameters ε^j [14.6].

A system of two identical spin $\frac{1}{2}$ particles can be in the singlet ($S = 0$) or triplet ($S = 1$) states, transitions between which are forbidden. For the singlet state, the angular momentum quantum number l is equal to the total moment quantum number j, so for a given j scattering is described by the single phase shift δ_0^j. In the triplet state, j can correspond to three values of l: $l = j - 1$, $l = j$, and $l = j + 1$. The states with $l = j$ and $l = \mp 1$ are of opposite parity. If the angular momentum corresponds to the values $l = j - 1$ and $l = j + 1$, then scattering is described by two proper scattering phase shifts δ_1^j and δ_2^j and the mixing parameter ε^j. If $l = j$, then scattering is described by the single phase shift which is usually denoted by δ_3^j.

The coefficients α, β, γ, δ, and ε entering the scattering amplitude (14.197) for two identical spin $\frac{1}{2}$ particles are given by the expressions

$$
\alpha = \frac{1}{4k} \sum_{l=0}^{\infty} \left[(l+2)\, a_l^{l+1} + (2l+1)\, a_l^l + (l-1)\, a_l^{l-1} \right.
$$
$$
\left. + (l+1)\, b_l^{l+1} + l b_l^{l-1} + (2l+1)\, c_l \right] P_l(\cos\vartheta) \quad ,
$$

$$
\beta = \frac{1}{4k} \left\{ \sum_{l=1}^{\infty} \left[(l+1)\, b_l^{l+1} + l b_l^{l-1} - (2l+1)\, c_l \right] P_l(\cos\vartheta) \right.
$$
$$
\left. - \sum_{l=2}^{\infty} \left[\frac{1}{l+1}\, a_l^{l+1} - \frac{2l+1}{l(l+1)}\, a_l^l + \frac{1}{l}\, a_l^{l-1} \right] P_l^2(\cos\vartheta) \right\} \quad ,
$$

$$
\gamma = \frac{1}{4k} \sum_{l=1}^{\infty} \left[\frac{l+2}{l+1}\, a_l^{l+1} - \frac{2l+1}{l(l+1)}\, a_l^l - \frac{l-1}{l}\, a_l^{l-1} \right.
$$
$$
\left. + b_l^{l+1} - b_l^{l-1} \right] P_l^1(\cos\vartheta) \quad , \tag{14.198}
$$

$$\delta = \frac{1}{4k \cos \vartheta} \left\{ \sum_{l=0}^{\infty} \left[\frac{1}{2} \left((l+2) a_l^{l+1} + (2l+1) a_l^l + (l-1) a_l^{l-1} \right) \right. \right.$$

$$\left. \times (\cos \vartheta - 1) + (l+1) b_l^{l+1} + l b_l^{l-1} - (2l+1) c_l \cos \vartheta \right] P_l(\cos \vartheta)$$

$$+ \frac{1}{2} \sum_{l=2}^{\infty} \left[\frac{1}{l+1} a_l^{l+1} - \frac{2l+1}{l(l+1)} a_l^l + \frac{1}{l} a_l^{l-1} \right]$$

$$\left. \times (1 + \cos \vartheta) P_l^2 (\cos \vartheta) \right\} \quad ,$$

$$\varepsilon = \frac{1}{4k \cos \vartheta} \left\{ \sum_{l=0}^{\infty} \left[\frac{1}{2} \left((l+2) a_l^{l+1} + (2l+1) a_l^l + (l-1) a_l^{l-1} \right) \right. \right.$$

$$\left. \times (1 + \cos \vartheta) - (l+1) b_l^{l+1} - l b_l^{l-1} - (2l+1) c_l \cos \vartheta \right] P_l(\cos \vartheta)$$

$$+ \frac{1}{2} \sum_{l=2}^{\infty} \left[\frac{1}{l+1} a_l^{l+1} - \frac{2l+1}{l(l+1)} a_l^l + \frac{1}{l} a_l^{l-1} \right] (\cos \vartheta - 1) P_l^2 (\cos \vartheta) \right\} \quad .$$

The quantities a_l^j and b_l^j for the triplet states ($S = 1$) are given by (14.178); c_l for the singlet states ($S = 0$) are

$$c_{l=j} = \sin \delta_0^j \exp(i \delta_0^j) \quad .$$

14.3   Derive the integral unitarity relation for the eleastic scattering amplitude for the case when all particles involved in the collision possess spins [14.12].

We assume that no processes other than elastic scattering occur in the system, and $f(k, k')$ is the matrix scattering amplitude. As follows from the unitarity condition for the S matrix, in the case of spin-possessing particles the scattering amplitude satisfies an integral relation of the type (5.9), i.e.,

$$f(k, k') - f^+(k, k') = \frac{ik}{2\pi} \int do'' \, f^+(k', k'') \, f(k, k'') \quad . \tag{14.199}$$

If $k = k'$, then (14.199) yields, along with the optical theorem, relations between the scattering amplitude matrix elements which do not vanish as $k' \to k$, and the spin characteristics of the system.

Let us consider in detail the integral unitarity relation (14.199) for scattering of two identical spin $\frac{1}{2}$ particles. The matrix scattering amplitude is given in this case by (14.197) with $\gamma' = 0$. We multiply (14.199) by the invariant spin matrices, in terms of which the amplitude (14.197) is expanded, and thus obtain five independent integral relations for the expansion coefficients α, β, γ, δ, and ε. In particular, if $k = k'$, the vector quantities σ_1, σ_2, and $l = k/k$ can form only three independent scalar matrices, i.e.,

$$1 , \quad l\sigma_1 \cdot l\sigma_2 , \quad \tfrac{1}{2}\left(\sigma_1\sigma_2 - l\sigma_1 \cdot l\sigma_2\right) . \tag{14.200}$$

(The three matrices are normalized in a unified manner for the sake of convenience.) Multiplying (14.199) by each matrix (14.200) and taking the traces of the left- and right-hand parts, we obtain three independent relations:

$$\int do\, \sigma(\vartheta) = \frac{4\pi}{k}\, \mathrm{Im}\,\{\alpha(0)\} ,$$

$$\sigma(\vartheta) \equiv \tfrac{1}{4}\,\mathrm{Tr}\,\{ff^+\} = |\alpha|^2 + |\beta|^2 + 2|\gamma|^2 + |\delta|^2 + |\varepsilon|^2 ; \tag{14.201}$$

$$\int do\, P_{ll}(\vartheta) = \frac{4\pi}{k}\,\mathrm{Im}\,\{\varepsilon(0)\} ,$$

$$P_{ll}(\vartheta) \equiv \mathrm{Tr}\,\{(l\sigma_1 \cdot l\sigma_2)\,ff^+\} = 2\,\mathrm{Re}\,\{\alpha\varepsilon^* - \beta\delta^*\} ; \tag{14.202}$$

$$\int do\, P_{tt}(\vartheta) = \frac{\cdots}{k}\,\mathrm{Im}\,\{\beta(0) + \delta(0)\} ,$$

$$P_{tt}(\vartheta) \equiv \mathrm{Tr}\,\{\sigma_1\sigma_2 - l\sigma_1 \cdot l\sigma_2)\,ff^+\}$$
$$= 2\left\{\mathrm{Re}\,\{(\alpha - \varepsilon)(\beta + \delta)^*\} + |\gamma|^2\right\} . \tag{14.203}$$

The first relation is the generalized optical theorem which gives the imaginary part of the central forward direction scattering amplitude as a function of the total scattering cross section. The second and third equations express the imaginary parts of the noncentral forward direction scattering amplitude in terms of the polarization tensor components given rise to by scattering.

14.3   Show that if the first Born approximation is applicable and the interaction is invariant with respect to time inversion, then scattering of an unpolarized beam by an unpolarized target does not produce polarization of the scattered beam [14.13].

We denote the matrix amplitude of spin s_1 – spin s_2 particle scattering as $f(k, k'; s_1\, s_2)$. If the interaction is invariant with respect to time inversion, then the amplitude $f(k, k'; s_1, s_2)$ satisfies the condition

$$f(k, k'; s_1, s_2) = \tilde{f}(-k', -k; -s_1^*, -s_2^*) , \tag{14.204}$$

where $\tilde{f}$ is the transposed matrix.

Indeed, according to (6.62), we have

$$\langle \chi_{s'\mu'}\,|f(k, k'; s_1, s_2)|\,\chi_{s\mu}\rangle$$
$$= \langle R\chi_{s\mu}\,|f(-k', -k; s_1, s_2)|\,R\chi_{s'\mu'}\rangle , \tag{14.205}$$

where R is the time inversion operator,

$$R = WK , \qquad W^+ = W^{-1} , \tag{14.206}$$

and

$$s'_\alpha = R\, s_\alpha\, R^{-1} = W\, s^*_\alpha\, W^{-1} = -s_\alpha \quad . \tag{14.207}$$

Substituting (14.206) into (14.205), we have

$$\langle \chi_{s'\mu'} \left| f(k, k'; s_1, s_2) \right| \chi_{s\mu} \rangle$$

$$= \langle \chi^*_{s\mu} \left| W^{-1} f(-k', -k; s_1, s_2)\, W \right| \chi^*_{s'\mu'} \rangle$$

$$= \langle \chi^*_{s\mu} \left| f(-k', -k; W^{-1} s_1 W, W^{-1} s_2 W) \right| \chi^*_{s'\mu'} \rangle$$

$$= \langle \chi^*_{s\mu} \left| f(-k', -k; -s^*_1, -s^*_2) \right| \chi^*_{s'\mu'} \rangle$$

$$= \langle \chi_{s'\mu'} \left| \tilde{f}(-k', -k; -s^*_1, s^*_2) \right| \chi_{s\mu} \rangle \quad .$$

Condition (14.204) follows from this relation.

Let us show that if the amplitude satisfies (14.204), then

$$\mathrm{Tr}\left\{ s_\alpha\, f(k, k'; s_1\, s_2)\, f^+(k, k'; s_1, s_2) \right\}$$

$$= \mathrm{Tr}\left\{ s_\alpha\, f^+(k, k'; s_1, s_2)\, f(k, k'; s_1, s_2) \right\} \quad , \quad \alpha = 1, 2 . \tag{14.208}$$

Note that $\mathrm{Tr}\left\{ s_\alpha\, f(k, k'; s_1, s_2)\, f^+(k, k'; s_1, s_2) \right\}$ is an axial vector which can depend only on the vectors k and k' which, in turn, can form the only pseudovector $k \times k'$. Therefore, (14.208) can be rewritten as

$$\mathrm{Tr}\left\{ s_\alpha\, f(k, k'; s_1\, s_2)\, f^+(k, k'; s_1, s_2) \right\} = (k \times k')\, F_\alpha(kk') \quad , \tag{14.209}$$

where $F_\alpha(kk')$ is a real fundtion which depends only on the scalar product kk'. It is clear that $F_\alpha(kk')$ is invariant with respect to the transformations

$$k \to -k \quad , \qquad k' \to -k' \quad ;$$

$$k \to k' \quad , \qquad k' \to k \quad . \tag{14.210}$$

We employ (14.207) and consider the combination

$$(k \times k')\, F_\alpha(kk') = -\mathrm{Tr}\left\{ W\, s^*_\alpha\, W^{-1} f(k, k'; s_1, s_2)\, f^+(k, k'; s_1, s_2) \right\}$$

or

$$(k \times k')\, F_\alpha(kk')$$

$$= -\mathrm{Tr}\left\{ s^*_\alpha\, f(k, k'; -s^*_1, -s^*_2)\, f^+(k, k'; -s^*_1, -s^*_2) \right\} \quad . \tag{14.211}$$

Making use of (14.210), we reduce (14.211) to

$$(k \times k')\, F_\alpha(kk')$$

$$= \mathrm{Tr}\left\{ s^*_\alpha\, f(-k', -k; -s^*_1, -s^*_2)\, f^+(-k', -k; -s^*_1, -s^*_2) \right\} \quad .$$

Using (14.204), (14.208) follows.

If the first Born approximation is applicable, then the transition matrix t is equal to the interaction potential V. If V is Hermitian, then the transition matrix is also Hermitian and hence

$$f(k, k'; s_1, s_2) = f^+(k', k; s_1 s_2) \quad . \tag{14.212}$$

Polarization of the particle α ($\alpha = 1, 2$) produced after unpolarized beam scattering by an unpolarized target, is given by

$$P = \frac{\text{Tr}\{s_\alpha f f^+\}}{\text{Tr}\{f f^+\}} \quad . \tag{14.213}$$

As follows from (14.212),

$$\text{Tr}\{s_\alpha f(k, k'; s_1, s_2) f^+(k, k'; s_1, s_2)\}$$
$$= \text{Tr}\{s_\alpha f^+(k', k; s_1, s_2) f(k', k; s_1, s_2)\} \quad . \tag{14.214}$$

Because the right-hand part of (14.209) changes its sign under the interchange $k \to k'$, we have

$$\text{Tr}\{s_\alpha f^+(k', k; s_1, s_2) f(k, k'; s_1, s_2)\}$$
$$= -\text{Tr}\{s_\alpha f^+(k, k'; s_1, s_2) f(k, k'; s_1, s_2)\} \quad . \tag{14.215}$$

Comparing (14.215) to (14.208), we see that polarization P_α is equal to zero. Thus we have proved that if the interaction potential is invariant with respect to time inversion, then in the first Born approximation scattering does not produce particle polarization.

Appendix

Here we give the explicit expressions for the functions $d^1_{MM'}(\vartheta)$ and $d^2_{MM'}(\vartheta)$ which were employed in the derivation of (14.188).

Table 1. $d^1_{MM'}(\vartheta)$

M \ M'	1	0	−1
1	$\frac{1}{2}(1+\cos\vartheta)$	$\frac{1}{\sqrt{2}}\sin\vartheta$	$\frac{1}{2}(1-\cos\vartheta)$
0	$-\frac{1}{\sqrt{2}}\sin\vartheta$	$\cos\vartheta$	$\frac{1}{\sqrt{2}}\sin\vartheta$
−1	$\frac{1}{2}(1-\cos\vartheta)$	$-\frac{1}{\sqrt{2}}\sin\vartheta$	$\frac{1}{2}(1+\cos\vartheta)$

Table 2. $d^2_{MM'}(\vartheta)$

$M' \backslash M$	2	1	0	-1	-2
2	$\frac{1}{4}(1+\cos\vartheta)^2$	$\frac{1}{2}\sin\vartheta\,(1+\cos\vartheta)$	$\sqrt{\frac{3}{8}}\sin^2\vartheta$	$\frac{1}{2}\sin\vartheta\,(1-\cos\vartheta)$	$\frac{1}{4}(1-\cos\vartheta)^2$
1	$-\frac{1}{2}\sin\vartheta\,(1+\cos\vartheta)$	$\frac{1}{2}(1+\cos\vartheta)(2\cos\vartheta-1)$	$\sqrt{\frac{3}{2}}\sin\vartheta\,\cos\vartheta$	$\frac{1}{2}(1-\cos\vartheta)(2\cos\vartheta+1)$	$\frac{1}{2}\sin\vartheta\,(1-\cos\vartheta)$
0	$\sqrt{\frac{3}{8}}\sin^2\vartheta$	$-\sqrt{\frac{3}{2}}\sin\vartheta\,\cos\vartheta$	$\frac{1}{2}(3\cos^2\vartheta-1)$	$\sqrt{\frac{3}{2}}\sin\vartheta\,\cos\vartheta$	$\sqrt{\frac{3}{8}}\sin^2\vartheta$
-1	$-\frac{1}{2}\sin\vartheta\,(1-\cos\vartheta)$	$\frac{1}{2}(1-\cos\vartheta)(2\cos\vartheta+1)$	$-\sqrt{\frac{3}{2}}\sin\vartheta\,\cos\vartheta$	$\frac{1}{2}(1+\cos\vartheta)(2\cos\vartheta-1)$	$\frac{1}{2}\sin\vartheta\,(1+\cos\vartheta)$
-2	$\frac{1}{4}(1-\cos\vartheta)^2$	$-\frac{1}{2}\sin\vartheta\,(1-\cos\vartheta)$	$\sqrt{\frac{3}{8}}\sin^2\vartheta$	$-\frac{1}{2}\sin\vartheta\,(1+\cos\vartheta)$	$\frac{1}{4}(1+\cos\vartheta)^2$

References[1]

Chapter 2

2.1   W. Heisenberg: Z. Phys. **120**, 513; 673 (1943)
2.2   F.J. Dyson: Phys. Rev. **75**, 486; 1736 (1949)

Chapter 3

3.1   B.A. Lippmann, J. Schwinger: Phys. Rev. **79**, 469 (1950)
3.2   T.A. Osborn: Ann. Phys. (NY) **58**, 417 (1970)
3.3   A.G. Sitenko: Ukr. Fiz. Zh. **4**, 152 (1959)
3.4   R. Glauber: *Lectures in Theoretical Physics*, Vol. 1 (Interscience, New York 1959) p. 315

Chapter 4

4.1   G.F. Drukarev: Zh. Eksp. Teor. Fiz. **19**, 247 (1949)
4.2   V.V. Babikov: *Phase Function Method in Quantum Mechanics* (Nauka, Moscow 1968)
4.3   F. Calogero: *Variable Phase Approach to Potential Scattering* (Academic, New York 1967)
4.4   E. Gerjuioy: Math. Phys. **6**, 993; 1396 (1965)
4.5   E.P. Wigner: Göttingen Nachrichten **31**, 546 (1932)
4.6   R. Jost: Helv. Phys. Acta **20**, 256 (1947)

Chapter 7

7.1   S.T. Ma: Phys. Rev. **71**, 195 (1947)
7.2   G.F. Drukarev: Zh. Eksp. Teor. Fiz. **21**, 59 (1951)
7.3   V. Bargmann: Rev. Mod. Phys. **21**, 488 (1949)
7.4   E.P. Wigner: Phys. Rev. **98**, 145 (1955)
7.5   R. Jost, W. Kohn: Phys. Rev. **87**, 977 (1952)

Chapter 8

8.1   N. Levinson: Danske Videnskab. Selskab, Mat.-Phys. Medd. **25**, 9 (1949)
8.2   N.N. Khuri: Phys. Rev. **107**, 1148 (1959)
8.3   L.D. Faddeev: Zh. Eksp. Teor. Fiz. **35**, 433 (1958)
8.4   H. Lehmann: Nuovo Cim. **10**, 579 (1958)
8.5   R. Blankenbecler, M.L. Goldberger: Phys. Rev. **126**, 766 (1962)

---

[1] The originals of many papers are written in Russian. The English translations are given where known.

Chapter 9

9.1   T. Regge: Nuovo Cim. **14**, 951 (1959); **17**, 947 (1960)
9.2   V. de Alfaro, T. Regge: *Potential Scattering* (North–Holland, Amsterdam 1965)
9.3   R.G. Newton: J. Math. Phys. **3**, 867; 1342 (1962)

Chapter 10

10.1   S. Mandelstam: Phys. Rev. **112**, 1344 (1958); **115**, 1741; 1759 (1959)
10.2   R. Blankenbecler, M.L. Goldberger, N.N. Khuri, S.B. Treiman: Ann. Phys. (NY) **10**, 62 (1960)
10.3   D.V. Shirkov, V.V. Serebriakov, V.A. Meshcheriakov: *Low Energy Dispersion Theories of Strong Interactions* (Nauka, Moscow 1967)

Chapter 11

11.1   V.A. Marchenko: Dok. Akad. Nauk SSSR **104**, 695 (1955)
11.2   Z.S. Agranovich, V.A. Marchenko: *Inverse Problem of Scattering Theory* (Kharkov University Press, Kharkov 1960)

Chapter 12

12.1   A.G. Sitenko, V.F. Kharchenko: Usp. Fiz. Nauk **103**, 469 (1971)
12.2   S. Weinberg: Phys. Rev. **131**, 440 (1963)
12.3   R. Jost, A. Pais: Phys. Rev. **82**, 840 (1951)
12.4   V.F. Kharchenko, N.M. Petrov: Nucl. Phys. A **137**, 417 (1969)
12.5   V.F. Kharchenko, S.A. Storozhenko: Nucl. Phys. A **137**, 437 (1969)
12.6   V.F. Kharchenko, S.A. Storozhenko: Ukr. Fiz. Zh. **15**, 1846 (1970)

Chapter 13

13.1   L.D. Faddeev: Zh. Eksp. Teor. Fiz. **39**, 1459 (1960)
13.2   L.D. Faddeev: in *Proc. V.A. Steklov Mathematical Institute* Vol. 69 (Nauka, Moscow 1963)
13.3   S.P. Merkuriev, L.D. Faddeev: *Quantum Scattering Theory of Few-Body Systems* (Nauka, Moscow 1985)
13.4   A.G. Sitenko, V.F. Kharchenko: Usp. Fiz. Nauk **103**, 469 (1971)
13.5   V.F. Kharchenko, N.M. Petrov: Nucl. Phys. A **137**, 417 (1969)
13.6   V.F. Kharchenko, S.A. Storozhenko: Ukr. Fiz. Zh. **15**, 1846 (1970)
13.7   A.G. Sitenko, V.F. Kharchenko: Nucl. Phys. **49**, 15 (1963)
13.8   G.V. Skorniakov, K.A. Ter–Martirosian: Zh. Eksp. Teor. Fiz. **31**, 775 (1956)
13.9   T.A. Osborn, D. Bolle: Phys. Rev. C **8**, 1198 (1973)

Chapter 14

14.1   L. Wolfenstein, J. Ashkin: Phys. Rev. **85**, 947 (1952)
14.2   J.M. Blatt, L.C. Biedenharn: Phys. Rev. **86**, 399 (1952)
14.3   A. Simon, T. Welton: Phys. Rev. **90**, 1036 (1953)
14.4   A. Simon: Phys. Rev. **92**, 1050 (1953); **93**, 1435 (1954)
14.5   L. Wolfenstein: Ann. Rev. Nucl. Sci. **6**, 43 (1956)
14.6   M.L. Goldberger, Y. Nambu, R. Oehme: Ann. Phys. (NY) **2**, 226 (1957)
14.7   L.D. Landau, E.M. Lifshitz: *Quantum Mechanics* (Pergamon, Oxford 1965)

14.8   E.P. Wigner: *Group Theory and its Applications to the Quantum Mechanics of Atomic Spectra* (Academic, New York 1959)

14.9   W. Lakin: Phys. Rev. **98**, 139 (1955)

14.10 H.P. Stapp: Phys. Rev. **107**, 607 (1957)

14.11 A.G. Sitenko, V.K. Tartakovsky: Ukr. Fiz. Zh. **5**, 581 (1960)

14.12 L.N. Puzikov, P.M. Ryndin, Y.A. Smorodinsky: Zh. Eksp. Teor. Fiz. **32**, 592 (1957)

14.13 M.L. Goldberger, K.M. Watson: *Collision Theory* (Wiley, New York 1964)

General Reading

Alfaro, V. de, Regge, T.: *Potential Scattering* (North–Holland, Amsterdam 1965)

Baldin, A.M., Goldanskii, V.I., Rosental', I.L.: *Kinematics of Nuclear Reactions* (Oxford University Press, Oxford 1961)

Baz', A.I., Zeldovich, Y.B., Perelomov, A.M.: *Scattering, Reactions, and Decays in Nonrelativistic Quantum Mechanics* (Israel Programme for Scientific Translations, Jerusalem 1969)

Blatt, J.M., Weisskopf, V.F.: *Theoretical Nuclear Physics* (Wiley, New York 1952)

Davydov, A.S.: *Quantum Mechanics* (Pergamon, Oxford 1965)

Goldberger, M.L., Watson, K.M.: *Collision Theory* (Wiley, New York 1964)

Joachain, C.J.: *Quantum Collision Theory* (North–Holland, Amsterdam 1975)

Landau, L.D., Lifshitz, E.M.: *Quantum Mechanics* (Pergamon, Oxford 1965)

Matthews, P.T.: *Relativistic Quantum Theory of Elementary Interactions* (Rochester University Press, New York 1957)

Mott, N.F., Massey, N.S.W.: *The Theory of Atomic Collisions* (Oxford University Press, Oxford 1965)

Newton, R.G.: *Scattering Theory of Waves and Particles* (McGraw–Hill, New York 1966)

Rodberg, L., Thaler, R.: *Introduction to the Quantum Theory of Scattering* (Academic , New York 1967)

Taylor, J.R.: *Scattering Theory* (Wiley, New York 1972)

Subject Index